Electron Transfer

Electron Transfer

Mechanisms and Applications

Edited by
Shunichi Fukuzumi

WILEY-VCH

Author

Shunichi Fukuzumi
Department of Material & Life Science
Osaka University
2-1 Yamada-oka, Suita
Osaka University
565-0871 Osaka
Japan

Library of Congress Card No.:
applied for

British Library Cataloguing-in-Publication Data
A catalogue record for this book is available from the British Library.

Bibliographic information published by the Deutsche Nationalbibliothek
The Deutsche Nationalbibliothek lists this publication in the Deutsche Nationalbibliografie; detailed bibliographic data are available on the Internet at <http://dnb.d-nb.de>.

Print ISBN: 978-3-527-32666-2
ePDF ISBN: 978-3-527-65180-1
ePub ISBN: 978-3-527-65179-5
oBook ISBN: 978-3-527-65177-1

Typesetting SPi Global, Chennai, India
Printing and Binding Markono Print Media Pte Ltd, Singapore

Printed on acid-free paper

10 9 8 7 6 5 4 3 2 1

Contents

Acknowledgments

The author gratefully acknowledges the contributions of his collaborators mentioned in the references. The author thanks Japan Science Technology Agency and the Ministry of Education, Culture, Sports, Science and Technology of Japan for the continuous support.

1
Introduction

The rapid consumption of fossil fuel has already caused unacceptable environmental problems such as the greenhouse effect by CO_2 emission, which is predicted to lead to disastrous climatic consequences [1]. Moreover, the consumption rate of fossil fuels is expected to increase further at least twofold relative to the present by midcentury because of population and economic growth, particularly in the developing countries. It is becoming more and more obvious that fossil fuels will run out eventually in the next century despite the recent shale gas revolution. Thus, renewable and clean energy resources are urgently required in order to solve global energy and environmental issues [2, 3]. Among renewable energy resources, solar energy is by far the largest exploitable resource [1–3]. Nature harnesses solar energy for its production by photosynthesis, and fossil fuels are the product of photosynthesis [4]. Fossil fuels range from volatile materials with low carbon:hydrogen ratios such as methane, to liquids such as petroleum, and to nonvolatile materials composed of almost pure carbon, such as anthracite coal. The age of the organisms and their resulting fossil fuels is typically millions of years, and sometimes exceeds 650 million years. The consumption rate of fossil fuels is becoming much faster than the production rate by nature. Thus, it is quite important to develop artificial photosynthetic systems for production of solar fuels, which are hopefully simpler and more efficient than natural systems. The conversion of photon energy to chemical energy in photosynthesis is achieved by electron transfer from the excited state of an electron donor (D^*: * denotes the excited state) to an electron acceptor (A) to produce the charge-separated state ($D^{\cdot+}$–$A^{\cdot-}$). The high oxidizing power of $D^{\cdot+}$ in $D^{\cdot+}$–$A^{\cdot-}$ is used for four-electron oxidation of water to produce dioxygen, whereas the high reducing power of $A^{\cdot-}$ is used to reduce nicotinamide adenine dinucleotide phosphate ($NADP^+$) coenzyme to NADPH [5]. NADPH reduces CO_2 by multi-electron reduction to produce sugar [4, 5]. Thus, electron transfer plays essential roles in photosynthesis. In order to develop artificial photosynthesis systems, it is quite important to control electron-transfer systems to maximize the energy conversion.

Electron transfer is the most fundamental chemical reaction in which only electron is removed or attached. However, it is quite important to recognize the fundamental difference of electron-transfer reactions from other chemical reactions in which chemical bonds are cleaved and formed during the reactions. Because

Electron Transfer: Mechanisms and Applications, First Edition. Shunichi Fukuzumi.

electron transfer occurs based on the Franck–Condon principle, nuclear configurations remain the same before and after the electron transfer [6]. In contrast, nuclear configurations are changed significantly in chemical reactions associated with the cleavage and formation of chemical bonds. The fastest electron transfer is achieved with no activation energy when the nuclear configurations are the same before and after the electron transfer with a large driving force of electron transfer. When the driving force of electron transfer is further increased, the nuclear configurations are not the same any more before and after the electron transfer. In such a case, activation energy is again required to change the nuclear configurations before the electron transfer to be the same as that after the electron transfer. This means that the rate of electron transfer is slowed down with increase in the driving force of electron transfer. This region is called the Marcus inverted region, which is discussed in detail later [6]. The importance of the Marcus inverted region is well recognized in the function of the photosynthetic reaction center (vide infra) [7–9].

The photoinduced electron-transfer processes occur in a membrane-bound protein in the photosynthetic reaction center, which contains a number of cofactors, including four bacteriochlorophylls (BChl) as shown in Figure 1.1 [10]. The central part is referred to as the special pair [$(BChl)_2$], while the other two bacteriochlorophylls (BChl) are referred to as "accessory" bacteriochlorophylls. There are also two bacteriopheophytins (BPhe), two ubiquinones (Q_A and Q_B), and a non-heme iron atom (not shown in Figure 1.1), which together with the special pair are organized in pseudo-C2 symmetry forming two branches (A and B). Light-initiated charge separation occurs between the special pair [$(BChl)_2$] and the neighboring pigments, leading to a radical cation [$(BChl)_2^{\cdot+}$]. Despite the quasi-symmetrical arrangement of the cofactors, the electrons are transported unidirectionally along the A-branch of the reaction center [11–13], which suggests that the symmetry-breaking specific interactions with the protein are fundamentally important. Light-initiated charge separation occurs between the special pair [$(BChl)_2$] and the neighboring pigment (BPhe), leading to the radical cation [$(BChl)_2^{\cdot+}$] and the radical anion ($BPhe^{\cdot-}$). Without Q_A, back electron transfer from $BPhe^{\cdot-}$ to $(BChl)_2^{\cdot+}$ occurs with the lifetime of 10 ns to regenerate $(BChl)_2$ and $BPhe^{\cdot-}$. In such a case there is no way for the energy conversion to occur when the photon energy is only converted to heat. In the presence of Q_A, electron transfer from $BPhe^{\cdot-}$ to Q_A occurs at a much faster rate than the back electron transfer from $BPhe^{\cdot-}$ to $(BChl)_2^{\cdot+}$ despite the smaller driving force of the electron transfer than that of the back electron transfer. This has been made possible because of the Marcus inverted region. The subsequent electron transfer from $Q_A^{\cdot-}$ to Q_B also occurs at a much faster rate than the back electron transfer from $Q_A^{\cdot-}$ to $(BChl)_2^{\cdot+}$ because of the much longer distance between $Q_A^{\cdot-}$ and $(BChl)_2^{\cdot+}$ than that between $Q_A^{\cdot-}$ and Q_B. The final charge-separated state [$(BChl)_2^{\cdot+}$ and $Q_B^{\cdot-}$] is obtained with a nearly 100% quantum yield, having a lifetime of seconds, which is long enough for further chemical reactions [14, 15]. Thus, the Marcus inverted region plays a pivotal role in the charge-separation processes in the photosynthetic reaction center.

As the reverse process of photosynthesis, in which the four-electron oxidation of water is achieved by using solar energy, the highly exergonic four-electron

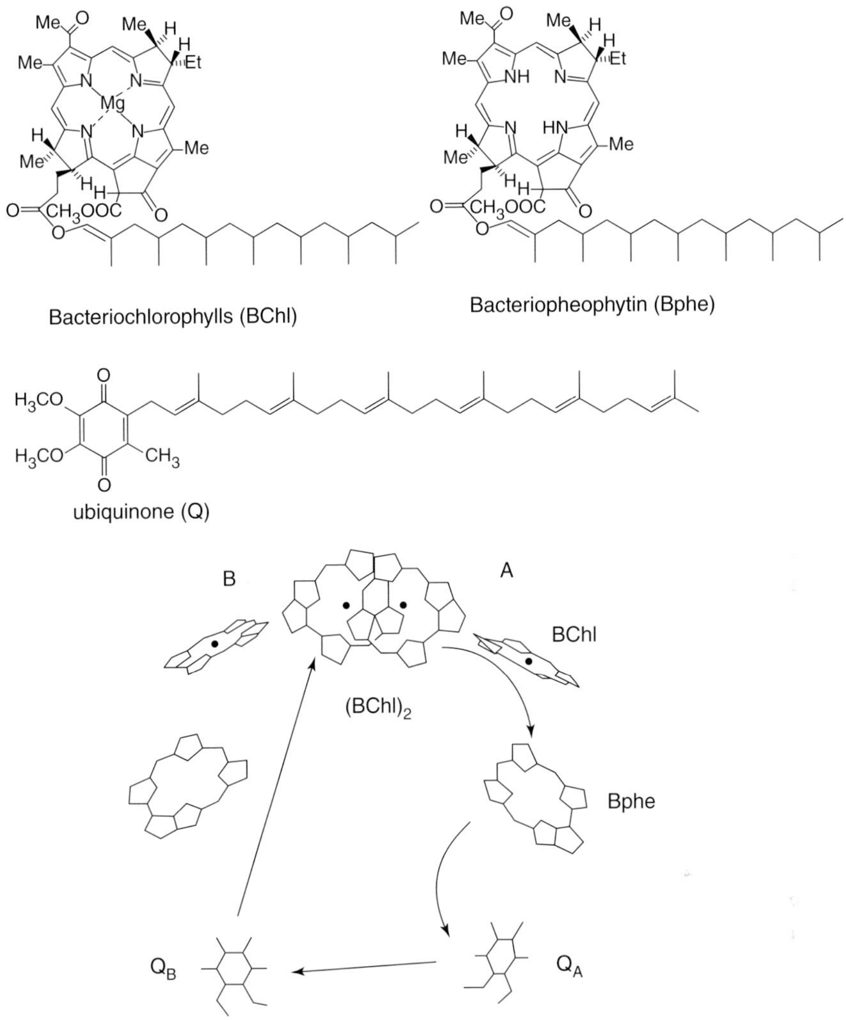

Figure 1.1 Cofactors and structure of photosynthetic reaction center of purple bacteria.

reduction of oxygen to water is essential to maintain the life of an aerobic organism by respiration [16–18]. Cytochrome *c* oxidases (C*c*Os) catalyze efficiently the reduction of molecular oxygen to water by the soluble electron carrier, cytochrome *c* [16–18]. In both the four-electron oxidation of water in photosynthesis and the four-electron reduction of O_2 in respiration, electron transfer is accompanied by proton transfer, which is referred to as proton-coupled electron transfer (PCET) [19–24]. Metal ions also play important roles in controlling electron transfer in metal ion-coupled electron transfer (MCET) [24–27].

This book describes mechanisms and applications of electron transfer inspired by electron-transfer processes in photosynthesis and respiration. The first rational design of a variety of donor–acceptor covalently linked ensembles including dyads, triads, tetrads, and pentads is described based on the Marcus theory of electron transfer, enabling to mimic the energy-transfer and electron-transfer processes in the photosynthetic reaction center. A specific challenge involves construction of simple donor–acceptor dyads, which afford longer lived and higher energy charge separated (CS) states than the natural system. The photosynthetic reaction center model compounds can be applied as effective redox catalysts in various catalytic chemical transformations. Then, the fundamental concepts of PCET and MCET are discussed and they are applied to developing efficient catalysts for multi-electron redox processes such as oxidation of water and reduction of dioxygen and carbon dioxide.

2

Marcus Theory of Electron Transfer

The dependence of rate constants of electron transfer on the driving force of electron transfer is well analyzed on the basis of the Franck–Condon principle (vide supra) by the Marcus theory of electron transfer, which provides basic principles to analyze the rate constant of electron transfer quantitatively [6]. According to the Marcus theory of electron transfer [6], the rate constant of nonadiabatic intramolecular electron transfer (k_{ET}) is given by Eq. (2.1):

$$k_{ET} = \left(\frac{4\pi^3}{h^2 \lambda k_B T}\right)^{1/2} V^2 \exp\left[-\frac{(\Delta G_{ET} + \lambda)^2}{4\lambda k_B T}\right] \tag{2.1}$$

where V is the electronic coupling matrix element, h is the Planck constant, T is the absolute temperature, ΔG_{ET} is the free energy change of electron transfer, and λ is the reorganization energy of electron transfer [6]. The ΔG_{ET} values are determined from the one-electron oxidation potentials (E_{ox}) of electron donors (D) and the one-electron reduction potentials (E_{red}) of electron acceptors by Eq. (2.2):

$$\Delta G_{ET} = e(E_{ox} - E_{red}) \tag{2.2}$$

where e is elementary charge. According to Eq. (2.1), the logarithm of the electron-transfer rate constant ($\log k_{ET}$) is related parabolically to the driving force of electron transfer from electron donors to acceptors ($-\Delta G_{ET}$) and the reorganization energy (λ) of electron transfer, that is, the energy required to structurally reorganize the donor, acceptor, and their solvation spheres upon electron transfer [6].

When the magnitude of the driving force of electron transfer becomes the same as the reorganization energy ($-\Delta G_{ET} = \lambda$), the electron-transfer rate reaches a maximum and is basically controlled by the magnitude of electronic coupling (V) between the donor and acceptor moieties (Eq. (2.1)). Upon passing this thermodynamic maximum, the highly exothermic region of the parabola ($-\Delta G_{ET} > \lambda$) is entered, in which an additional increase of the driving force results in an actual slowdown of the electron-transfer rate, due to an increasingly poor vibrational overlap of the product and reactant wave functions. This highly exergonic range is generally referred to as the Marcus inverted region [6–8, 28, 29]. In such a case, the magnitude of the reorganization energy is the key parameter to control the electron-transfer process. The smaller the reorganization energy, the faster is the forward photoinduced charge-separation

Electron Transfer: Mechanisms and Applications, First Edition. Shunichi Fukuzumi.

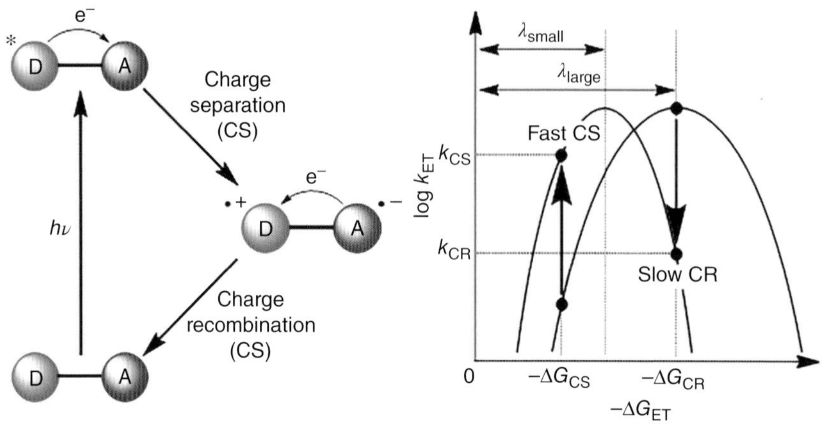

Figure 2.1 (a) Schematic diagram of photoinduced electron transfer of an electron donor–acceptor (D–A) dyad. (b) Driving force dependence of log k_{ET} with different λ values (for the meaning of λ, see text).

(CS) process, but the charge-recombination (CR) process becomes slower when the driving force for back electron transfer ($-\Delta G_{ET}$) is larger than the reorganization energy (λ) of electron transfer as shown in Figure 2.1. Among the key parameters that govern electron-transfer reactions (V, ΔG_{ET}, and λ), the reorganization energy (λ) imposes the most important impact. The primary electron-transfer processes of photosynthesis are, for example, characterized by an extremely small reorganization energy ($\lambda \sim 0.2$ eV), attained by the protein environment. This aspect is essential to achieving ultrafast charge separation and retarding the energy-wasting charge recombination, which is highly exergonic ($-\Delta G = 1.2$ eV) [11–15]. Thus, the components of artificial photosynthesis, which are not necessarily natural components in photosynthesis, must have small λ values of electron transfer.

3

Photosynthetic Reaction Center Models

The first candidates of artificial photosynthetic reaction centers are porphyrins, which have been involved in a number of important biological electron-transfer systems including the primary photochemical reactions of chlorophylls in the photosynthetic reaction centers. Porphyrins contain an extensively conjugated two-dimensional π-system and are thereby suitable for efficient electron transfer, because the uptake or release of electrons results in minimal structural and solvation change upon electron transfer, resulting in a small λ value of electron transfer [30]. In addition, rich and extensive absorption features of porphyrinoid systems guarantee increased absorption cross sections and efficient use of the solar spectrum [30]. In contrast with the two-dimensional porphyrin π-system, fullerenes contain an extensively conjugated three-dimensional π-system [31]. Electron transfer to C_{60} is highly efficient because of the minimal changes of structure and solvation associated with the electron-transfer reduction [31–33]. Thus, a series of porphyrin–fullerene linked molecules have been designed and synthesized to mimic the charge-separation processes in the photosynthetic reaction center, starting from a zinc porphyrin–C_{60} dyad (ZnP–C_{60}), a ferrocene–zinc porphpyrin–C_{60} triad (Fc–ZnP–C_{60}), a zinc porphyrin-free base porphyrin–C_{60} triad, up to a tetrad (Fc–ZnP–H_2P–C_{60}) [34–39].

The driving force dependence of the electron-transfer rate constants (k_{ET}) of these dyad, triad, and tetrad molecules for CS and CR processes is shown in Figure 3.1, where log k_{ET} is plotted against the driving force ($-\Delta G_{ET}$) [39]. The lines in Figure 3.1 represent the best fit to Eq. (2.1) (ZnP–C_{60}: $\lambda = 0.66$ eV, $V = 3.9\ \text{cm}^{-1}$; Fc–ZnP–C_{60}, Fc–H_2P–C_{60}, and ZnP–H_2P–C_{60}: $\lambda = 1.09$ eV, $V = 0.019\ \text{cm}^{-1}$; Fc–ZnP–$H_2$P–$C_{60}$: $\lambda = 1.32$ eV, $V = 0.00017\ \text{cm}^{-1}$) [39]. The λ-value increases, whereas the V value decreases with increasing edge-to-edge distance in the order of the dyad ($R_{ee} = 11.9$ Å), the triad ($R_{ee} = 30.3$ Å), and the tetrad ($R_{ee} = 48.9$ Å). Such an increase in the λ value with increasing distance results from an increase in the solvent reorganization energy (λ_s), because the λ_s value is known to increase with increasing distance between an electron donor and an acceptor as given by Eq. (3.1):

$$\lambda_s = e^2\,[(2r_1)^{-1} + (2r_2)^{-1} - (r_{12})^{-1}][n^{-2} - (e_s)^{-1}] \tag{3.1}$$

where r_1 and r_2 are the radii of the donor and acceptor, r_{12} is the donor–acceptor distance, ε is the dielectric constant, and n is the refractive index [6, 40].

Electron Transfer: Mechanisms and Applications, First Edition. Shunichi Fukuzumi.

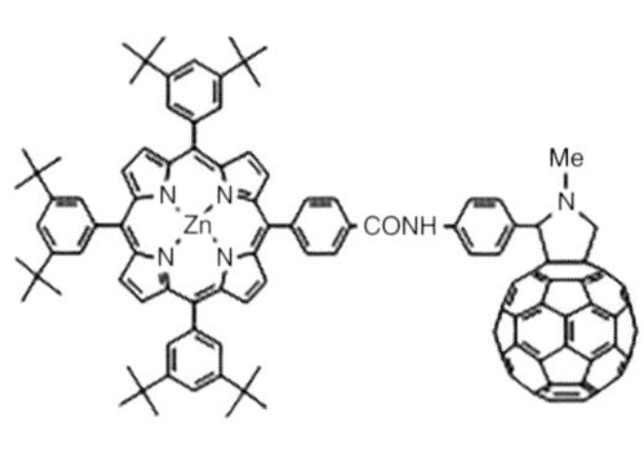

Fc–C_{60}

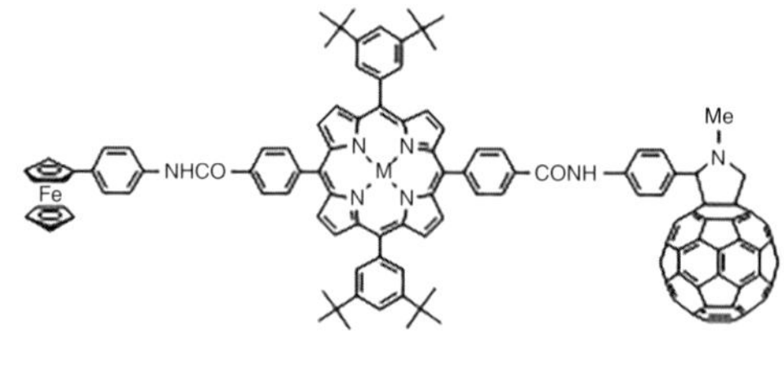

Fc–ZnP–C_{60}(M=Zn) Fc–H_2P–C_{60}(M=H_2)

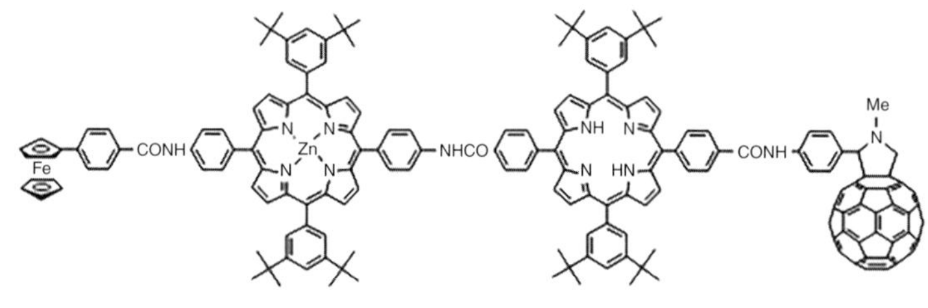

Fc–ZnP–H_2P–C_{60}

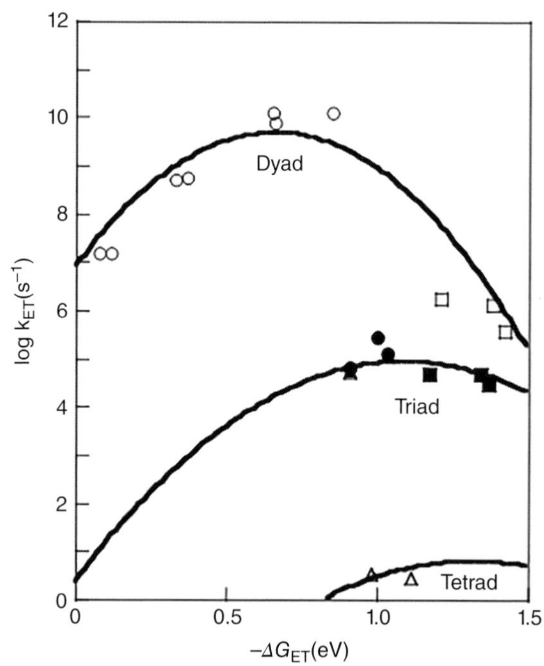

Figure 3.1 Driving force ($-\Delta G_{ET}$) dependence of intramolecular electron-transfer rate constants (k_{ET}) in ZnP–C_{60} (CS [○]; CR [□]), Fc–ZnP–C_{60} (●), Fc–H_2P–C_{60} (▲), ZnP–H_2P–C_{60} (■), and Fc–ZnP–H_2P–C_{60} (△). Source: Imahori et al. 2001 [39]. Reproduced with permission of American Chemical Society.

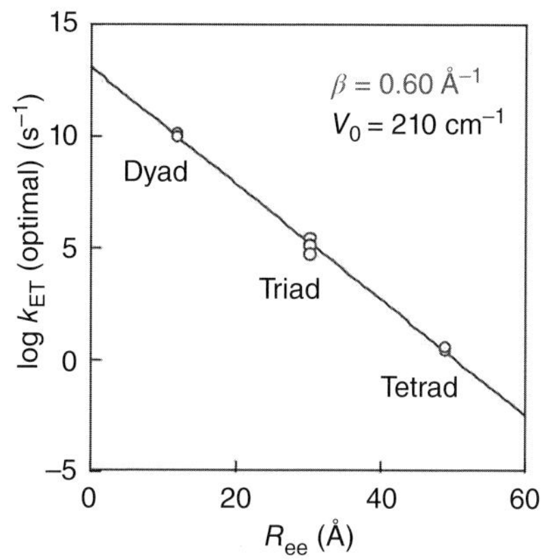

Figure 3.2 Edge-to-edge distance (R_{ee}) dependence of optimal electron-transfer rate constants ($\log[k_{ET}(\text{optimal})/s^{-1}]$) in ZnP–$C_{60}$ dyad, Fc–ZnP–C_{60} triad, Fc–H_2P–C_{60} triad, ZnP–H_2P–C_{60} triad, and Fc–ZnP–H_2P–C_{60} tetrad. The line represents the best fit to Eq. (2.1) ($\beta = 0.60$ Å^{-1}, $V_0 = 210\,cm^{-1}$).

In the case of the dyad, the CR rate in $ZnP^{\cdot+}$–$C_{60}^{\cdot-}$ in the Marcus inverted region is much slower than the CS rate constants from both the singlet and triplet excited states in the Marcus normal region (Figure 3.1). This enables subsequent electron transfer from Fc to $ZnP^{\cdot+}$ in the triad (Fc–$ZnP^{\cdot+}$–$C_{60}^{\cdot-}$) and from ZnP to $H_2P^{\cdot+}$ in ZnP–$H_2P^{\cdot+}$–$C_{60}^{\cdot-}$ to produce the final CS state, Fc^+–ZnP–$C_{60}^{\cdot-}$ and $ZnP^{\cdot+}$–H_2P–$C_{60}^{\cdot-}$, in competition with the back electron transfer in the initial CS states. In the case of Fc^+–ZnP–H_2P–$C_{60}^{\cdot-}$, the charges are separated at a long distance (R_{ee} = 48.9 Å) [39]. The lifetime of the resulting CS state at such a long distance in a frozen benzonitrile (PhCN) has been determined as 0.38 seconds [39]. Similar tetrad and pentad compounds were reported to attain long-lived CS states up to 1.7 seconds [40–42]. It should be noted that the CS lifetime is temperature independent, since the CR process is at the Marcus top region (Figure 3.1) [39]. This is the first example to achieve a CS lifetime that is comparable to that observed for the bacterial photosynthetic reaction center. However, such an extremely long CS lifetime could only be determined in frozen media, since in condensed media bimolecular back electron transfer between two Fc^+–ZnP–H_2P–$C_{60}^{\cdot-}$ is much faster than the unimolecular CR process [39]. The maximum k_{ET} values (k_{ETmax}) of dyads, triads, and tetrads in the Marcus plot (Figure 3.2) are correlated with the edge-to-edge distance (R_{ee}), separating the radical ions, according to Eq. (3.2) [39]:

$$\ln\, k_{ETmax} = \ln\left(\frac{2\pi^{3/2}V_0^{\,2}}{h(\lambda k_B)^{1/2}}\right) - \beta R_{ee} \tag{3.2}$$

where V_0 refers to the maximal electronic coupling element, while β is the decay coefficient factor (damping factor), which depends primarily on the nature of the bridging molecule. From the linear plot of $\ln k_{ETmax}$ vs. R_{ee} the β value is obtained as 0.60 Å^{-1} [39]. This β value is located within the boundaries of nonadiabatic electron transfer reactions for saturated hydrocarbon bridges (0.8–1.0 Å^{-1}) and unsaturated phenylene bridges (0.4 Å^{-1}) [43].

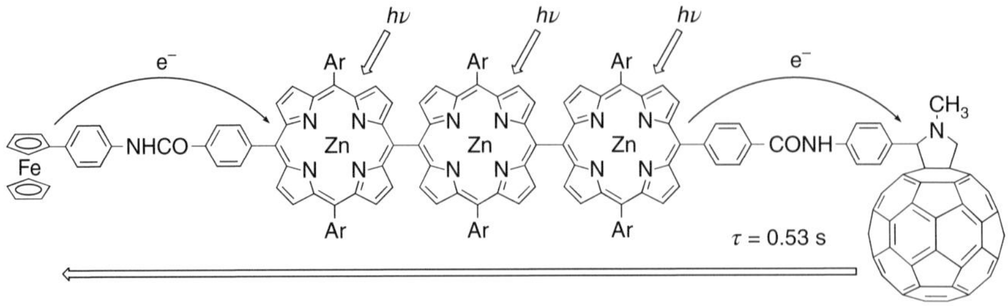

Scheme 3.1 Multistep photoinduced electron transfer in a ferrocene-*meso*, *meso*-linked porphyrin trimer–fullerene pentad (Fc–$(ZnP)_3$–C_{60}); Ar = 3,5-$Bu^t{}_2C_6H_3$. Source: Imahori et al. 2004 [44]. Reproduced with permission of John Wiley & Sons.

The best molecule mimicking multistep electron-transfer processes in the photosynthetic reaction center so far reported is a ferrocene-*meso*, *meso*-linked porphyrin trimer–fullerene pentad (Fc–$(ZnP)_3$–C_{60} in Scheme 3.1), where the C_{60} and the ferrocene (Fc) are tethered at both the ends of $(ZnP)_3$ (R_{ee} = 46.9 Å) [44]. The lifetime of the final CS state (0.53 seconds at 163 K) has been attained without lowering the CS efficiency (Φ = 0.83) [44].

4

Electron Donor–Acceptor Dyads

Multistep electron-transfer processes have been utilized to attain a long distance charge separation (CS), mimicking the natural photosynthetic reaction center (vide supra). However, a significant amount of energy is lost during the multistep electron-transfer processes to reach the final CS state [11–15]. In photosynthesis a two–step photoexcitation, the so-called "Z-scheme," is thereby required to recover the energy loss via the multistep electron-transfer processes and to gain strong oxidizing power to oxidize water as well as high reducing power to reduce NAD^+ coenzyme [4]. The design and synthesis of molecular machinery mimicking such an elaborated "Z-scheme" in nature seems far beyond our capability, and even if it could be done, the synthetic cost would certainly preclude any type of practical application. Thus, it is highly desired to design simple molecular electron donor–acceptor dyads that are capable of fast CS but can retain slow charge recombination (CR). Theoretically, it is possible to obtain such an electron donor–acceptor dyad, because the CS lifetime increases with increasing driving force of electron transfer in the Marcus inverted region (vide supra). However, the driving force of electron transfer should be lower than the triplet excited state of one of the components of donor–acceptor dyads. Otherwise, the CS state would decay rapidly to the triplet excited state in the Marcus normal region rather than to the ground state in the Marcus inverted region [35].

A number of simple donor–acceptor dyads have been designed and synthesized to attain long-lived CS state, where the donor and acceptor molecules are linked with a short spacer to minimize the solvent reorganization energy [45–50]. Efficient photoinduced electron transfer occurs in a zinc imidazoporphyrin–C_{60} dyad (ZnImP–C_{60}) with a short linkage to form the CS state ($ZnImP^{\cdot+}$–$C_{60}{}^{\cdot-}$) with the rate constant of 1.4×10^{10} s^{-1} (Scheme 4.1) [45]. The CS state (1.34 eV) is lower in energy than both the triplet excited states of C_{60} (1.50 eV) and ZnImP (1.36 eV) [45]. The CS state, produced upon photoexcitation of ZnImP–C_{60}, is detected by the transient absorption spectrum, which has absorption bands at 700 and 1040 nm due to $ZnImP^{\cdot+}$ and $C_{60}{}^{\cdot-}$, respectively [45]. The CS state decays by back electron transfer to the ground state, obeying first-order kinetics with a rate constant of 3.9×10^3 s^{-1} (the lifetime is 260 μs) at 298 K [45]. At 278 K the lifetime of the CS state was determined as 310 μs, which is much longer than those of conventional donor–acceptor dyads with longer spacers [7–9].

Electron Transfer: Mechanisms and Applications, First Edition. Shunichi Fukuzumi.

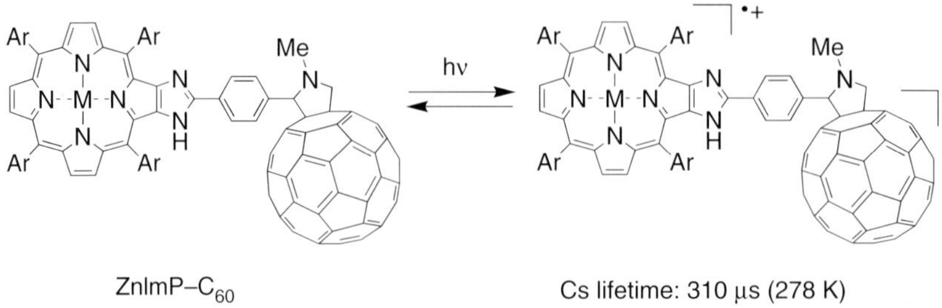

Scheme 4.1 Formation of a long-lived CS state of a zinc imidazoporphyrin–C_{60} dyad (ZnImP–C_{60}) with a short linkage (Ar = 3,5-$Bu^t{}_2C_6H_3$). Source: Kashiwagi et al. 2003 [45]. Reproduced with permission of American Chemical Society.

An electron donor–acceptor dyad linked with a short spacer containing Au(III) and Zn(II) porphyrins ($ZnPQ–AuPQ^+$ in Scheme 4.2) also affords a long-lived electron-transfer state with a lifetime of 10 μs in nonpolar solvents such as cyclohexane [46]. The introduction of quinoxaline to the gold porphyrin results in a lowering of the electron-transfer state energy. In contrast to the case of neutral donor–acceptor dyads, the energy of the electron-transfer state ($ZnPQ^{\cdot+}$–AuPQ) becomes smaller in a less polar solvent, which is lower than the energies of the triplet excited states of ZnPQ (1.32 eV) and $AuPQ^+$ (1.64 eV) [46]. Photoinduced electron transfer occurs from the singlet excited state of the ZnPQ ($^1ZnPQ^*$) to the metal center of the $AuPQ^+$ moiety to produce $ZnPQ^{\cdot+}–Au^{II}PQ$. The observed long lifetime of $ZnPQ^{\cdot+}–Au^{II}PQ$ results from a small reorganization energy for the metal-centered electron transfer of $AuPQ^+$ in nonpolar solvents due to the small change in solvation upon electron transfer as compared with that in polar solvents [46]. In a polar solvent such as benzonitrile (PhCN), no CS state was observed, but instead only the triplet–triplet absorption due to $^3ZnPQ^*–AuPQ^+$ was observed [46]. The absence of an observable CS state in PhCN is ascribed to the much slower photoinduced electron transfer due to the large reorganization

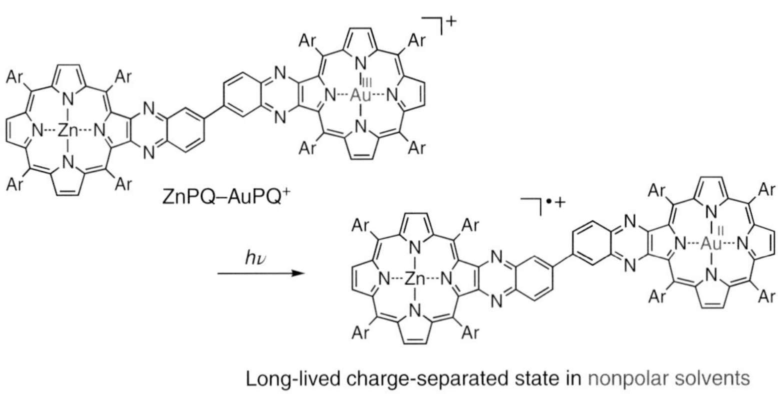

Scheme 4.2 Formation of a long-lived CS state of $ZnPQ–AuPQ^+$ in nonpolar solvents (Ar = 3,5-$Bu^t{}_2C_6H_3$). Source: Fukuzumi et al. 2003 [46]. Reproduced with permission of American Chemical Society.

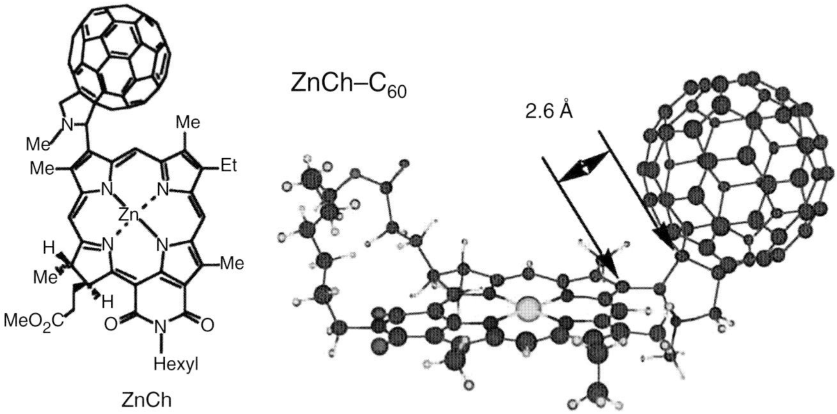

Figure 4.1 Structure of a closely linked ZnCh–C_{60} dyad. Source: Ohkubo et al. 2004 [47]. Reproduced with permission of John Wiley & Sons.

energy as compared with that in nonpolar solvents allowing an efficient intersystem crossing process in the ZnPQ–AuPQ$^+$ dyad to produce the triplet excited state ^{3}ZnPQ*–AuPQ$^+$ [46].

A closely linked zinc chlorin–fullerene dyad (ZnCh–C_{60} in Figure 4.1) affords a longer CS lifetime as compared with other zinc chlorin–fullerene dyads with longer spacers [47–51]. A deoxygenated PhCN solution containing ZnCh–C_{60} gives rise upon a 388 nm laser pulse to a transient absorption maximum at 460 nm due to the singlet excited state of ZnCh [47]. The decay rate constant was determined as 1.0×10^{11} s^{-1}, which agrees with the value determined from the fluorescence lifetime measurements [47]. The decay of absorbance at 460 nm due to ^{1}ZnCh* is accompanied by an increase in absorbance at 590 nm due to ZnCh$^{\cdot+}$ [47]. This indicates that electron transfer from ^{1}ZnCh* to C_{60} occurs rapidly to form the CS state, ZnCh$^{\cdot+}$–$C_{60}{}^{\cdot-}$. The CS state decays via back electron transfer to the ground state rather than to the triplet excited state, because the CS state is lower in energy (1.26 eV) than the triplet excited states of both C_{60} (1.50 eV) and ZnCh (1.36–1.45 eV) [47]. The lifetime of the CS state is determined as 230 μs at 298 K. The large temperature dependence of the CS lifetime is observed and the lifetime of the CS state at 123 K becomes as long as 120 seconds [47].

Covalently and non-covalently linked porphyrin–quinone dyads constitute one of the most extensively investigated photosynthetic reaction center models, in which the fast photoinduced electron transfer from the porphyrin singlet excited state to the quinone occurs to produce the CS state, mimicking well the photosynthetic electron transfer [52–54]. Unfortunately, the CR rates of the CS state of porphyrin–quinone dyads are also fast and the CS lifetimes are mostly on the order of picoseconds or subnanoseconds in solution [52–54]. In general, a three-dimensional C_{60} is superior to a two-dimensional quinone in terms of the smaller reorganization of electron transfer of C_{60} as compared with quinone (vide supra) to attain the long-lived CS state [31–33, 55]. When the geometry between a porphyrin ring and quinone is optimized by using hydrogen bonds, which can also control the redox potentials of quinones, however, a surprisingly long lifetime up to one microsecond has been attained [56]. In a series of ZnP–n–Q ($n = 3$, 6, 10) in Scheme 4.3, the hydrogen bond between two amide groups provides a structural scaffold to assemble the donor (ZnP) and the acceptor (Q) moiety, leading to attaining the long-lived CS state [56].

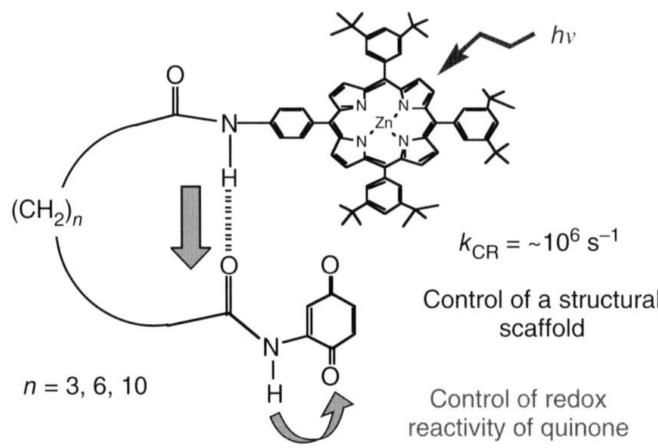

Scheme 4.3 Zinc porphyrin–quinone linked dyads (ZnP–n–Q; $n = 3$, 6, 10) with hydrogen bonds. Source: Okamoto and Fukuzumi 2005 [56]. Reproduced with permission of American Chemical Society.

As described above, the closely linked donor–acceptor dyads afford long-lived CS states. As long as porphyrins and C_{60} are used as components of donor–acceptor dyads, however, the low lying triplet energies of porphyrins and C_{60} have precluded to attain the long-lived CS states with a higher energy than the triplet energies [35]. In such a case, it is highly desired to find a chromophore that has a high triplet energy and a small λ value of electron transfer. Among many choromophores, acridinium ion is the best candidate for such a purpose, since the λ value for the electron self-exchange between the acridinium ion and the corresponding one-electron reduced radical (acridinyl radical) is the smallest (0.3 eV) among the redox-active organic compounds [57]. Another important property of acridinium ion is a high triplet excited energy [58, 59]. Thus, an electron donor moiety (mesityl group) is directly connected at the 9-position of the acridinium ion to yield 9-mesityl-10-methylacridinium ion (Acr^+–Mes) [60], in which the solvent reorganization of electron transfer is minimized because of the short linkage between the donor and acceptor moieties. The X-ray crystal structure of Acr^+–Mes is shown in Figure 4.2a [60]. The dihedral angle made by aromatic ring planes is perpendicular and therefore there is no π conjugation between the donor and acceptor moieties. Indeed, the absorption and fluorescence spectra of Acr^+–Mes are superpositions of the spectra of each component, i.e. mesitylene and 10-methylacridinium ion. The highest occupied molecular orbital (HOMO) and lowest unoccupied molecular orbital (LUMO) orbitals of Acr^+–Mes calculated by a density functional theory (DFT) method with Gaussian 98 (B3LYP/6-31G* basis set) are localized on mesitylene and acridinium moieties (Figure 4.2b,c), respectively [60]. The energy of the electron-transfer state ($Acr^{\cdot}$–$Mes^{\cdot+}$) in PhCN is determined by the redox potentials of each component of Acr^+–Mes as 2.37 eV [60].

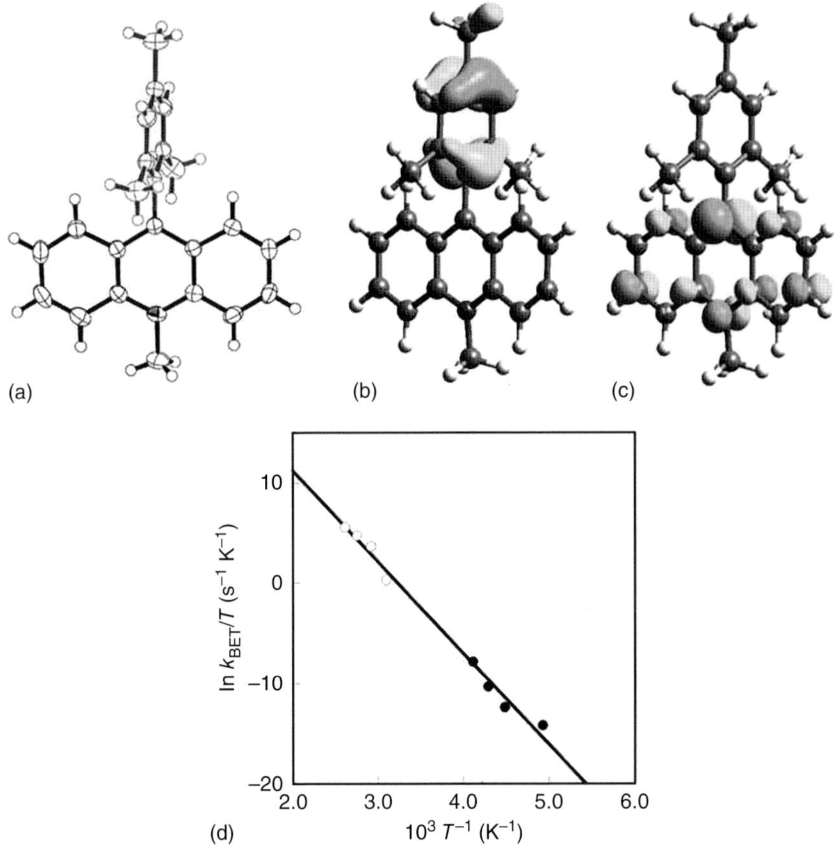

Figure 4.2 (a) X-ray crystal structure of Acr^+–Mes. (b) HOMO and (c) LUMO orbitals calculated by DFT method with Gaussian 98 (B3LYP/6-31G* basis set). (d) Plot of k_{BET}/T vs. T^{-1}. Source: Fukuzumi et al. 2004 [60]. Reproduced with permission of American Chemical Society.

Photoirradiation of a deaerated PhCN solution of Acr^+–Mes by a nanosecond laser light at 430 nm results in the formation of $Acr^{\cdot}$–$Mes^{\cdot+}$ with a quantum yield close to unity (98%) via photoinduced electron transfer from the mesitylene moiety to the singlet excited state of the acridinium ion moiety ($^1Acr^{+*}$–Mes) [60]. The intramolecular back electron transfer from the $Acr^{\cdot}$ moiety to the $Mes^{\cdot+}$ moiety in $Acr^{\cdot}$–$Mes^{\cdot+}$ was too slow to compete with the intermolecular transfer (k_{BET}) in Figure 4.2d, agreeing with the Marcus equation in the deeply inverted region (Eq. (2.1)). The lifetime of the electron-transfer state in frozen medium becomes longer with decreasing temperature to approach a virtually infinite value at 77 K [60]. However, the decay time profile of $Acr^{\cdot}$–$Mes^{\cdot+}$ in solution obeyed second-order kinetics (NOT first-order kinetics) [60]. This is the same as the case of Fc^+–ZnP–H_2P–$C_{60}^{\cdot-}$, in which bimolecular back electron transfer predominates due to the slow intramolecular back electron transfer (vide supra) [39]. In contrast, the decay of $Acr^{\cdot}$–$Mes^{\cdot+}$ obeys first-order kinetics in PhCN at high temperatures [60]. This indicates that the rate of the intramolecular back electron transfer of $Acr^{\cdot}$–$Mes^{\cdot+}$ becomes much faster than the rate of the intermolecular back electron transfer at higher temperatures because of the larger activation energy of the former than that of the latter. Such a remarkable result has sparked a flurry of work by others in the field of artificial photosynthesis [61].

Benniston et al. claimed that the triplet excited state of the acridinium ion moiety ($^3Acr^{+*}$–Mes) might be formed rather than the electron-transfer state ($Acr^{\cdot}$–$Mes^{\cdot+}$) and that the energy of $^3Acr^{+*}$–Mes is lower than that of $Acr^{\cdot}$–$Mes^{\cdot+}$ [62]. They reported that the triplet excitation energy of Acr^+–Mes was 1.96 eV based on the phosphorescence spectrum [62]. If this value were correct, the one-electron oxidation potential (E_{ox}) of $^3Acr^{+*}$–Mes would be −0.08 V vs. saturated calomel electrode (SCE), which is determined from the one-electron oxidation potential of the Mes moiety (1.88 V) [60] and the triplet excitation energy (1.96 V). In such a case, electron transfer from the triplet excited state of Acr^+–Mes to *N*,*N*-dihexylnaphthalenediimide (NIm: E_{red} = −0.46 V vs. SCE) would be energetically impossible judging from the positive free energy change of electron transfer (0.38 eV). However, the addition of NIm (1.0×10^{-3} M) to a PhCN solution of Acr^+–Mes and laser photoexcitation results in the formation of $NIm^{\cdot-}$ as detected clearly by the well-known absorption bands at 480 and 720 nm [63, 64], accompanied by the decay of transient absorption at 510 nm due to the $Acr^{\cdot}$ moiety of the electron-transfer (ET) state as shown in Figure 4.3a [65]. Similarly, the addition of aniline (3.0×10^{-5} M) to a PhCN solution of Acr^+–Mes results in the formation of aniline radical cation (λ_{max} = 430 nm) [66], accompanied by decay of the $Mes^{\cdot+}$ moiety at 500 nm as shown in Figure 4.3b [65]. The rate constant of formation of aniline radical cation was determined to be $5.6 \times 10^9\ M^{-1}\ s^{-1}$, which is close to the diffusion rate constant in PhCN [60]. Thus, the photogenerated state of Acr^+–Mes has both the reducing and oxidizing abilities to reduce NIm and to oxidize aniline, respectively. Only the electron-transfer state ($Acr^{\cdot}$–$Mes^{\cdot+}$) has such a dual ability, which has now been well confirmed for electron-transfer oxidation of many electron donors with $Acr^{\cdot}$–$Mes^{\cdot+}$ and electron-transfer reduction of many electron acceptors such as hexyl viologen, *p*-benzoquinone, and Selectfluor (fluorinating reagent) with $Acr^{\cdot}$–$Mes^{\cdot+}$ [67–70]. However, this conclusion is contradictory to the reported triplet energy (1.96 eV), which is lower in energy than the ET state [62]. This contradiction comes from an acridine impurity, which may be left in the preparation of by Benniston et al. who synthesized the compound via methylation of the corresponding acridine [62]. The yield of acridinium ion is about 50–70% after reflux at high temperature for a few days [62]. In such a case, acridine may remain as an impurity even after purification of the acridinium ion by recrystallization. When Acr^+–Mes was prepared by the Grignard reaction of 10-methyl-9(10*H*)acridone with 2-mesitylmagnesium bromide, there was no acridine [60]. Thus, Acr^+–Mes without acridine afforded no phosphorescence spectrum in both deaerated glassy 2-MeTHF and ethanol at 77 K. It is well known that acridine derivatives exhibit phosphorescence at 15650–15850 cm^{-1} [71]. It was confirmed that the phosphorescence maximum of 9-phenylacridine in glassy 2-MeTHF at 77 K afforded the same spectrum reported by Benniston et al. [62] Thus, the reported low triplet energy of Acr^+–Mes, which contradicts our results on the long-lived electron-transfer state, results from the acridine

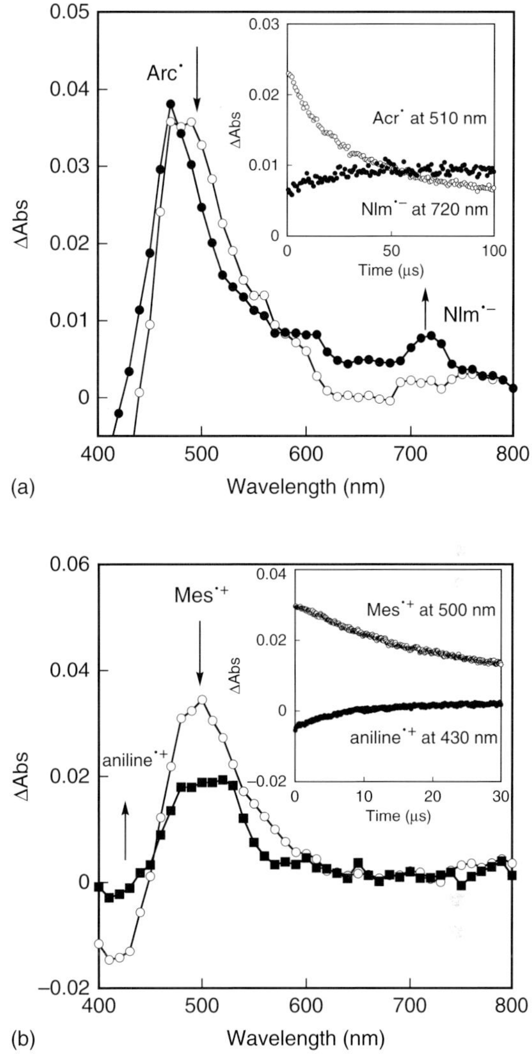

Figure 4.3 Transient absorption spectra of Acr^+–Mes (5.0×10^{-5} M) in deaerated MeCN at 298 K taken at 2 and 20 μs after laser excitation at 430 nm in the presence of (a) *N*,*N*-dihexylnaphthalenediimide (1.0×10^{-3} M) or (b) aniline (3.0×10^{-5} M). Inset: Time profiles of the absorbance decay at 510 nm and the rise at 720 nm and (b) the decay at 500 nm and the rise at 430 nm. Source: Fukuzumi and coworkers 2005 [65]. Reproduced with permission of Royal Society of Chemistry.

impurity contained in Acr^+–Mes used by Benniston et al. who also reported that photoirradiation of a PhCN solution of Acr^+–Mes results in the formation of the acridinyl radical ($Acr^{\cdot}$–Mes) [62]. They implied that this stable radical species could be mistaken as a long-lived electron-transfer state [62]. When PhCN is purified, however, no change in the absorption spectrum is observed [60, 65]. The formation of $Acr^{\cdot}$–Mes results from electron transfer from a donor impurity contained in unpurified PhCN (e.g. aniline) to the $Mes^{\cdot+}$ moiety of $Acr^{\cdot}$–$Mes^{\cdot+}$ as indicated in Figure 4.3b. Even an extremely small amount (5.0×10^{-5} M) of aniline is enough to react with $Acr^{\cdot}$–$Mes^{\cdot+}$ to produce $Acr^{\cdot}$–Mes, which is stable due to the bulky Mes substituent, because the lifetime of $Acr^{\cdot}$–$Mes^{\cdot+}$ is long enough to react with such a small concentration of an electron donor. It should be noted that no net photochemical reaction occurs without a donor impurity because the long-lived Acr$^{\bullet}$–Mes$^{\bullet+}$ decays via bimolecular back electron transfer to the ground state [60, 65]. Thus, misleading effects of impurities indeed result from the long-lived electron-transfer state, which has both oxidizing and reducing abilities.

In contrast to the photoirradiation of a purified PhCN solution of Acr$^+$–Mes at 298 K, which results in no change in the absorption spectrum (Figure 4.4a), when the photoirradiation of the same solution was performed at low temperatures (213–243 K) with a 1000 W high-pressure mercury lamp through the UV light cutting filter (>390 nm) and the sample was cooled to 77 K, the color of the frozen sample at 77 K was clearly changed as shown in the inset of Figure 4.4b. When a glassy 2-methyltetrahydrofuran (2-MeTHF) is employed for the photoirradiation of Acr$^+$–Mes at low temperature, the resulting glassy solution measured at 77 K affords the absorption spectrum due to the electron-transfer state, which consists of the absorption bands of the Acr$^{\bullet}$ moiety (500 nm) and the Mes$^{\bullet+}$ moiety (470 nm) as shown in Figure 4.4b. No decay of the absorption due to the electron-transfer state in Figure 4.2b was observed until liquid nitrogen ran out [65].

The long lifetime of the ET state of Acr$^+$–Mes has allowed observing the structural change in the Acr$^+$–Mes(ClO_4^-) crystal upon photoinduced ET directly by using laser pump and X-ray probe crystallographic analysis (Figure 4.5) [72]. Upon photoexcitation of the crystal of Acr$^+$–Mes(ClO_4^-), the *N*-methyl group of the Acr$^+$ moiety was bent and its bending angle was 10.3(16)° when the *N*-methyl carbon moved 0.27(4) Å away from the mean plane of the ring as shown in Figure 4.5 [72]. This bending is caused by the photoinduced electron transfer from the Mes moiety to the Acr$^+$ moiety to produce Acr$^{\bullet}$–Mes$^{\bullet+}$, because the sp^2 carbon of the *N*-methyl group of Acr$^+$ is changed to the sp^3 carbon in the one-electron reduced state (Acr$^{\bullet}$) [72]. The bending of the *N*-methyl group

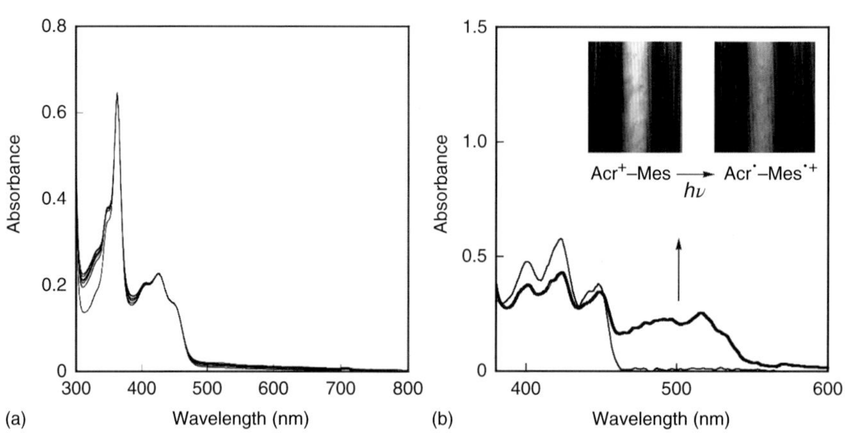

Figure 4.4 (a) UV–vis spectral change in the steady-state photolysis of a deaerated PhCN solution of Acr$^+$–Mes (3.3×10^{-5} M). Spectra were recorded at 90-second interval. (b) UV–vis absorption spectra obtained by photoirradiation with high-pressure mercury lamp of deaerated 2-MeTHF glasses of Acr$^+$–Mes at 77 K. Inset: picture images of frozen PhCN solutions of Acr$^+$–Mes before and after photoirradiation at low temperatures and taken at 77 K.

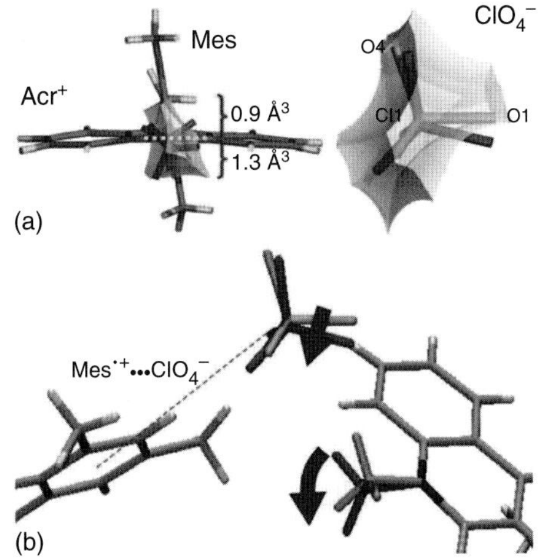

Figure 4.5 (a) Diagram of the reaction cavity: (left) diagram around the *N*-methyl group, with numbers indicating the volumes of the divided cavity formed by the dotted line; (right) drawing around ClO_4^-. (b) Cooperative photoinduced geometrical changes. The dashed line indicates the suggested $Mes^{\cdot+}\cdots ClO_4^-$ electrostatic interaction. Source: Hoshino et al. 2012 [72]. Reproduced with permission of American Chemical Society.

by photoexcitation was accompanied by the rotation and movement of the ClO_4^- by the electrostatic interaction with the $Mes^{\cdot+}$ moiety (Figure 4.5) [72]. Thus, the observed bending of the *N*-methyl group and the movement of ClO_4^- provide strong evidence for the generation of the ET state of Acr^+–Mes upon photoexcitation. In contrast to the case of Acr^+–Mes, no geometrical difference was observed upon photoexcitation of Acr^+–Ph, which does not afford the ET state [72].

Immobilization of Acr^+–Mes has also been achieved by incorporating Acr^+–Mes cation into nanosized mesoporous silica–alumina (AlMCM-41), which has cation exchange sites to obtain a nanocomposite (Acr^+–Mes@AlMCM-41) [73]. The shape and size of nanosized AlMCM-41 were controlled by changing the preparation conditions as shown in Figure 4.6, where TEM images reveal a tubular or rod-like (tAlMCM-41) morphology in the diameter of 50–100 nm with the length of 0.2–2 μm array (part a) and also a sphere morphology (sAlMCM-41, part b) [73]. The X-ray powder pattern of tAlMCM-41 exhibited a well-resolved pattern with a prominent peak (100) observed at c. $2\theta = 2.56°$, indicating a highly ordered material with a hexagonal array [73]. Uniform channels c. 4 nm in diameter exist in a tube. Because the Acr^+–Mes molecular size is small enough as compared with the pore size of mesoporous silica with its diameter of more than 3 nm, cation exchange with Acr^+–Mes occurs spontaneously upon mixing Na^+–exchanged AlMCM-41 with Acr^+–Mes in acetonitrile [73]. The cation exchange percentages of tAlMCM-41 and sAlMCM-41 by Acr^+–Mes were determined to be 16% and 18%, respectively [73]. The Acr^+–Mes incorporated into AlMCM-41 is stable without leaching out in acetonitrile at room temperature [73].

Upon photoexcitation of Acr^+–Mes@tAlMCM-41 suspended in MeCN, photoinduced ET from the Mes moiety to the singlet excited state of the Acr^+ moiety occurred within 10 ps to produce the ET state as detected by laser flash photolysis and electron paramagnetic resonance (EPR) measurements [60, 67]. In contrast to the case in solution (vide supra), no bimolecular decay of the ET

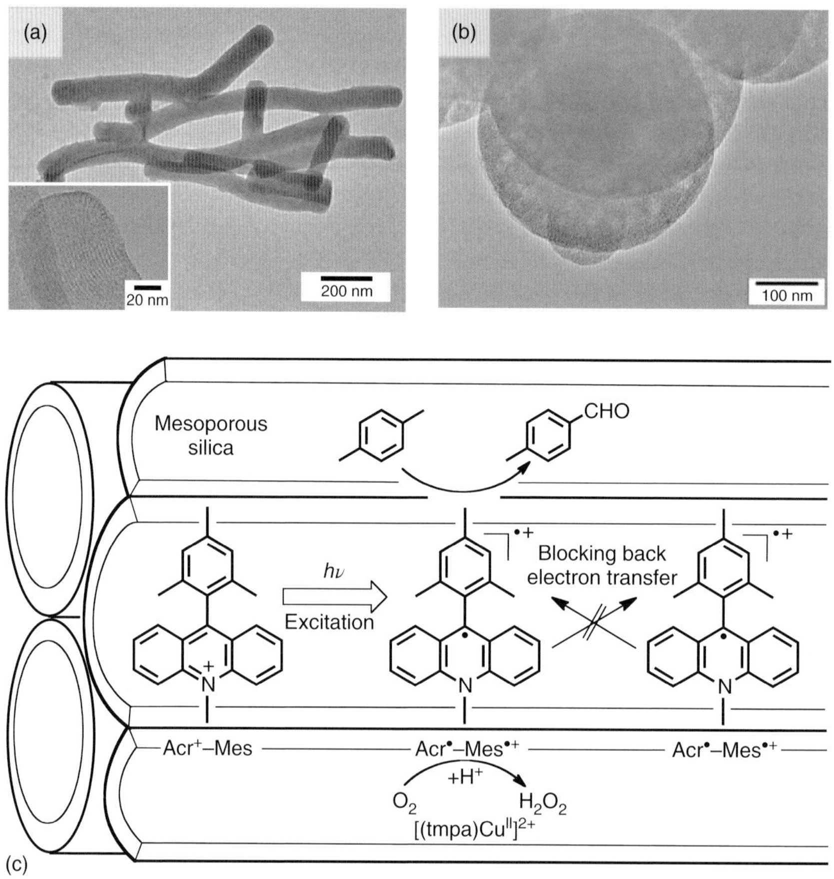

Figure 4.6 Transmission electron microscope (TEM) images of (a) tAlMCM-41 and (b) sAlMCM-41 (the high-resolution image of tAlMCM-41 is inserted in (a)). (c) Reaction scheme of photocatalytic oxygenation of *p*-xylene with Acr^+–Mes and $[(tmpa)Cu^{II}]^{2+}$ incorporated into sAlMCM-41. Source: Fukuzumi et al. 2012 [73]. Reproduced with permission of PNAS.

state occurs because each Acr^+–Mes molecule is isolated inside AlMCM-41 [73]. The lifetime of the ET state of Acr^+–Mes@tAlMCM-41 suspended in acetonitrile was determined to be 2.3 seconds at 198 K, which is much longer than that in solution because of the inhibition of bimolecular BET in AlMCM-41 as illustrated in Figure 4.6 [73]. Thus, incorporation of a simple electron donor–acceptor dyad into AlMCM-41 has made it possible to elongate the lifetime of the charge-separated state, which is longer than that of the bacterial photosynthetic reaction center (one second) [74].

The triplet ET state of $Acr^{\bullet}$–$Mes^{\bullet +}$@tAlMCM-41 was detected by an EPR spectrum measured at 4 K, which exhibited a fine structure together with a strong sharp signal at $g = 4.0$ [73]. The distance between two electron spins was determined from the zero-field splitting parameters to be 7.7 Å, which agrees with the expected distance of 7.2 Å between an sp^2 carbon atom at the 4 position of the mesityl moiety and sp^2 carbon atoms at the 3 and 6 positions of the acridinyl moiety [73]. Polycondensation of Acr^+–Mes-bridged organosilane in the presence of a nonionic surfactant is also reported to yield a mesostructured organosilica solid with a functional framework that exhibited long-lived photoinduced CS [75].

Nano-sized charge-separated molecules can also be obtained by using single-walled carbon nanotubes (SWNTs) [76], which exhibit excellent chemical and physical properties as revealed by various potential applications [77–81]. Extensive efforts have so far been devoted to assemble electron donor and acceptor molecules on SWNTs [82–88]. However, the fine control of size (i.e. length) of SWNTs remains a formidable challenge, because SWNTs have seamless cylindrical structures made up of a hexagonal carbon network, which leads to the difficulty of solubilization/functionalization without treatment with strong acid or vigorous sonication [89–92]. On the other hand, the cup-stacked carbon nanotubes (CSCNTs) that consist of cup-shaped nanocarbon (CNC) units, which stack via van der Waals attractions, have merited special attention from the viewpoint of the conventional carbon nanotube alternatives [93–96]. The tube–tube van der Waals energy between CNCs has been counterbalanced by the thermal or photoinduced electron transfer multi-electron reduction due to electrostatic repulsion, resulting in the highly dispersible CNCs with size homogeneity [97, 98].

The CNCs with controlled size have been functionalized with a large number of porphyrin molecules [99]. The general procedure for the synthesis of the porphyrin-functionalized cup-shaped nanocarbons [CNC–$(H_2P)_n$] is shown in Figure 4.7a [99]. The CNCs are first functionalized with aniline as the precursor for further functionalization with porphyrins. The aniline-functionalized nanocarbons react with the porphyrin derivatives to construct the nanohybrids.

The structure of the CNCs of the CNC–$(H_2P)_n$ nanohybrids is shown by the TEM in Figure 4.7b, which reveals a CNC with a hollow core along the length of the nanocup with well-controlled diameter (c. 50 nm) and size (c. 100 nm) [99]. The weight percentage of porphyrins attached to the CNCs was determined by thermogravimetric analysis (TGA) and elemental analysis to be ca. 20% [99]. This corresponds to one functional group per 640 carbon atoms of the nanocup framework for CNC–$(H_2P)_n$ nanohybrid. Thus, the π-framework of the CNC is not destroyed despite attachment of a large number of porphyrin molecules on the CNC.

Spectroscopic evidence for the covalent functionalization of CNC–$(H_2P)_n$ nanohybrid was obtained by an intensity increase of the Raman signal at 1353 cm^{-1} (D band) in the functionalized CNC as compared with the pristine CSCNTs [99], because the D band has been used for monitoring the process of functionalization that transforms sp^2 to sp^3 sites [99]. The UV–vis absorption spectrum of CNC–$(H_2P)_n$ nanohybrid agreed with that of the superposition of ref-H_2P [tetrakis(*N*-octadecyl-4-aminocarboxyphenyl)porphyrin] and CNCs, indicating that there is no significant interaction between attached porphyrins and CSCNTs in the ground states [99].

The fluorescence lifetime of CNC–$(H_2P)_n$ was determined to be 3.0 ± 0.1 ns, which is much shorter than that of ref-H_2P (14.1 ± 0.1 ns) [99]. The fluorescence

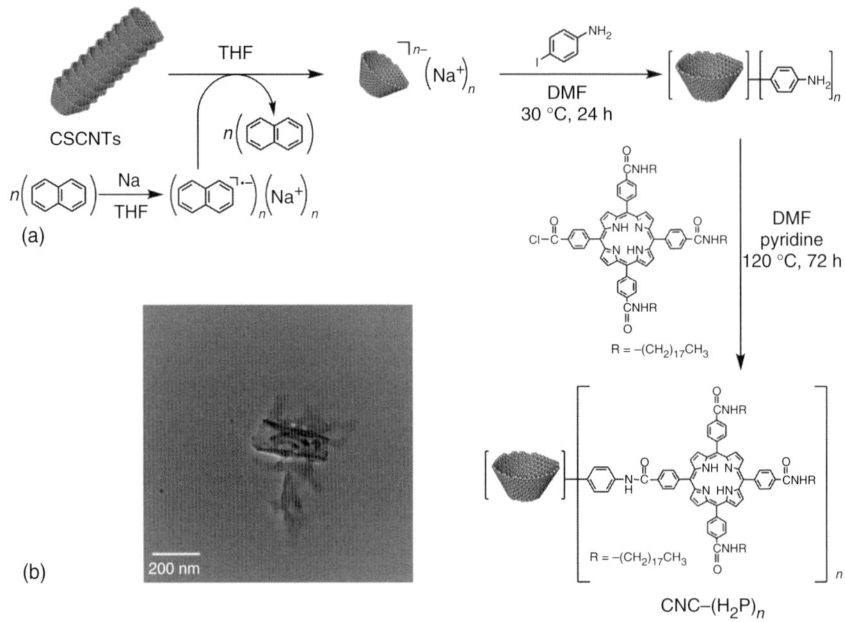

Figure 4.7 (a) Synthetic procedure of CNC–$(H_2P)_n$. (b) TEM image of CNC–$(H_2P)_n$. Source: Ohtani et al. 2009 [99]. Reproduced with permission of John Wiley & Sons.

emission at 650 nm was also quenched in CNC–$(H_2P)_n$ [99]. The short fluorescence lifetime of CNC–$(H_2P)_n$ and an efficient fluorescence quenching of porphyrins in CNC–$(H_2P)_n$ as compared to the ref-H_2P may result from the photoinduced electron transfer from the singlet excited state of H_2P ($^1H_2P^*$) to CNC in CNC–$(H_2P)_n$. The occurrence of photoinduced electron transfer to afford the charge-separated (CS) state of CNC–$(H_2P)_n$ was confirmed by nanosecond laser flash photolysis measurements in Figure 4.8, where the absorption bands in the visible and near infrared (NIR) regions are attributed to $H_2P^{\cdot+}$, which are clearly different from the triplet–triplet absorption of ref-H_2P [99]. The formation of the CS state was also confirmed by EPR measurements under photoirradiation of CNC–$(H_2P)_n$ in frozen *N*,*N*-diemthylformamide (DMF) at 153 K. The observed isotropic EPR signal at $g = 2.0044$ agrees with that of ref-$H_2P^{\cdot+}$ produced by one-electron oxidation with $[Ru(bpy)_3]^{3+}$ (bpy = 2,2′-bipyridine) in deaerated $CHCl_3$ [99]. The EPR signal corresponding to the reduced carbon-based nanomaterials was too broad to be detected, probably due to delocalization of electrons in CNC [99].

The CS state of CNC–$(H_2P)_n$ detected in Figure 4.8a decays obeying clean first-order kinetics: the first-order plots at different initial CS concentrations afford linear correlations with the same slope (Figure 4.8b) [99]. Thus, the decay of the CS state results from back electron transfer in the nanohybrid rather than intermolecular back electron transfer from $CNC^{\cdot-}$ to $H_2P^{\cdot+}$. The CS lifetime was

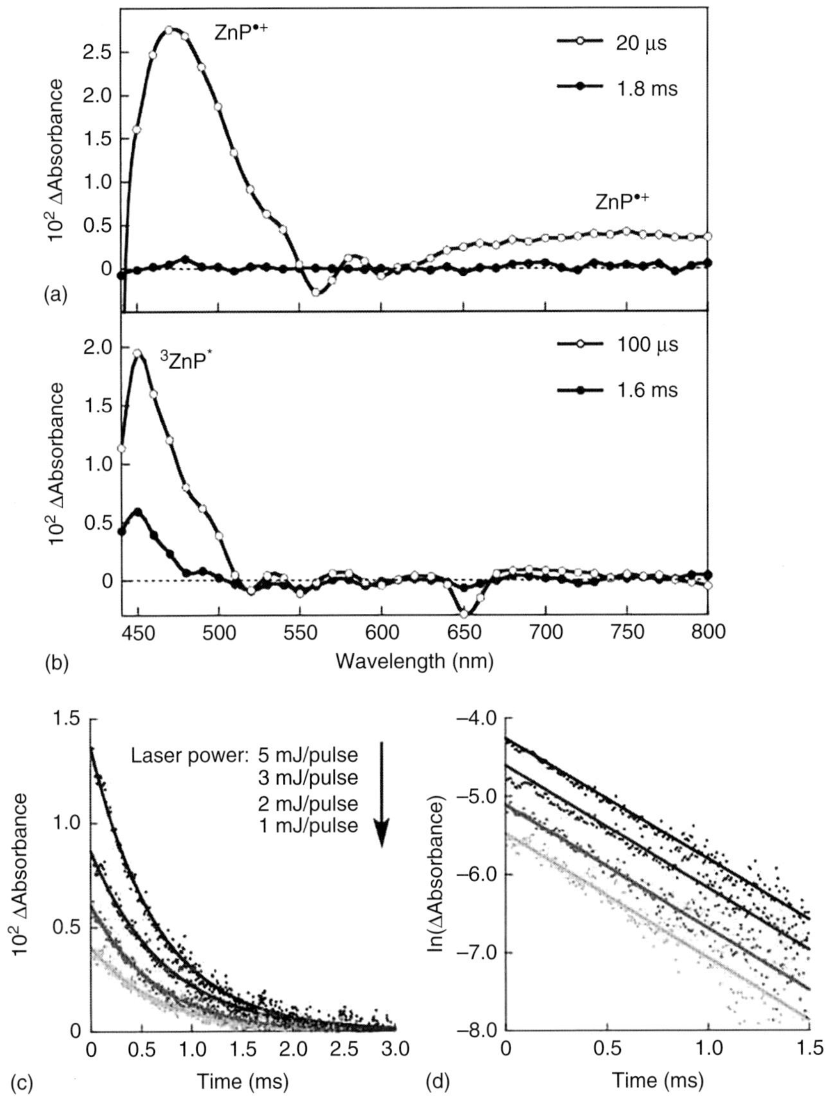

Figure 4.8 (a) Transient absorption spectra of (a) CNC–$(H_2P)_n$ taken at 20 and 1.8 ms after laser excitation at 426 nm and (b) ref-H_2P in deaerated DMF at 298 K taken at 100 ms and 1.6 ms after laser excitation at 426 nm. (c) Decay time profiles and (d) first-order plots at 470 nm with different laser powers (5, 3, 2, and 1 mJ/pulse). Source: Ohtani et al. 2009 [99]. Reproduced with permission of John Wiley & Sons.

determined from the first-order plots in Figure 4.8b to be 0.64 ± 0.01 ms, which is the longest lifetime ever reported for electron donor-attached nanocarbon materials [99]. Such a long CS lifetime may be ascribed to the efficient electron migration in the CNCs following CS.

5

Supramolecular Electron Transfer

The use of non-covalent interactions, such as metal–ligand coordination, electrostatic effects, hydrogen bonds, and rotaxane formation, provides a much easier means of constructing electron donor–acceptor supramolecular ensembles that in the limit could mimic the electron-transfer events of photosynthesis [100–105]. Among the various non-covalent interactions that might prove useful for the assembly of electron-transfer systems, anion binding appears particularly attractive. Recently, anion recognition has emerged as an important sub-discipline within the broader field of supramolecular chemistry [106–116]. A variety of anion receptors have been synthesized and their use in stabilizing supramolecular complexes has been reported [106–116]. In these latter complexes, the individual components (receptor, anion) are in equilibrium with the corresponding supramolecular ensembles.

5.1 Cation–Anion Binding

When pyrene carboxylate anion (as the TBA [tetrabutylammonium] salt) was mixed with C8·2HCl, spectral changes were seen to indicate the formation of the supramolecular complex [Py–C8] (Figure 5.1a) [117]. The binding constant (K_a) corresponding to the formation of Py–C8 was determined from these spectral changes (cf. Figure 5.1b) to be $(2.6 \pm 0.3) \times 10^5\ M^{-1}$ in MeCN at 298 K [117]. Transient absorption spectra obtained 10 μs after laser excitation at 355 nm (Figure 5.1) indicate the formation of $C8^{\cdot+}$ (740 and 820 nm) and $Py^{\cdot-}$ (480 nm) rather than $C8^{\cdot-}$ and $Py^{\cdot+}$ [118]. Thus, photoinduced electron transfer occurs from C8 to the singlet excited state of Py ($^1Py^*$), producing the charge-separated (CS) state, $C8^{\cdot+}$–$Py^{\cdot-}$; this happens even though the energy of the putative $C8^{\cdot+}$–$Py^{\cdot-}$ state (2.58 eV) is higher than that of $C8^{\cdot-}$–$Py^{\cdot+}$ (1.31 eV) [117].

The reason why $C8^{\cdot+}$–$Py^{\cdot-}$ is formed instead of $C8^{\cdot-}$–$Py^{\cdot+}$ is well rationalized by the Marcus theory of electron transfer [6], which predicts that the charge separation to produce the lower energy CS state ($C8^{\cdot-}$–$Py^{\cdot+}$) with the large driving force of 2.15 eV occurs deeply in the Marcus inverted region. This makes the CS rate much slower than the CS rate required to produce the high energy CS state of $C8^{\cdot+}$–$Py^{\cdot-}$ in the Marcus normal region for which the driving force would be 0.88 eV [117].

Electron Transfer: Mechanisms and Applications, First Edition. Shunichi Fukuzumi.

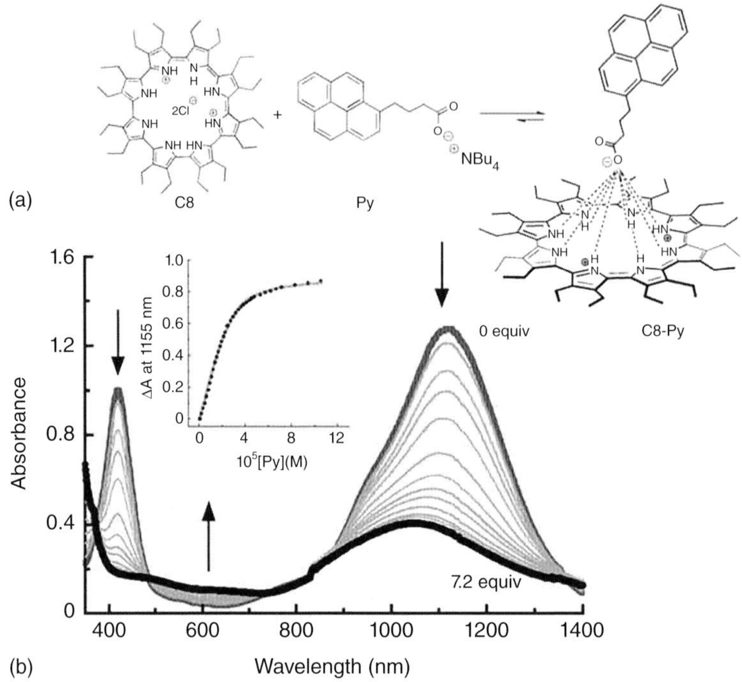

Figure 5.1 (a) Proposed complex formation. (b) UV–vis absorption titration of tetra-*n*-butylammonium 1-pyrenebutyrate, Py, into cyclo[8]pyrrole, C8, at 1.5×10^{-5} M in MeCN at 298 K. Inset: Curve fit (line) to a 1 : 1 binding isotherm produced from the change in absorbance at 1155 nm (points). Source: Sessler et al. 2008 [117]. Reproduced with permission of American Chemical Society.

The charge reversal seen in the "umpolung" system defined by C8 and Py has also been reported for electron transfer from electron donors to $^1Py^*$ [119]. In the present instance, the CS rate constant was determined to be $5.1 \times 10^6\ s^{-1}$ from the decrease in the fluorescence lifetime of C8–Py (140 ns) as compared to that of the reference pyrene (480 ns). The slow CS was confirmed by femtosecond laser flash photolysis, where little change was observed in the transient absorption spectrum due to the dynamic behavior of $^1Py^*$ in C8–Py.

The CS state decays to the triplet excited state ($^3C8^*$) rather than to the ground state, as inferred from the fact that the transient spectrum of the ensemble recorded at 900 μs in Figure 5.2 agrees with that observed for $^3C8^*$ species produced upon photoexcitation of C8 alone [117]. This mode of recombination is also rationalized in terms of Marcus theory. Specifically, the slower back electron transfer to the ground state as compared to the triplet excited state reflects a process that lies in the Marcus inverted region, wherein electron transfer becomes slower as the driving force of electron transfer increases [6–8, 28, 29]. Fitting the decay at both 470 and 740 nm to a single exponential results in a charge-recombination (CR) rate constant of $(3.3 \pm 0.1) \times 10^3\ s^{-1}$, a value that corresponds to a lifetime of the CS state of 300 μs [117]. Changing the laser power has no effect on the rates, as underscored by the invariance of

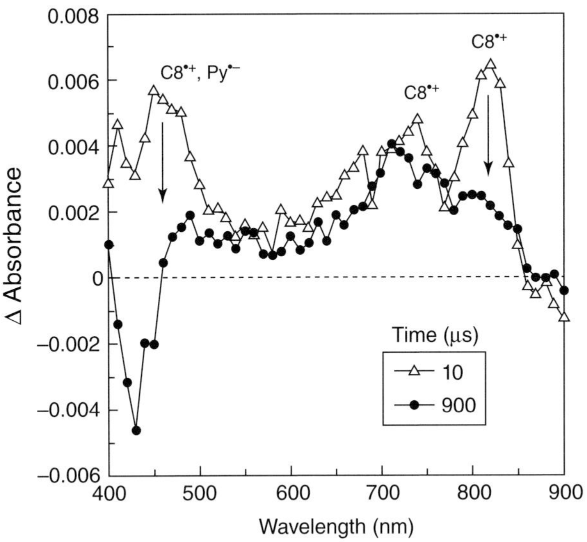

Figure 5.2 Transient absorption spectra of **C8** and **Py**, in a 1 : 1 ratio at 5.0×10^{-5} M in deaerated MeCN at 298 K taken at 10 μs (Δ) and 900 μs (•) after laser photoexcitation (355 nm, 25 mJ/pulse). Source: Sessler et al. 2008 [117]. Reproduced with permission of American Chemical Society.

the first-order plots. This indicates that intramolecular electron transfer occurs within the supramolecular complex. As expected, the use of pyrene-1-butyric acid, instead of the base (i.e. pyridine) in conjunction with C8 under otherwise identical conditions, yielded none of the peaks characteristic of $C8^{\cdot+}$ following photoirradiation because the anion binding is essential for the formation of the supramolecular complex [117]. Thus, anion binding provides a convenient way to construct an electron donor–acceptor supramolecular complex that gives rise to long-lived CS states.

A different electron-transfer complex stabilized by hydrogen bonds is prepared from ferrocenecarboxylic acid (FcCOOH) and H_2DPP (dodecaphenylporphyrin) [H_4DPP$(FcCOO)_2$]. The X-ray crystal structure of H_4DPP$(FcCOO)_2$ is shown in Figure 5.3, where H_4DPP^{2+} and two ferrocene-carboxylate ($FcCOO^-$) anions are linked by hydrogen bonds, albeit in an asymmetric manner [120]. On one side of the saddle-distorted H_4DPP^{2+} macrocycle, two-point hydrogen bonds, involving two of the pyrrole N–H protons and two of the oxygen atoms in the bound $FcCOO^-$ carboxylate anion, are seen. These bonds are characterized by interatomic (N· · ·O) distances of 2.726(5) and 2.587(4) Å, respectively. On the other side of the ring, hydrogen bond interactions, involving two of the four N–H pyrrolic protons and one of the $FcCOO^-$ carboxylate oxygen atoms, are seen. These interactions give rise to interatomic distances of 2.773(4) and 2.763(5) Å. Both $FcCOO^-$ anions reside in a cleft-like environment created by the two phenyl groups attached to the pyrroles of the porphyrins (cf. Figure 5.3) [121–123].

A 1 : 2 supramolecular complex between H_4DPP^{2+} and $FcCOO^-$ [H_4DPP $(FcCOO)_2$] was also formed in PhCN as indicated by the UV–vis titration shown

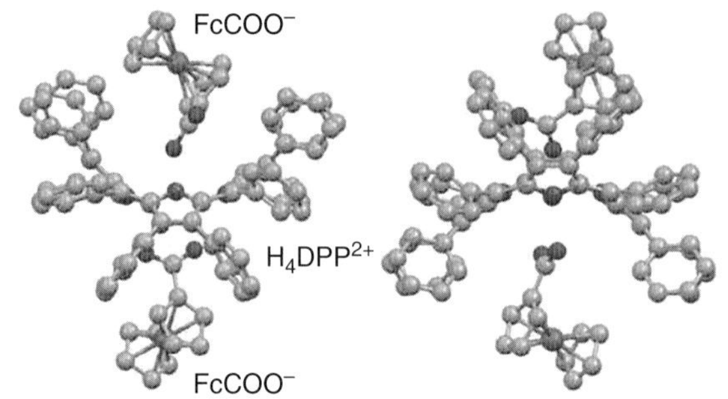

Figure 5.3 Crystal structure of $H_4DPP(FcCOO)_2$ viewed from different directions. Hydrogen atoms and solvent molecules are omitted for clarity. Source: Honda et al. 2010 [120]. Reproduced with permission of American Chemical Society.

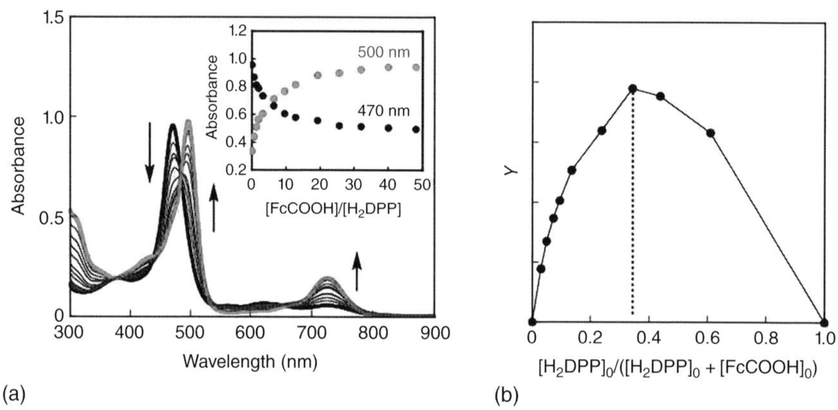

Figure 5.4 (a) Absorption spectral changes observed over the course of a titration of H_2DPP with FcCOOH in PhCN at room temperature. Inset: absorbance change at 470 nm (black line) and 500 nm (grey line). (b) Job's continuous variation plot to determine the stoichiometry of the complex formation between H_2DPP and FcCOOH: Molar fraction of H_2DPP is varied while keeping the total concentration constant. Source: Honda et al. 2010 [120]. Reproduced with permission of American Chemical Society.

in Figure 5.4a [120]. The Job's plot in Figure 5.4b confirms the 1 : 2 binding stoichiometry. The observed one-step spectral change in the course of the titration with FcCOOH is consistent with the presence of hydrogen-bonding interactions in solution that mirror those observed in the crystalline state. This is an important observation because formation of hydrogen bonds between the N–H protons of pyrroles and carboxylate anions in solution is known to facilitate diprotonation of the core H_2DPP moiety [124]. However, this fate is avoided in the present instance.

The formation constant of $H_4DPP(FcCOO)_2$ was determined from the spectral change to be $9.2 \times 10^8\ M^{-2}$ [120]. The presence of hydrogen bonds was further supported by the ^1H NMR spectral changes observed over the course of the FcCOOH addition [120]. Specifically, after adding up to 2 equiv of FcCOOH, peaks assigned to the *meso*-phenyl protons of H_4DPP^{2+} were all downfield-shifted, which is consistent with the formation of H_4DPP^{2+} [124–126]. No further shift was observed upon the addition of more than 2 equiv of FcCOOH. In contrast, upon the addition of increasing quantities of FcCOOH, the signals ascribed to the ferrocene moiety exhibited upfield shifts, a finding that is ascribed to shielding caused by the ring current of the porphyrin ring. In $CDCl_3$, the largest upfield shift was seen in the case of the proton adjacent to the carboxylate moiety; this is consistent with the presence of a hydrogen bonding interaction between the pyrrole NH protons and the carboxylate group of $FcCOO^-$ [120].

Rate constants of photoinduced electron transfer and back electron transfer in $H_4DPP(FcCOO)_2$ were determined by femtosecond laser flash photolysis measurements [120]. After femtosecond laser excitation at 430 nm, the transient absorption bands at 550 and 1030 nm due to the singlet excited state of the H_4DPP^{2+} moiety, $^1[H_4DPP^{2+}]^*$, were observed at 1.5 ps. The spectrum then evolved to that assigned to the one-electron reduced species of H_4DPP^{2+}, $H_4DPP^{\cdot+}$, in the electron-transfer state corresponding to the self-associated ensemble $H_4DPP(FcCOO)_2$. From the rise and the decay of absorbance at 545 nm the rate constants for the photoinduced electron transfer and back electron transfer were determined to be $5.0 \times 10^{11}\ s^{-1}$ and $6.1 \times 10^{10}\ s^{-1}$, respectively [120]. Similarly, the rate constants (k_{ET}) of photoinduced electron transfer and back electron transfer were determined for the supramolecular complexes of H_4DPP^{2+} with other ferrocene derivatives, which likewise contained carboxylate anions as the anion linking motif [120]. The curve fit of the kinetic results for a variety of electron donors, obtained using Eq. (2.1), is shown in Figure 5.5. Inspection of Figure 5.5 reveals a small reorganization energy ($\lambda = 0.68 \pm 0.03$ eV) and a large electronic coupling matrix element ($V = 43 \pm 7\ cm^{-1}$) [120]. The λ value is comparable to those of covalently linked donor–acceptor systems consisting of electrically neutral porphyrins as electron donors (0.41 – 0.66 eV) [39, 127]. The small λ value has the consequence that the back electron transfer falls in the Marcus inverted region, where the CS lifetime becomes longer with increasing driving force. The large V value indicates that the electrostatic and hydrogen-bonding interactions in the supramolecular complex established as a

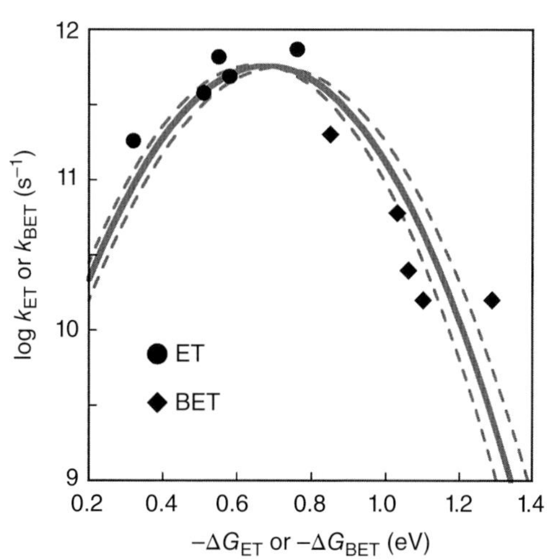

Figure 5.5 Driving force dependence of log k_{ET} (•) or k_{BET} (♦) for intrasupramolecular photoinduced electron transfer and back electron transfer in supramolecular complexes of H_4DPP^{2+} with hydrogen-bonded electron donors in PhCN at room temperature. Fits to the Marcus equation for electron transfer (Eq. (2.1)) are shown by the solid line with $\lambda = 0.68$ eV and $V = 43\ cm^{-1}$ and the dotted lines with $\lambda = 0.65$ and 0.71 eV and $V = 43\ cm^{-1}$, respectively. Source: Honda et al. 2010 [120]. Reproduced with permission of American Chemical Society.

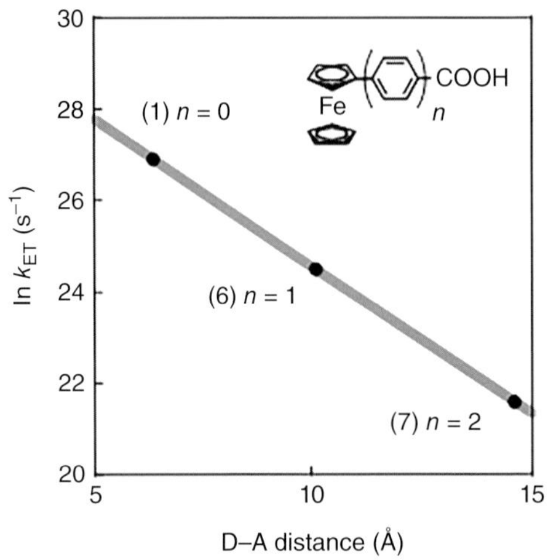

Figure 5.6 Dependence of ln k_{ET} on distance for intrasupramolecular electron transfer in ensembles derived from the electron donors ((1) $FcCOO^-$, (6) $FcPhCOO^-$, and (7) $FcbphCOO^-$) and the excited state species $^1[H_4DPP^{2+}]^*$ produced as the result of photoexcitation. Here, the distance between an electron donor and an electron acceptor is defined as that between the Fe atom of the ferrocene moiety and the center of a mean plane of the porphyrin ring. Source: Honda et al. 2010 [120]. Reproduced with permission of American Chemical Society.

result of anion binding are strong enough to establish strong electronic coupling between the electron donor and acceptor moieties.

The dependence of k_{ET} on distance within intrasupramolecular photoinduced electron-transfer ensembles was also examined by using ferrocenecarboxylic acid derivatives having linear phenylene linker(s) between the ferrocene moiety and the carboxyl group. This arrangement is shown in Figure 5.6, where the distance between electron donor and acceptor is defined as that between the Fe atom of ferrocene and the center of the mean plane of the porphyrin ring in H_4DPP^{2+} [120]. For these systems, the dependence of the rate constant for intrasupramolecular electron transfer (k_{ET}) on distance is given by Eq. (5.1) [6]:

$$\ln k_{ET} = \ln k_0 - \beta r \tag{5.1}$$

where k_0 is the rate constant for adiabatic intrasupramolecular electron transfer, r is the donor–acceptor center-to-center distance and β is the decay coefficient factor (damping factor), which depends primarily on the nature of the bridging molecule [28]. A plot of $\ln k_{ET}$ vs. r is shown in Figure 5.6. From the slope of this linear plot, the β value was determined to be 0.64 Å^{-1}, which is comparable to those of covalently phenylene-bridged multi-porphyrin systems [128]. Thus, the electron donor–acceptor linkage by anion binding in this series of supramolecular complexes provides electron coupling interactions between the electron donor and acceptor moieties, which are as strong as those present in comparable covalently linked systems.

Anion binding has also been utilized to form supramolecular donor–acceptor systems composed of anionic sulfonated porphyrins (H_2TPPS^{4-} and $ZnTPPS^{4-}$), which function as electron donors, and a cationic lithium ion encapsulated fullerene ($Li^+@C_{60}$) as an electron acceptor (Scheme 5.1) [129, 130]. In PhCN, the UV–vis spectral features of free base tetraphenylporphyrin tetrasulfonate $[(Bu_4N^+)_4H_2TPPS^{4-}]$ at 298 K undergo a discernible change upon addition of $Li^+@C_{60}$. Specifically, the Soret band undergoes a red shift to 427 nm with an isosbestic point at 430 nm (Figure 5.7a) [130]. The absorbance change exhibits saturation behavior with increasing $Li^+@C_{60}$ concentration (Figure 5.7b) [130].

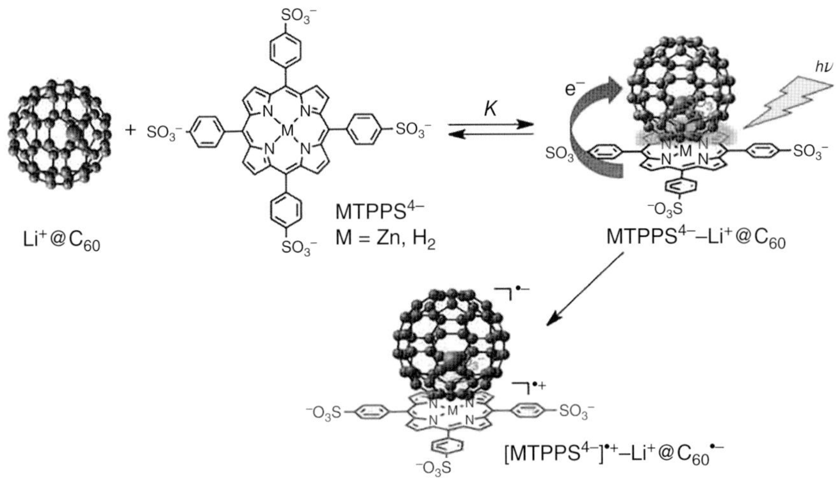

Scheme 5.1 Supramolecular complex formation and photoinduced charge separation of $MTPPS^{4-}$ (M = Zn and H_2) with $Li^+@C_{60}$ in PhCN. Source: Ohkubo et al. 2012 [130]. Reproduced with permission of Royal Society of Chemistry.

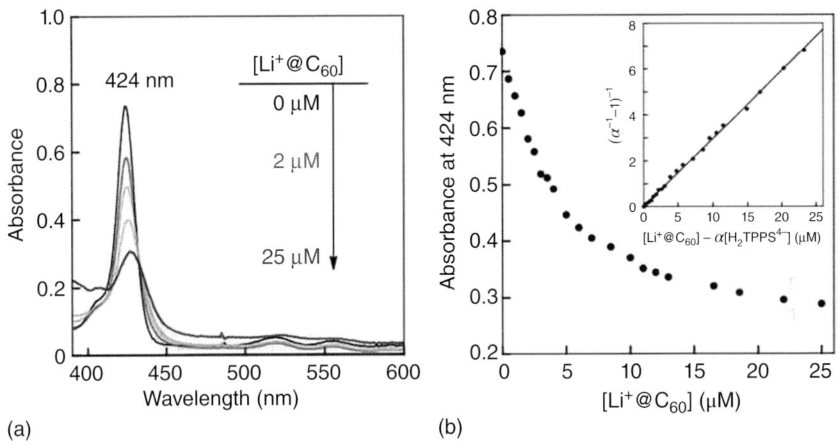

Figure 5.7 UV–vis spectra of H_2TPPS^{4-} (2.0×10^{-6} M) in the presence of various concentrations of $Li^+@C_{60}$ (0 to 2.5×10^{-5} M) in PhCN. (b) Absorption change at 424 nm. Inset: Plot of $(\alpha^{-1} - 1)^{-1}$ vs. $[Li^+@C_{60}] - \alpha[H_2TPPS^{4-}]$; $\alpha = (A - A_0)/(A_\infty - A_0)$. Source: Ohkubo et al. 2012 [130]. Reproduced with permission of Royal Society of Chemistry.

From these concentration-induced spectral changes in PhCN, the formation constant (K) corresponding to the $H_2TPPS^{4-}/Li^+@C_{60}$ complex was determined to be 3.0×10^5 M^{-1}. A Job plot provided support for the expectation that the binding stoichiometry was 1 : 1. When H_2TPPS^{4-} was replaced by $ZnTPPS^{4-}$, the K value was determined to be 1.6×10^5 M^{-1}. Concordant K values were obtained from fluorescence quenching experiments [130].

The fluorescence quenching of H_2TPPS^{4-} and $ZnTPPS^{4-}$ by $Li^+@C_{60}$ occurs via energy transfer from the singlet excited states of H_2TPPS^{4-} and $ZnTPPS^{4-}$

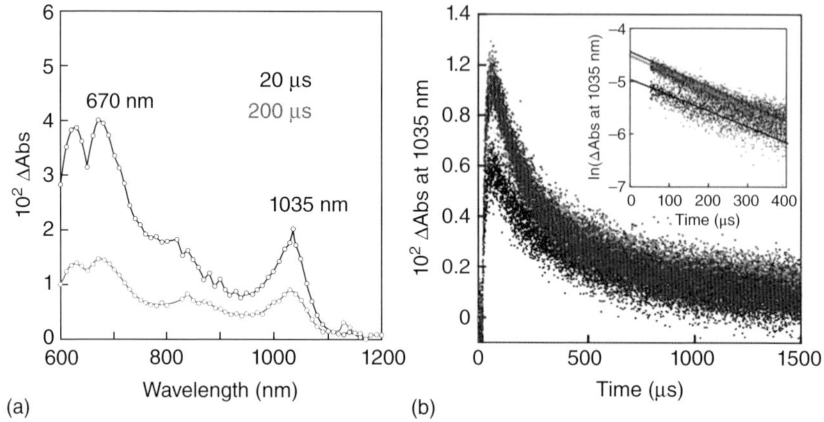

Figure 5.8 (a) Transient absorption spectra of H_2TPPS^{4-} (2.5×10^{-5} M) in the presence of $Li^+@C_{60}$ (5.0×10^{-5} M) in deaerated PhCN at 298 K taken at 20 and 200 μs after nanosecond laser excitation at 520 nm. (b) Decay time profiles at 1035 nm with different laser intensities (1, 3, 6 mJ/pulse). Source: Ohkubo et al. 2012 [130]. Reproduced with permission of Royal Society of Chemistry.

to $Li^+@C_{60}$ as revealed by the femtosecond laser flash photolysis measurements [130]. Nanosecond laser flash photolysis measurements have revealed that electron transfer from H_2TPPS^{4-} and $ZnTPPS^{4-}$ to the triplet excited state of $Li^+@C_{60}$ occurs to produce the CS state as shown in Figure 5.8, where the absorption band of $[H_2TPPS^{4-}]^{\cdot+}$ (λ_{max} = 670 nm) [131] and that of $Li^+@C_{60}$ radical anion (λ_{max} = 1035 nm) [129] are observed. The decay of the CS state $[(H_2TPPS^{4-})^{\cdot+}–Li^+@C_{60}{}^{\cdot-}]$ in PhCN was monitored using different laser intensities. Various first-order plots for the decay time profiles of the CS state at different concentrations are shown in Figure 5.8b. The first-order plots afford approximately linear correlations with the same slope irrespective of the difference in concentration of the CS state. Thus, there is no or little contribution from the bimolecular CR process of free $[H_2TPPS^{4-}]^{\cdot+}$ and $Li^+@C_{60}{}^{\cdot-}$ molecules [130]. The lifetimes of the triplet CS state of the supramolecular complex was determined to be 310 μs for H_2TPPS^{4-} and 300 μs for $ZnTPPS^{4-}$ at 298 K [130]. This is the longest lifetime of the CS state ever reported for monomer porphyrin/fullerene systems linked non-covalently in solution. The quantum yield of the CS state is determined to be 0.39 using the absorption of the CS state ($Li^+@C_{60}{}^{\cdot-}$: $\varepsilon_{1035} = 7300\ M^{-1}\ cm^{-1}$) [130].

The activation enthalpies of the CR processes were determined to be 3.0 kcal mol^{-1} for $ZnTPPS^{4-}–Li^+@C_{60}$ and 5.4 kcal mol^{-1} for $H_2TPPS^{4-}–Li^+@C_{60}$ [130]. This indicates that there is a significant energy difference between the singlet and triplet CS states and that the CR processes may occur through the thermally activated singlet CS state. The lifetime of the CS state at 77 K is estimated to be as long as 60 hours for $H_2TPPS^{4-}–Li^+@C_{60}$ [130]. Such a long-lived triplet CS state was detected by EPR measurements by photoirradiation of the $H_2TPPS^{4-}–Li^+@C_{60}$ complex in frozen PhCN as shown in Figure 5.9. The spin–spin interaction in the triplet radical ion pair of the supramolecular

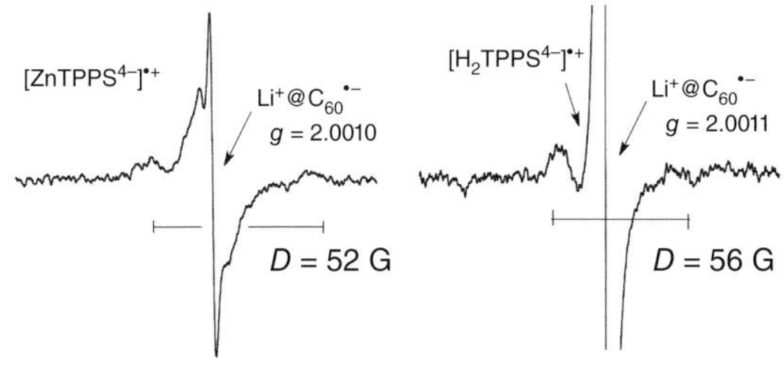

Figure 5.9 EPR spectra of (a) $(ZnTPPS^{4-})^{\cdot+}$–$Li^+@C_{60}^{\cdot-}$ and (b) $(H_2TPPS^{4-})^{\cdot+}$–$Li^+@C_{60}^{\cdot-}$ in PhCN generated by photoirradiation with a high-pressure Hg lamp (1000 W) at 77 K. Source: Ohkubo et al. 2012 [130]. Reproduced with permission of Royal Society of Chemistry.

complex is clearly shown at 77 K, where the fine structure due to the triplet CS state is clearly observed at $g = 2$. From the zero-field splitting values ($D = 52$ G for $ZnTPPS^{4-}$ and 56 G for H_2TPPS^{4-}) the distances (r) between two electron spins were estimated using the relation $D = 27\,800/r^3$ to be 8.1 and 7.9 Å, respectively [130]. These r values agree with the center-to-center distance of a reported crystal structure of porphyrin/C_{60} [130].

In solution, fast bimolecular charge recombination processes of the charge-separated states of electron donor–acceptor linked molecules usually predominate over much slower intramolecular charge recombination pathways (vide supra) [7, 8]. Such a bimolecular charge recombination process is retarded in a supramolecular electron donor–acceptor complex composed of one tetra-anionic porphyrin (**1**-M^{4-}: M = H_2 and Zn) and two dicationic porphyrins (**2**-Zn^{2+}) produced by the two-electron oxidation of a π-extended 1,3-dithiol-2-ylidene quinoidal porphyrin (porphyrin-bridged TTF) in Figure 5.10 [132]. **2**-Zn^{2+} forms a 2 : 1 supramolecular complex with **1**-M^{4-} in PhCN, wherein both the top and bottom faces are covered by large charged molecules as shown in the X-ray crystal structure in Figure 5.11 [132]. The binding constants of **1**-Zn^{4-}/(**2**-Zn^{2+})$_2$ and **1**-H_2^{4-}/(**2**-Zn^{2+})$_2$ in PhCN at 298 K were determined from the spectral titrations to be $(5.9 \pm 0.5) \times 10^{11}$ and $(6.6 \pm 0.5) \times 10^{12}$ M^{-2}, respectively [132]. The large binding constants result from the strong electrostatic interaction between the cationic and anionic porphyrins.

Photoinduced electron transfer from **1**-M^{4-} to **2**-Zn^{2+} occurs via the triplet excited state to afford the CS state [132]. The transient absorption associated with the CS state was surprisingly persistent [132]. The spin state of the CS state was determined to be triplet as detected by EPR measurements [132]. The decay of the triplet CS state obeyed first-order kinetics and the slope of the first-order plots proved invariant to laser intensity and therefore the decay rate is independent from the concentration of the photoinduced CS state. This indicates that the charge recombination occurs within the supramolecular complex and there is no contribution from intermolecular charge recombination between supramolecular complexes [132]. The intermolecular charge recombination processes are prohibited as a result of electrostatic repulsion. The two positively charged moieties, **2**-Zn^{2+} and **2**-$Zn^{\cdot+}$, between which the radical trianionic porphyrin **1**-$M^{\cdot3-}$ is sandwiched, provide a positively charged buffer layer that prevents close contacts with either another supramolecular ensemble or a free electron donor moiety (i.e. **2**-$Zn^{\cdot+}$). The lifetimes of the CS states of **1**-$H_2^{\cdot3-}$/(**2**-$Zn^{\cdot+}$)(**2**-Zn^{2+}) and

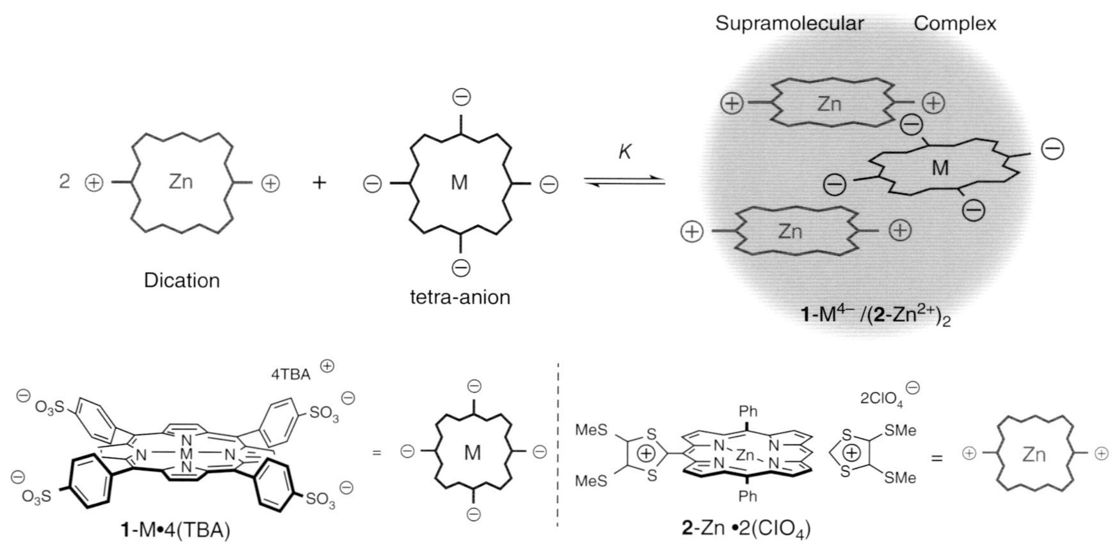

Figure 5.10 Formation of supramolecular porphyrin complexes, **1**-M^{4-}/(**2**-Zn^{2+})$_2$; M = H_2 or Zn. Source: Bill et al. 2014 [132]. 2014. https://pubs.rsc.org/en/content/articlelanding/2014/sc/c4sc00803k#!divAbstract and https://creativecommons.org/licenses/by/3.0/. Licensed under CCBY 3.0.

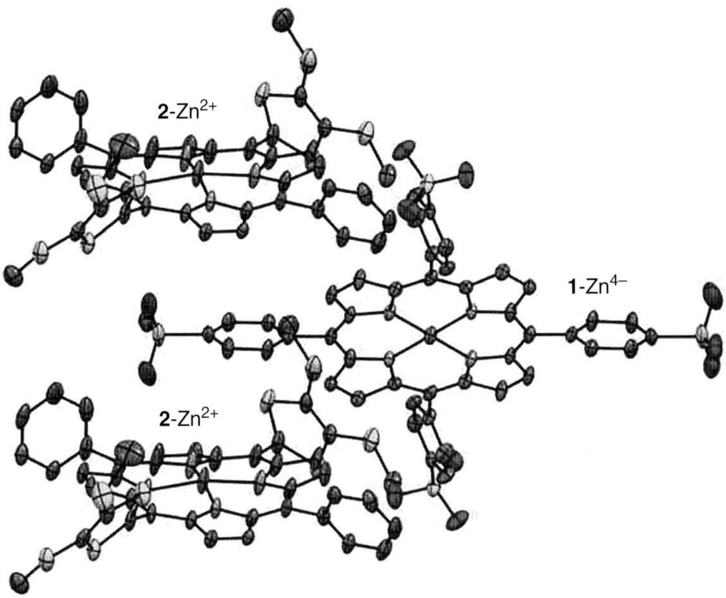

Figure 5.11 X-ray crystal structure of the supramolecular complex **1**-Zn^{4-}/(**2**-Zn^{2+})$_2$. Thermal ellipsoids are scaled to the 50% level. Hydrogen atoms and solvent molecules are omitted for clarity. Source: Bill et al. 2014 [132]. https://pubs.rsc.org/en/content/articlelanding/2014/sc/c4sc00803k#!divAbstract and https://creativecommons.org/licenses/by/3.0/. Licensed under CCBY 3.0.

1-$Zn^{\cdot 3-}$/(**2**-$Zn^{\cdot +}$)(**2**-Zn^{2+}) in PhCN at 298 K were determined to be 83 and 43 ms, respectively, which are the longest CS lifetimes ever reported for supramolecular π-electron donor–acceptor complexes in solution [180]. The temperature dependence of the rate constant of intramolecular back electron transfer in the CS state of the supramolecular complex [**1**-$H_2^{\cdot 3-}$/(**2**-$Zn^{\cdot +}$)(**2**-Zn^{2+})] was analyzed based on Eq. (2.1) to afford the λ and V values as 0.24 eV and 0.16 cm^{-1}, respectively [132]. The small V value results from the spin-forbidden back electron transfer of the triplet CS state and small orbital interactions between **1**-$H_2^{\cdot 3-}$ and **2**-$Zn^{\cdot +}$ arising from the slipped-sandwich arrangement (Figure 5.11) [132].

5.2 π-Complexes

As compared to porphyrin monomers, porphyrin dimers with appropriate linkage can accommodate electron acceptor guest molecules by π–π interactions to form sandwich complexes [133–144]. Fox example, a cyclic Ni porphyrin dimer (Ni-CPD_{Py}) linked by butadiyne moieties bearing 4-pyridyl groups (Figure 5.12) forms a sandwich complex with C_{60} ($C_{60}\subset Ni_2$-CPD_{Py}) as shown in the X-ray crystal structure (Figure 5.13), where the dimer bites a C_{60} molecule by tilting the porphyrin rings with respect to each other and there are strong π–π interactions between the porphyrin rings and C_{60} [144]. The adjacent dimers are linked

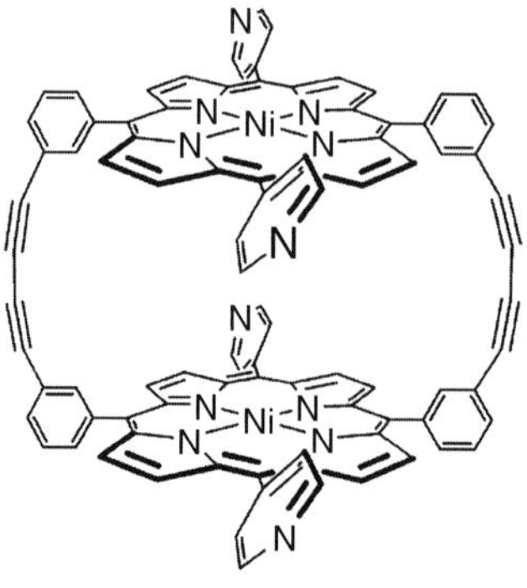

Figure 5.12 Chemical structure of Ni$_2$-CPD$_{Py}$. Source: Nobukuni et al. 2009 [144]. Reproduced with permission of American Chemical Society.

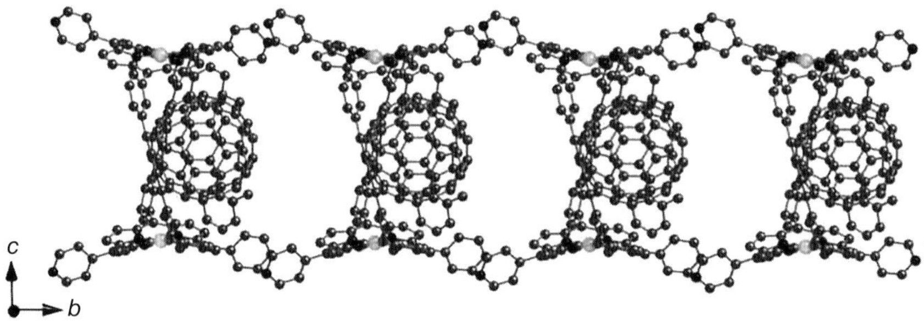

Figure 5.13 Crystal structures of tubular assemblies of C_{60}⊂Ni$_2$-CPD$_{Py}$. Hydrogen atoms are omitted for clarity. Source: Nobukuni et al. 2009 [144]. Reproduced with permission of American Chemical Society.

by hydrogen bonds and π–π interactions [144]. The C_{60} molecules are linearly arranged in the inner channel to give a supramolecular peapod [144].

The linear arrangement of C_{60} in C_{60}⊂Ni$_2$-CPD$_{Py}$ results in high electron mobilities of $\sum\mu = 0.72$ and $0.20\,cm^2\,V^{-1}\,s^{-1}$ along the *b* and *c* axes, respectively, which were determined by flash-photolysis time-resolved microwave conductivity (FP-TRMC) measurements [144]. The TRMC technique can evaluate the intrinsic mobility without being affected by chemical or physical defects in the material and/or the organic/metal–electrode interfaces [145–147]. The observed value along the *b* axis of the single crystal of C_{60}⊂Ni$_2$-CPD$_{Py}$ is comparable to that of the single crystal of C_{60} ($\sum\mu = 0.50\,cm^2\,V^{-1}\,s^{-1}$ measured by TOF) [148]. The observed high electron mobility along the *b* axis results from the well-ordered linear arrangement of C_{60} in the porphyrin nanotube. However, the expected charge-separated state could not be observed in the time-resolved transient absorption spectra of C_{60}⊂Ni$_2$-CPD$_{Py}$, because the singlet excited state of the nickel porphyrin immediately changes to the triplet excited state by intersystem crossing, and the low energy triplet excited state of C_{60} ($^3C_{60}{}^*$) is formed by energy transfer [144]. The estimated energy level of the charge-separated state (1.98 eV) is higher than that of $^3C_{60}{}^*$ (1.60 eV) [144]. When Ni$_2$-CPD$_{Py}$ was replaced by a free base porphyrin dimer (H$_4$-CPD$_{Py}$), complete charge-separated

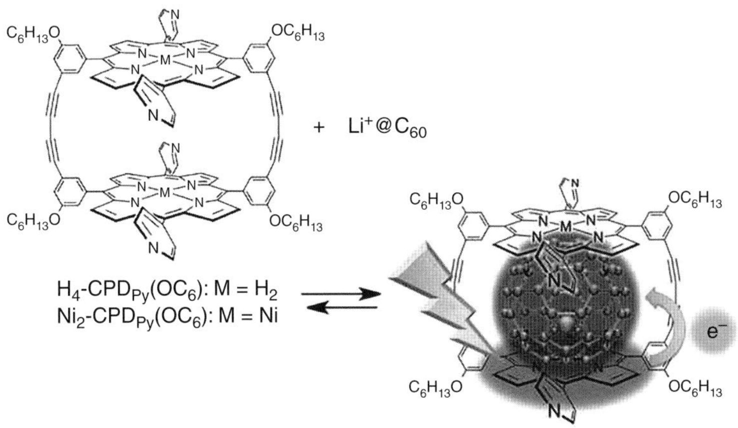

Figure 5.14 Supramolecular formation and photoinduced charge separation between $MCPD_{Py}(OC_6)$ and $Li^+@C_{60}$. Source: Kamimura et al. 2013 [150]. https://pubs.rsc.org/en/content/articlelanding/2013/sc/c3sc22065f#!divAbstract and https://creativecommons.org/licenses/by/3.0/. Licensed under CCBY 3.0.

state $\{H_4\text{-}CPD_{Py}{}^{\cdot+} + C_{60}{}^{\cdot-}\}$ was observed by femtosecond laser flash photolysis of $C_{60}\subset H_4\text{-}CPD_{Py}$ in the solid state with a lifetime of 470 ps [149]. The photovoltaic activity of $C_{60}\subset Ni_2\text{-}CPD_{Py}$ and $C_{60}\subset H_4\text{-}CPD_{Py}$ was evaluated by using solar cells composed of modified electrodes and $I^-/I_3{}^-$ solution [149]. The $C_{60}\subset H_4\text{-}CPD_{Py}$-modified electrode exhibited IPCE of 17% and a power conversion efficiency (η) of 0.33%, which was more than 16 times larger than that of $OTE/SnO_2/(C_{60}\subset Ni_2\text{-}CPD_{Py})_n$ (0.02%) [149]. Such a significant enhancement of the η value demonstrates that the formation of highly ordered clusters and the efficient charge separation of $(C_{60}\subset H_4\text{-}CPD_{Py})_n$ contributes to the improvement of light energy conversion properties [149].

When C_{60} is replaced by $Li^+@C_{60}$, porphyrin dimers with four long alkoxy substituents on the *meso*-phenyl groups ($MCPDPy(OC_6)$) in Figure 5.14 form strong supramolecular complexes in even a polar solvent such as PhCN [150]. The association constants (K_{assoc}) of $Li^+@C_{60}\subset MCPD_{Py}(OC_6)$ in PhCN at 298 K were determined to be $2.6 \times 10^5\ M^{-1}$ for $Li^+@C_{60}\subset H_4\text{-}CPD_{Py}(OC_6)$ and $3.5 \times 10^5\ M^{-1}$ for $Li^+@C_{60}\subset Ni_2\text{-}CPD_{Py}(OC_6)$ [150].

Upon laser excitation of $Li^+@C_{60}\subset Ni_2\text{-}CPD_{Py}(OC_6)$, transient absorption bands due to $Ni_2\text{-}CPD_{Py}(OC_6)^{\cdot+}$ and $Li^+@C_{60}{}^{\cdot-}$ were observed as shown in Figure 5.15a [150]. In this case, electron transfer occurs from $Ni_2\text{-}CPD_{Py}(OC_6)$ to the triplet excited state of $Li^+@C_{60}$ ($^3Li^+@C_{60}{}^*$) rather than from $^3[Ni_2\text{-}CPD_{Py}(OC_6)]^*$ to $Li^+@C_{60}$ as indicated by the disappearance of the absorption band at 750 nm due to $^3Li^+@C_{60}{}^*$, accompanied by appearance of the absorption band at 1035 nm due to $Li^+@C_{60}{}^{\cdot-}$ (Figure 5.15b) [150]. The rate constant of electron transfer from $Ni_2\text{-}CPD_{Py}(OC_6)$ to $^3Li^+@C_{60}{}^*$ to produce the CS state was determined from the rise in the absorbance at 1035 nm due to $Li^+@C_{60}{}^{\cdot-}$ to be $5.7 \times 10^7\ s^{-1}$ [150]. The absorbance at 1035 nm due to $Li^+@C_{60}{}^{\cdot-}$ in the CS state decayed obeying first-order kinetics with the same slope irrespective of the difference in the laser intensity

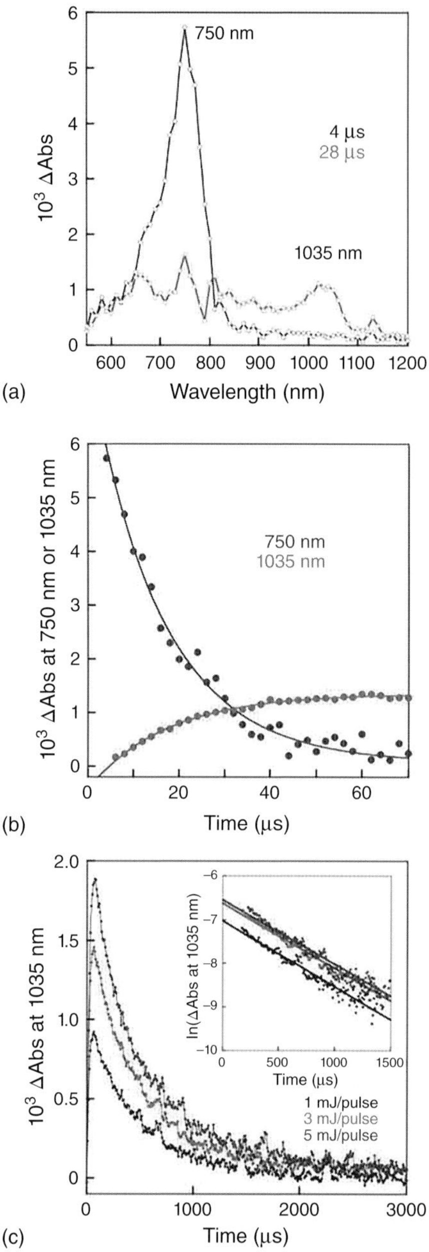

Figure 5.15 (a) Transient absorption spectra of Ni_2-$CPD_{Py}(OC_6)$ with $Li^+@C_{60}$ in deaerated PhCN at room temperature taken at 4 and 28 μs after nanosecond laser excitation at 520 nm. $[Ni_2\text{-}CPD_{Py}(OC_6)] = 2.5 \times 10^{-5}$ M, $[Li^+@C_{60}] = 5.0 \times 10^{-5}$ M. (b) Rise and (c) decay time profiles at 1035 nm with different laser intensities (1, 3, 5 mJ/pulse). Inset: First-order plots. Source: Kamimura et al. 2013 [150]. https://pubs.rsc.org/en/content/articlelanding/2013/sc/c3sc22065f#!divAbstract. https://creativecommons.org/licenses/by/3.0/. Licensed under CCBY 3.0.

(Figure 5.15c) [150]. This clearly indicates that the decay of the CS state occurs via intrasupramolecular back electron transfer rather than a bimolecular back electron-transfer reaction between the CS states. The CS lifetime was determined from the slope of the first-order plots in Figure 5.13c to be 0.67 ms, which is the longest value ever reported for non-covalent monomer dimer porphyrin–fullerene supramolecules in solution [150]. The CS state was also observed for $Li^+@C_{60}⊂H_4$-$CPD_{Py}(OC_6)$. The quantum yields of the CS states were estimated to be 0.13 for $Li^+@C_{60}⊂Ni_2$-$CPD_{Py}(OC_6)$ and 0.32 for $Li^+@C_{60}⊂H_4$-$CPD_{Py}(OC_6)$ and by means of the comparative method with the absorption intensities of the CS states ($Li^+@C_{60}{}^{\bullet-}$: $\varepsilon(1035\,nm) = 7300\,M^{-1}\,cm^{-1}$) [150]. When $Li^+@C_{60}$ was replaced by pristine C_{60}, no CS states were produced as predicted by their higher energy levels than those of the triplet excited states of $CPD_{Py}(OC_6)$ and C_{60} [150].

The mechanisms of intrasupramolecular photoinduced charge separation in $Li^+@C_{60}⊂Ni_2$-$CPD_{Py}(OC_6)$ is shown in Scheme 5.2 [150]. The singlet excited state of Ni_2-$CPD_{Py}(OC_6)$ ($^1[H_4$-$CPD_{Py}(OC_6)]^*$) is generated upon photoexcitation of $Li^+@C_{60}⊂Ni_2$-$CPD_{Py}(OC_6)$ at 420 nm, where the porphyrin moiety is exclusively excited. The excited state energy of Ni_2-$CPD_{Py}(OC_6)$ (E_s =1.97 eV) is slightly higher than that of $^1[Li^+@C_{60}]^*$ (E_s = 1.94 eV) [151]. Although electron transfer from $^1[Ni_2$-$CPD_{Py}(OC_6)]^*$ to $Li^+@C_{60}$ is energetically possible (Scheme 5.2), the intersystem crossing to generate $^3[Ni_2$-$CPD_{Py}(OC_6)]^*$ occurs with the rate constant of $8.3 \times 10^8\,s^{-1}$ [150]. Then, electron transfer occurs from $^3[Ni_2$-$CPD_{Py}(OC_6)]^*$ to $Li^+@C_{60}$ with the driving force of 0.44 eV to produce the CS state. The CS state decays slowly via intrasupramolecular BET with the lifetime of 0.67 ms (Scheme 5.2) [150].

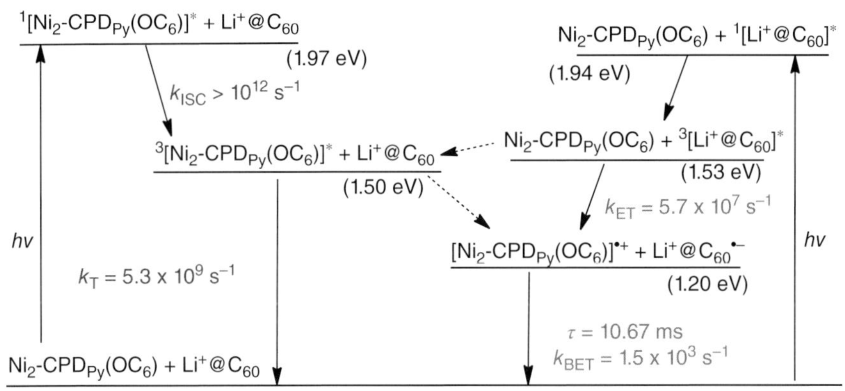

Scheme 5.2 Energy diagram for $Li^+@C_{60}⊂Ni_2$-$CPD_{Py}(OC_6)$; broken arrow: minor pathway. Source: Kamimura et al. 2013 [150]. https://pubs.rsc.org/en/content/articlelanding/2013/sc/c3sc22065f#!divAbstract. https://creativecommons.org/licenses/by/3.0/. Licensed under CCBY 3.0.

When porphyrin dimers are replaced by a porphyrin trimer, the tripod conformation of a porphyrin trimer ($TPZn_3$) in Figure 5.16 makes it possible to capture a fullerene derivative containing a pyridine moiety (PyC_{60}) [152] inside the cavity strongly by π–π interactions together with the coordination bond between Zn^{2+} and the pyridine moiety (Scheme 5.3) [153–156]. The association constant of $TPZn_3$ with PyC_{60} ($1.1 \times 10^5\,M^{-1}$ in toluene) determined from the UV–vis absorption spectral titration (Figure 5.17a) is much larger as compared with those of the corresponding monomer (MPZn) and dimer porphyrin ($DPZn_2$) [156].

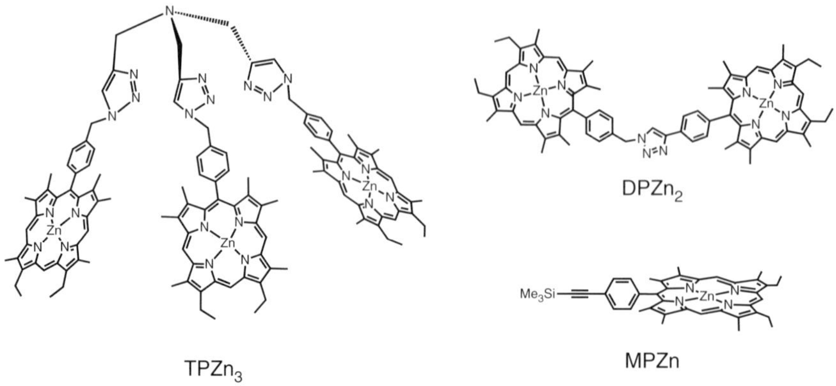

Figure 5.16 A porphyrin tripod and the reference dimer and monomer. Source: Takai et al. 2010 [156]. Reproduced with permission of American Chemical Society.

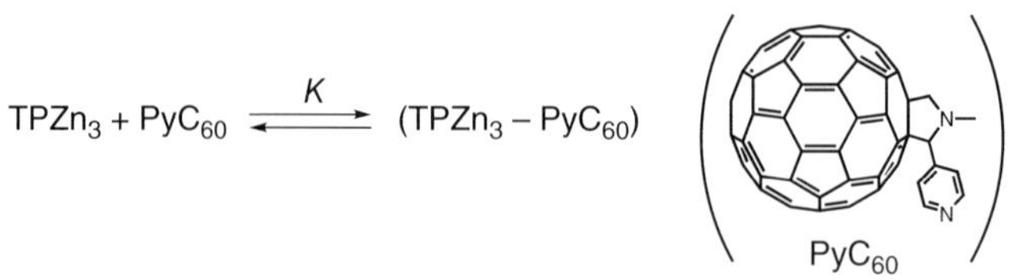

Scheme 5.3 Formation of a supramolecular complex between TPZn$_3$ and PyC$_{60}$. Source: Takai et al. 2010 [156]. Reproduced with permission of American Chemical Society.

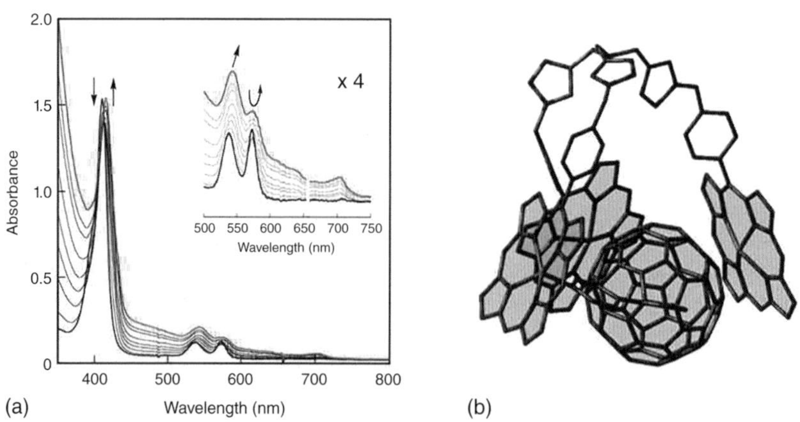

Figure 5.17 (a) UV–vis spectral changes upon addition of PyC$_{60}$ (0–48 μM) to a *o*-DCB solution of TPZn$_3$ (3 μM) at 298 K. (b) The structure of the TPZn$_3$–PyC$_{60}$ complex was optimized by DFT at the B3LYP/3-21G(*) level. Source: Takai et al. 2010 [156]. Reproduced with permission of American Chemical Society.

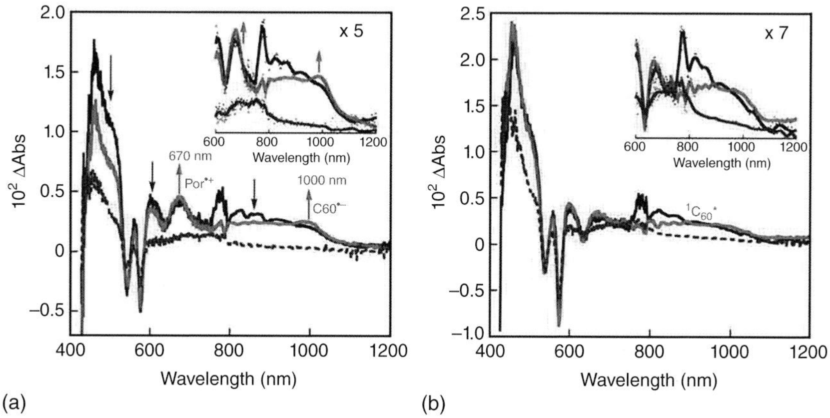

Figure 5.18 Differential transient absorption spectra of (a) $TPZn_3$ (7.0×10^{-6} M) and (b) MPZn (1.1×10^{-5} M) in the presence of PyC_{60} (2.3×10^{-5} M) obtained at 2 ps, 62 ps, and 2800 ps after femtosecond laser pulse irradiation at 410 nm in deaerated toluene at 298 K. Source: Takai et al. 2010 [156]. Reproduced with permission of American Chemical Society.

The ^1H NMR signals of $TPZn_3$ exhibit downfield shifts upon complexation with PyC_{60}, whereas the pyridyl protons of PyC_{60} exhibit large upfield shifts by the complexation, which is ascribed to the influence of the large porphyrin aromatic ring current [156]. This result clearly shows that the pyridyl group of PyC_{60} coordinates to the central zinc ions of $TPZn_3$. The encapsulation of PyC_{60} into the cavity of $TPZn_3$ was supported by the DFT-optimized structure (B3LYP/3-21G(*) basis set) in Figure 5.17b [156].

The occurrence of photoinduced electron transfer from $^1TPZn_3{}^*$ to PyC_{60} was confirmed by femtosecond laser flash photolysis measurements in Figure 5.18a, where the transient absorption spectrum due to $^1TPZn_3{}^*$ changes as time elapses to afford the absorption bands at $\lambda_{max} = 1000$ nm due to the monofunctionalized fullerene radical anion [157, 158] and at 670 nm due to the one-electron oxidized species of $TPZn_3$ ($TPZn_3{}^{\cdot+}$) [156, 159, 160].

In sharp contrast to the $TPZn_3$–PyC_{60} complex, the transient absorption spectrum of monomer porphyrin (MPZn) in the presence of PyC_{60} (Figure 5.18b) exhibits absorbance change due to the energy transfer from $^1MPZn^*$ to PyC_{60} to give the singlet excited state $^1PyC_{60}{}^*$ (1.76 eV), followed by conversion to the triplet excited states $^3MPZn^*$ and $^3PyC_{60}{}^*$ at (Figure 5.16b), accompanied by the recovery of the ground state [156].

The energy diagrams of photodynamics for $TPZn_3$ and MPZn in the presence of PyC_{60} in toluene are shown in Scheme 5.4a,b, respectively [156]. The energy level (1.49 eV) of the CS state ($TPZn_3{}^{\cdot+}$–$PyC_{60}{}^{\cdot-}$) is lower than the energy level of the triplet excited state of PyC_{60} moieties (1.56 eV). The rate constant (k_{ET}) of photoinduced electron transfer from $^1TPZn_3{}^*$ to PyC_{60} is larger than the rate constant of intersystem crossing. From the rate constant of back electron transfer ($k_{BET} = 1.9 \times 10^9\ s^{-1}$), the lifetime of the CS state is determined to be $\tau_{CS} = 0.53$ ns. In contrast, only energy transfer from $^1MPZn^*$ to PyC_{60} occurs to produce $^1PyC_{60}{}^*$, in competition with intersystem crossing to $^3MPZn^*$ [156].

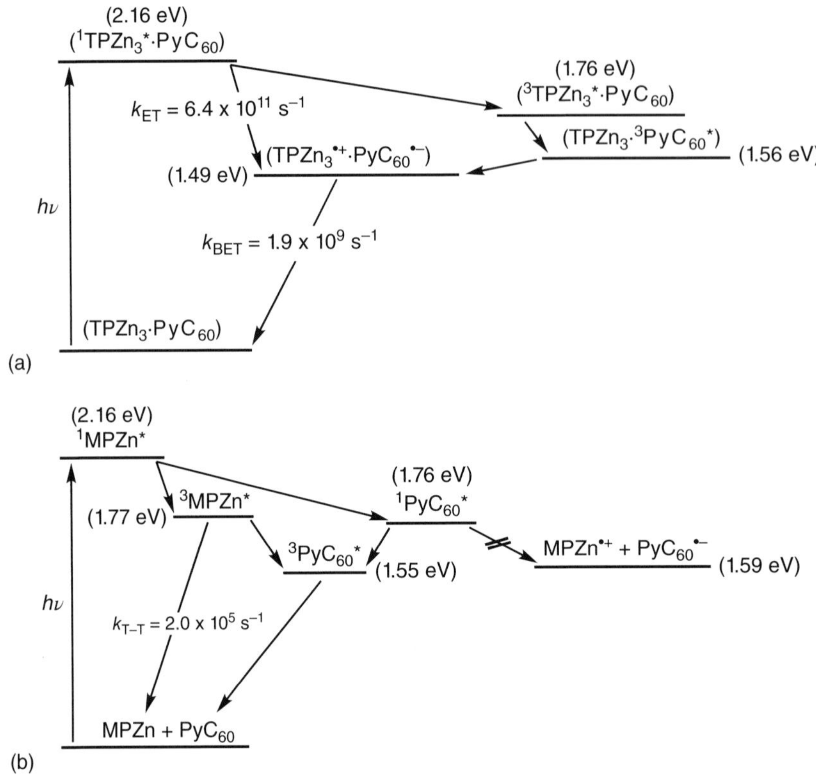

Scheme 5.4 Energy diagram for photodynamics of (a) $TPZn_3$ and (b) MPZn in the presence of PyC_{60} in toluene. Source: Takai et al. 2010 [156]. Reproduced with permission of American Chemical Society.

$TPZn_3$ also forms a stable 1 : 1 complex with gold(III) tetra(4-pyridyl)porphyrin ($AuTPyP^+$) in nonpolar solvents [161]. The strong binding of $TPZn_3$ with $AuTPyP^+$ results from the encapsulation of $AuTPyP^+$ inside the cavity of $TPZn_3$ through multiple coordination bonds. The efficient quenching of the singlet excited state of $TPZn_3$ occurs via a photoinduced electron-transfer pathway in the $TPZn_3$–$AuTPyP^+$ complex as in the case of the $TPZn_3$–PyC_{60} complex [161].

Besides coordination bonds, π–π interaction has merited increasing attention because of its important role in biological systems, such as for π-stacking of double-strand DNA. Acridinium ion is suitable as a component of an artificial photosynthetic reaction center because of the small reorganization energy (λ) of electron transfer, which results in fast photoinduced electron transfer but extremely slow back electron transfer (vide supra) [60]. A free base cofacial bisporphyrin, H_4DPOx, forms a π-complex with acridinium ion (AcH^+) by π–π interaction in PhCN (Scheme 5.5) [135]. Formation of the H_4DPOx–AcH^+ π-complex was probed by UV–vis and NMR spectra [135]. The binding constant between AcH^+ and H_4DPOx is determined as 9.7×10^4 M^{-1} [135]. The optimized structure of the π-complex between H_4DPOx and AcH^+ by DFT calculation is shown in Figure 5.19a [135]. The HOMO (highest occupied molecular orbital)

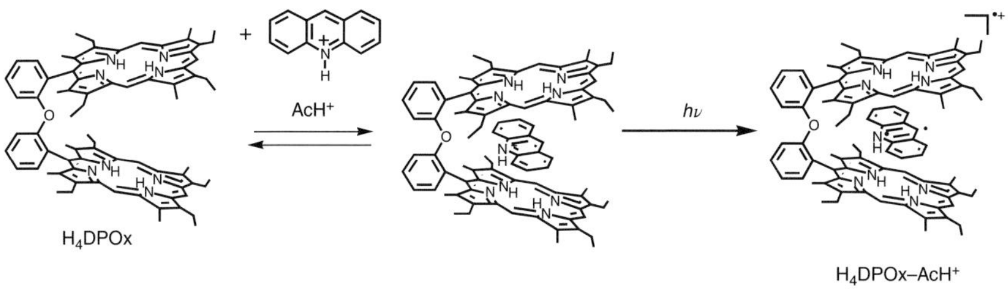

Scheme 5.5 Photoinduced electron transfer in a π-complex between a free base cofacial bisporphyrin (H_4DPOx) and acridinium ion (AcH^+). Source: Tanaka et al. 2006 [135]. Reproduced with permission of American Chemical Society.

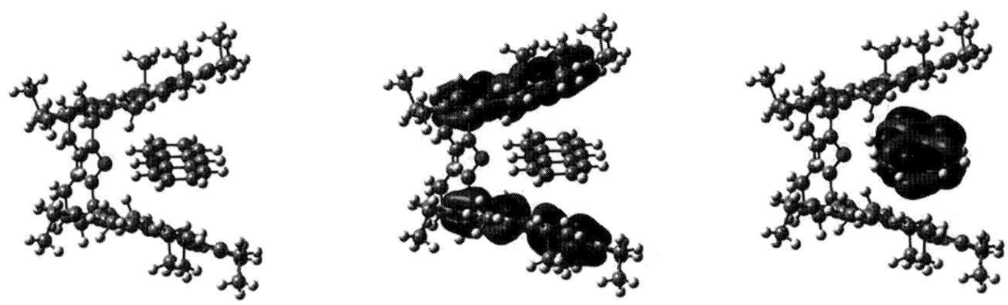

Figure 5.19 (a) Optimized structure of $H_4DPOx–AcH^+$ by DFT calculation. (b) HOMO and (c) LUMO orbitals calculated by a DFT method at B3LYP/6-31G*//B3LYP/3-21G level. Source: Tanaka et al. 2006 [135]. Reproduced with permission of American Chemical Society.

and LUMO (lowest unoccupied molecular orbital) of $H_4DPOx^{\cdot+}–AcH^{\cdot}$ are localized on the porphyrin and AcH^+ moieties as shown in Figure 5.19b,c, respectively. Thus, photoinduced electron transfer from the singlet excited state ($^1H_4DPOx^*$) to the AcH^+ moiety in the π complex occurs efficiently to produce the ET state ($H_4DPOx^{\cdot+}–AcH^{\cdot}$) upon photoexcitation of the H_4DPOx moiety (Scheme 5.5) [135].

The occurrence of photoinduced electron transfer in the π-complex was confirmed by nanosecond laser flash photolysis [135]. The quantum yield of the ET state at 298 K is determined as 0.90 ± 0.05 [135]. The energy diagram of photoinduced electron transfer in the $H_4DPOx–AcH^+$ complex is summarized in Scheme 5.6 [135]. The energy level of the ET state ($-\Delta G_{ET}$) in the π complex is determined as 1.28 eV from the redox potentials of the π-complex, which is lower than that of the triplet states of H_4DPOx (1.59 eV) and AcH^+ (2.01 eV) [135]. This is the reason why the transient absorption spectra due to the ET state can be observed instead of the T–T absorption spectrum. The rate constant (k_{ET}) of electron transfer from the singlet excited state of H_4DPOx ($^1H_4DPOx^*$) to AcH^+ in the $H_4DPOx–AcH$ complex was determined as $2.9 \times 10^8\ s^{-1}$ by time-resolved fluorescence measurements [135].

The lifetime of the ET state exhibits large temperature dependence. The temperature dependence of the rate constant of back electron transfer (k_{ET}) is in accordance with the Marcus equation (Eq. (2.1)) as shown in Figure 5.20a [135].

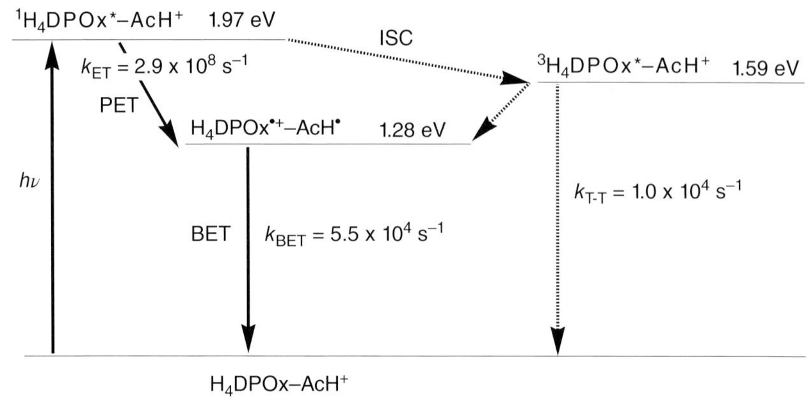

Scheme 5.6 Energy diagram of photoinduced electron transfer of the $H_4DPOx–AcH^+$ complex. Source: Tanaka et al. 2006 [135]. Reproduced with permission of American Chemical Society.

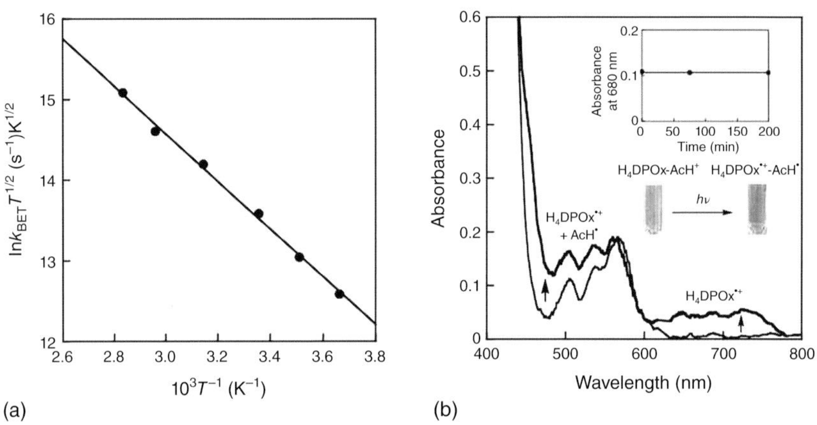

Figure 5.20 (a) Plot of $\ln(k_{BET}T^{1/2})$ vs. T^{-1} for the intramolecular BET in $H_4DPOx^{\cdot+}–AcH^{\cdot}$ in PhCN determined by laser flash photolysis. (b) Visible absorption spectra obtained by photoirradiation with high-pressure mercury lamp of deaerated 2-MeTHF/butyronitrile (9 : 1, v/v) glasses of H_4DPOx (1.8×10^{-5} M) and AcH^+ (3.6×10^{-4} M) at low temperature and measured in liquid N_2 dewar at 77 K. Inset: Time profile at 680 nm and picture images before and after photoirradiation at low temperatures measured at 77 K. Source: Tanaka et al. 2006 [135]. Reproduced with permission of American Chemical Society.

The plot of $\ln(k_{ET}T^{1/2})$ vs. T^{-1} affords a linear curve in Figure 5.20a. From the slope and the intercept, the λ and V values are determined as 0.54 ± 0.1 eV and 1.6 ± 0.3 cm^{-1}, respectively [135].

The extrapolated ET state lifetime at 77 K from the linear plot in Figure 5.20a is estimated as 355 days. Such an extremely long-lived ET state is indeed shown by the steady-state photoirradiation of glassy 2-MeTHF containing 10% butyronitrile of $H_4DPOx–AcH^+$ by a 1000 W high-pressure mercury lamp at low temperature [135]. The new absorption bands due to $H_4DPOx^{\cdot+}$ (450–540 nm

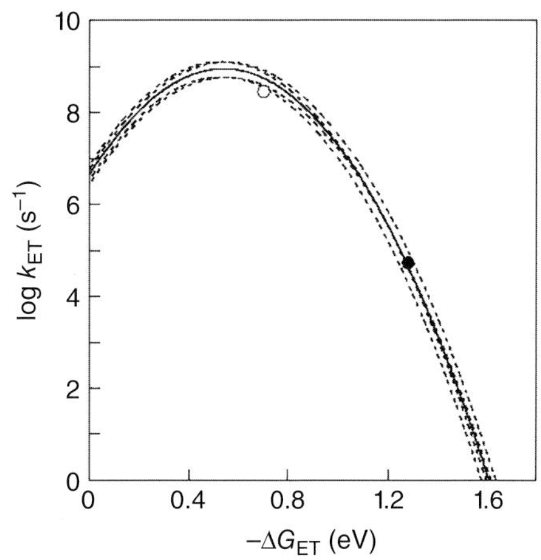

Figure 5.21 Marcus driving force ($-\Delta G_{ET}$) dependence of the ET rate constants (log k_{ET}) using Eq. (2.1) with the λ (0.54 ± 0.01 eV) and V (1.6 ± 0.3 cm^{-1}) values, derived from the temperature dependence of k_{BET} in Figure 5.20a, including the experimental values for photoinduced ET (○) and BET (•) in deaerated PhCN at 298 K. The broken curves indicate the error limit derived from uncertainty of the λ value and the V value. Source: Tanaka et al. 2006 [135]. Reproduced with permission of American Chemical Society.

and 620–800 nm) and AcH$^{\cdot}$ (450–540 nm) are clearly observed as shown in Figure 5.20b [135]. The observed ET state exhibits no decay for 200 minutes at 77 K (inset of Figure 5.20b) [135]. The color change from the ground state of the H_4DPOx–AcH$^+$ complex to the ET state is also shown in the inset of Figure 5.20b [135]. It is important to note that the color and the absorption spectrum of the ET state is back to the original one when temperature is increased to 298 K and such a change can be repeated many times [135].

Since the λ and V values are determined from the temperature dependence of k_{BET}, the driving force dependence of the ET rate constant in the π-complex at 298 K is obtained according to Eq. (2.1) as shown in Figure 5.21 [135]. The k_{ET} value of electron transfer from $^1H_4$DPOx* to AcH$^+$ in the H_4DPOx–AcH complex (2.9×10^8 s^{-1}) agrees with the predicted value from the Marcus curve within the experimental error (open circle in Figure 5.21) [135]. The forward ET is nearly on the Marcus top region, whereas the BET process is deeply in the Marcus inverted region. This is the reason why an extremely long-lived ET state of the π-complex is formed in PhCN, in particular at low temperature. Formation of the simple π-complex formed between an electron donor and an acceptor with a long ET lifetime can be utilized to obtain more sophisticated electron donor–acceptor ensembles for the construction of efficient energy conversion systems as described later.

The π–π interaction also plays an important role in diameter-selective separation of SWNTs from starting purified SWNTs by forming supramolecular nanohybrids between porphyrin–peptide hexadecamer [P(H_2P)$_{16}$] [162, 163] and SWNTs as shown in Figure 5.22a [164]. The supramolecular complex formation is made possible by wrapping of peptidic backbone in P(H_2P)$_{16}$ and π–π interaction between porphyrins and nanotubes to extract the large-diameter nanotubes (c. 1.3 nm) as revealed by ultraviolet–visible–near infrared and Raman spectroscopy as well as high-resolution transmission electron microscopy [164]. Laser photoexcitation of P(H_2P)$_{16}$/SWNTs in DMF results in the appearance of prominent transient absorption bands at 450 nm together with a broad absorption band in the 500–800 nm region, which overlaps with a number of bleaching

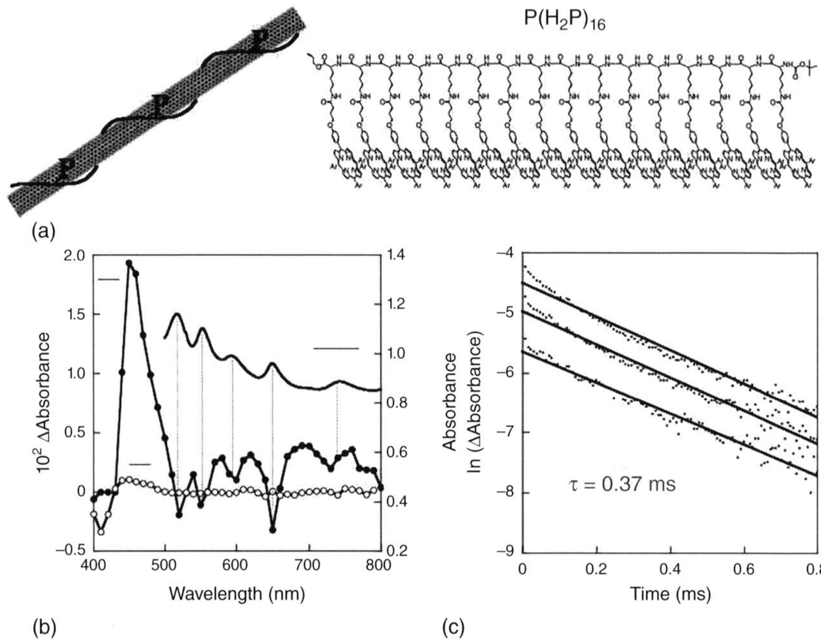

Figure 5.22 (a) Supramolecular nanohybrids of porphyrin–peptide hexadecamer [$P(H_2P)_{16}$] with SWNTs. (b) Transient absorption spectra of $P(H_2P)_{16}$/SWNTs (solid line with black circles) and $P(H_2P)_{16}$ (solid line with white circles) in DMF at 298 K taken at 6 μs after laser excitation at 440 and 427 nm, respectively. Inset: UV–vis–NIR absorption spectra of $P(H_2P)_{16}$/SWNTs. (c) First-order plots at 480 nm with different laser powers (7, 3, and 1 mJ, respectively). Source: Saito et al. 2007 [164]. Reproduced with permission of American Chemical Society.

bands due to the porphyrin (518, 553, 597, and 649 nm) and SWNTs (750 nm) as shown in Figure 5.22b (closed circles) [164]. By taking into account the bleaching bands, the observed transient absorption spectrum agrees with that of free base porphyrin radical cation [164]. In contrast, no transient absorption is observed upon photoexcitation of $P(H_2P)_{16}$ (open circles in Figure 5.22b) [164]. Thus, the observed transient absorption spectrum is assigned to the CS state of $P(H_2P)_{16}$/SWNTs, which is produced by photoinduced electron transfer from the singlet and triplet excited states of $P(H_2P)_{16}$ to SWNTs. The CS lifetime is determined as 0.37 ± 0.03 ms from the first-order decay of the absorbance due to the CS state (Figure 5.22c) [164]. This method opens a new strategy toward the extraction of large-diameter SWNTs without destruction of the π-conjugated system within SWNTs and development of an efficient light energy conversion system.

5.3 Electron-Transfer Switching

As described above, anion binding results in a negative shift in the one-electron oxidation potentials of the electron donors, leading to stabilization of the CS state of the anion-bound electron donor–acceptor dyads. Such a change in the one-electron oxidation potential of the electron donors by anion binding can also be used to control the thermal (i.e. ground state) electron transfer from an electron donor to an electron acceptor, provided these moieties are part of an appropriately designed supramolecular electron-transfer complex. Substrate binding can also be used to modulate the redox potentials of the ensemble and help switch electron transfer "on" or "off." As a proof of these design principles, we created a non-covalent ensemble based on a tetrathiafulvalene-calix[4]pyrrole (TTF-C4P) donor [165] and a dicationic mesityl quinone (BIQ^{2+}) acceptor [166]. Using this system, it was found that the addition of selected anions or cations could be used to control the direction of electron transfer [167].

When TTF-C4P (30 μM) is mixed with 1 M equiv of $BIQ^{2+}\cdot 2PF_6^-$, no reaction was observed [167]. Under these conditions the TTF-C4P exists in its so-called 1,3-alternate conformation whose size and shape preclude effective supramolecular interactions with BIQ^{2+}. When tetra-*n*-hexylammonium chloride (THACl) was added to the solution of TTF-C4P and $BIQ^{2+}\cdot 2PF_6^-$, however, new absorption bands at 379, 751, and 1995 nm appeared at the expense of the original TTF-C4P absorption band (λ_{max} = 329 nm). A clear isosbestic point at 340 nm is seen, as shown in Figure 5.23a [167]. Under these conditions, switching of the TTF-C4P unit to its so-called cone conformation is expected. The absorption bands at c. 751 and 1995 nm are ascribed to $TTF^{\cdot +}$ and $[TTF]_2^{\cdot +}$ radical cations derived from TTF-C4P, whereas the absorption band at 379 nm is assigned to the reduced radical $BIQ^{\cdot +}$. Thus, electron transfer from TTF-C4P to $BIQ^{2+}\cdot 2PF_6^-$ occurs in the presence of THACl and the yield of the electron-transfer products increases with increasing concentration of THACl (Figure 5.23a) [167]. Once produced, no further changes in these optical features

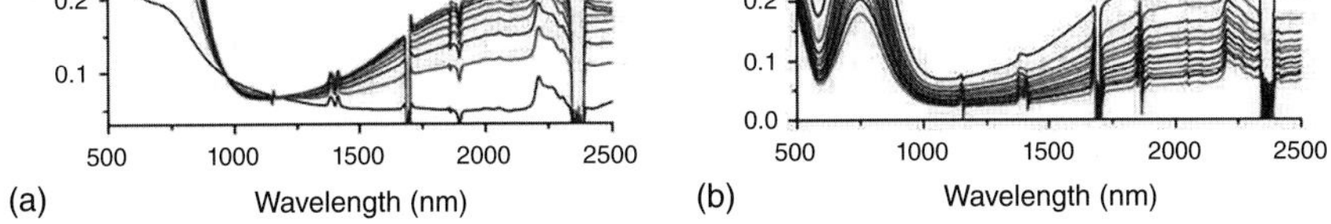

Figure 5.23 (a) Spectroscopic changes observed when TTF-C4P (30 μM) is treated first with 1 M equiv of $BIQ^{2+}\cdot 2PF_6^-$ and then titrated with increasing quantities of THACl in $CHCl_3$. The final spectrum was recorded in the presence of 10 equiv. The inset shows the EPR spectra of TTF-C4P (1.1×10^{-4} M) in $CHCl_3$ recorded at room temperature in the presence of 1 M equiv of $BIQ^{2+}\cdot 2PF_6^-$ and after incremental addition of THACl. (b) Cation induced reverse electron transfer seen upon the step-wise addition of up to 15 equiv of TEACl to a 1 : 1 solution of TTF-C4P and $BIQ^{2+}\cdot 2Cl^-$ in $CHCl_3$. The inset shows the EPR spectra of TTF-C4P, in the presence of 1 equiv of $BIQ^{2+}\cdot 2Cl^-$ at 1.1×10^{-4} M in $CHCl_3$ recorded at room temperature upon incremental addition of TEACl up to 5 M equiv. Source: Park et al. 2010 [167]. Reproduced with permission of The American Association for the Advancement of Science.

were observed under ambient conditions. However, addition of TEACl to a $CHCl_3$ solution of TTF-C4P$^{\cdot+}$ and BIQ$^{\cdot+}$ resulted in back electron transfer from BIQ$^{\cdot+}$ to TTF-C4P$^{\cdot+}$ as indicated by a decrease in absorbance intensity at c. 751 and 1995 nm, accompanied by restoration of the absorbance feature at 329 nm ascribed to the TTF-C4P receptor (Figure 5.23b) [167]. These optical changes are consistent with "switched on" anion-induced electron transfer from TTF-C4P to $BIQ^{2+}\cdot 2PF_6^-$ and its subsequent cation-triggered reversal ("switching off") as shown in Scheme 5.7 [167].

Structural analysis of the crystals of the electron-transfer product produced by mixing TTF-C4P to $BIQ^{2+}\cdot 2PF_6^-$ in the presence of the chloride anion revealed a supramolecular donor–acceptor ensemble, wherein two anion-bound bowl-like TTF-C4P moieties encapsulate a bis-imidazolium quinone guest in a 2 : 1 manner (Figure 5.24a). Based on the structural parameters, this capsule was thought to consist of a tightly coupled biradical species best described as [TTF-C4P]$^{\cdot 2+}\cdot$BIQ$^{\cdot+}\cdot 2(Cl^-)$ [167].

The existence of a tightly coupled biradical "capsule" was further confirmed by low temperature (4 K) EPR analyses of single crystals of the [TTF-C4P]$^{\cdot 2+}\cdot$BIQ$^{\cdot+}\cdot 2(Cl^-)$ salt [167]. A typical triplet signal with zero-field splitting was observed at $g = 4.08$ and 2.007, respectively (Figure 5.24b). The zero field splitting parameter (D) when plotted against the angle of rotation for the single crystal resulted in a sine curve, a finding that is consistent with the presence of an intramolecular radical ion pair [167].

The electron acceptor, the BIQ^{2+} cation, used in the supramolecular donor–acceptor complex with TTF-C4P can be replaced by a different electron

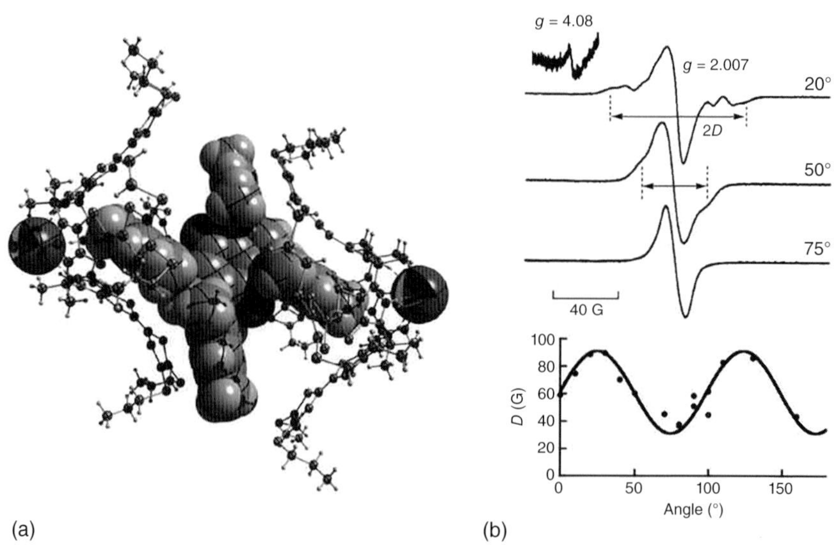

Figure 5.24 (a) Single crystal X-ray structure of the supramolecular donor–acceptor complex of net stoichiometry $[TTF\text{-}C4P]_2\cdot[BIQ^{2+}]\cdot 2(Cl)$. (b) ESR spectra of a crystal of TTF-PQMes recorded at different crystal angles at 4 K (upper panel). The angle variation of the D value. Source: Park et al. 2010 [167]. Reproduced with permission of The American Association for the Advancement of Science.

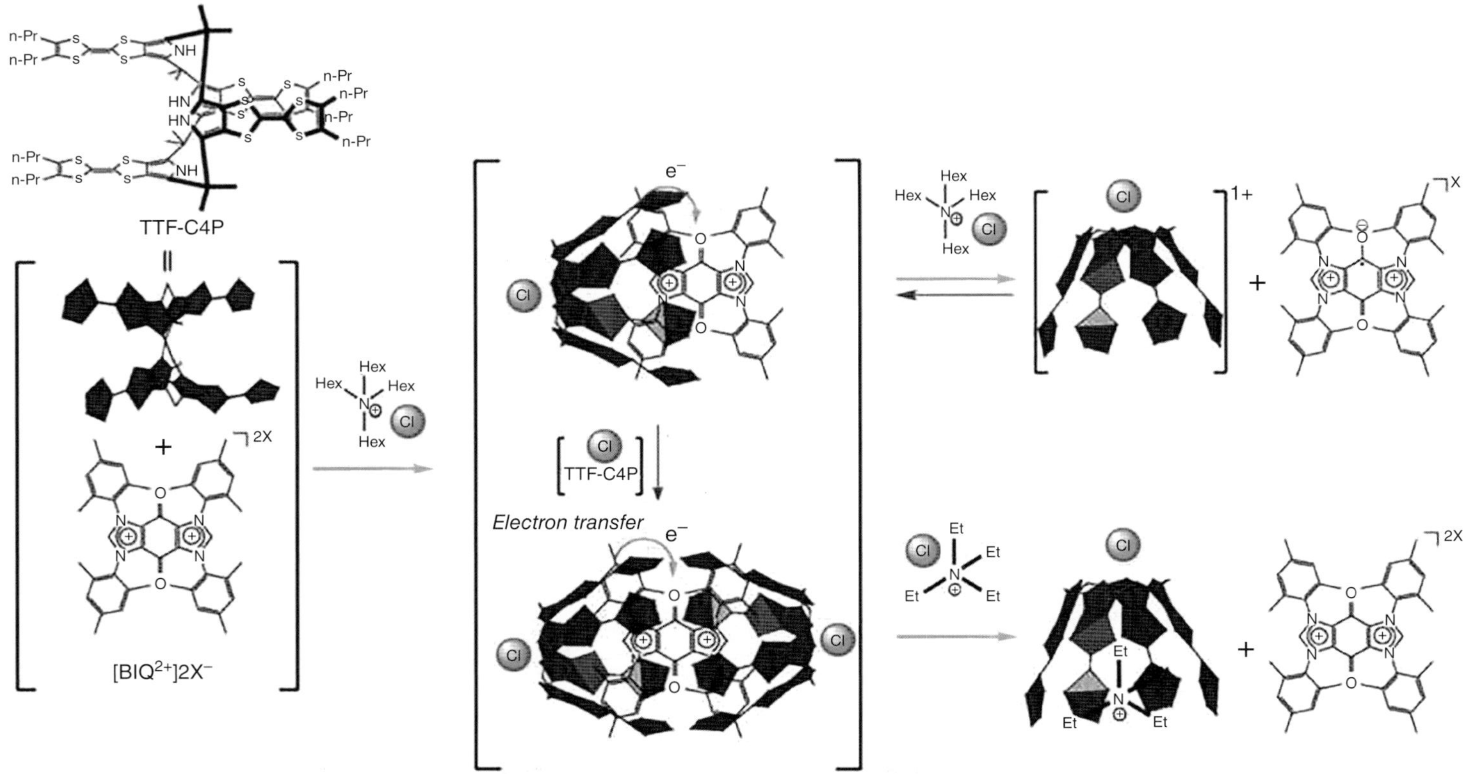

Scheme 5.7 Chemical structures of TTF-C4P and BIQ^{2+} salts, and their proposed ion-mediated electron-transfer reactions. Source: Park et al. 2010 [167]. Reproduced with permission of The American Association for the Advancement of Science.

acceptor that has a similar one-electron reduction potential. Fullerenes are attractive in this context. To date, supramolecular complexes of C_{60} with electron donor hosts have attracted significant attention because of their ability to promote efficient photoinduced electron-transfer reactions [168–171]. However, the ground state C_{60} is not a sufficiently strong electron acceptor to support a thermal ET reaction in the case of TTF-C4P [167]. Lithium ion encapsulated C_{60} ($Li^+@C_{60}$) is known to act as a more effective electron acceptor than pristine C_{60} [129, 130]. When $Li^+@C_{60}$ is employed as an electron acceptor in conjunction with TTF-C4P, a supramolecular ET complex is formed in the presence of a triggering anion (vide infra) [172].

Upon mixing $[Li^+@C_{60}]PF_6$ and TTF-C4P in PhCN, no evidence of electron transfer is observed as expected from the higher one-electron oxidation potential of the TTF-C4P receptor (E_{ox} vs. SCE = 0.51 V) than the one-electron reduction potential of the putative guest, $Li^+@C_{60}$ (E_{red} vs. SCE = 0.14 V) [172]. However, the addition of tetra-*n*-hexylammonium chloride (THACl) to this solution of TTF-C4P and $Li^+@C_{60}$ induced electron transfer from the TTF-C4P to $Li^+@C_{60}$, presumably triggered by binding of the Cl^- anion to the calix[4]pyrrole moiety. Electron transfer was inferred from the appearance of an absorption band at 1035 nm ascribable to the one-electron reduced species produced from $Li^+@C_{60}$ ($Li^+@C_{60}{}^{\cdot-}$) as shown in Figure 5.25a [172]. The absorbance at 1035 nm was also seen to increase with increasing THACl concentration to reach a constant value at the point where electron transfer is complete (Figure 5.25b) [129, 130]. As described above, Cl^- binds to TTF-C4P to induce a conformation change from the so-called 1,3-alternate to the cone conformation due to concerted NH–anion hydrogen bonding interactions, a change that allows initial substrate binding and stabilization of the radical ion pair consisting of the TTF-C4P$^{\cdot+}$(Cl^-) host and the bound $Li^+@C_{60}{}^{\cdot-}$ guest (Scheme 5.8) [172]. In PhCN, Cl^- interacts with TTF-C4P in a 1 : 1 manner with an effective binding constant of c. 1.9×10^4 M^{-1} (Figure 5.25b) [172].

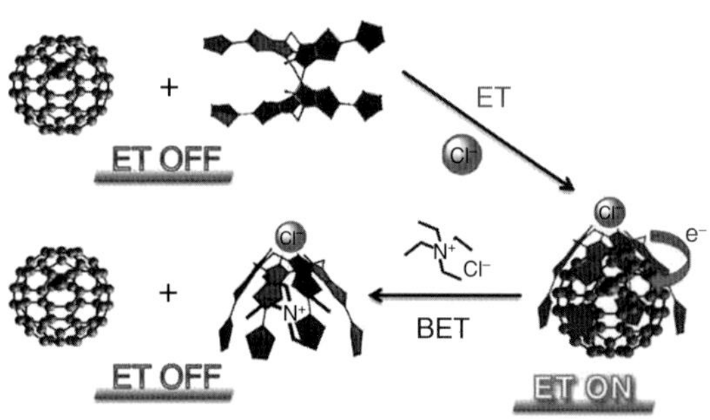

Scheme 5.8 Ion-mediated electron-transfer reactions of a supramolecule between $Li^+@C_{60}$ and TTF-C4P. Source: Fukuzumi et al. 2011 [172]. Reproduced with permission of American Chemical Society.

The radical ion pair between TTF-C4P$^{\cdot+}$(Cl^-) and $Li^+@C_{60}{}^{\cdot-}$ was detected by EPR (Figure 5.26). Here, a four-line signal due to TTF-C4P$^{\cdot+}$(Cl^-) in which the electron spin is localized at only one TTF moiety (a_N = 0.91 G; a_H = 1.10 G),

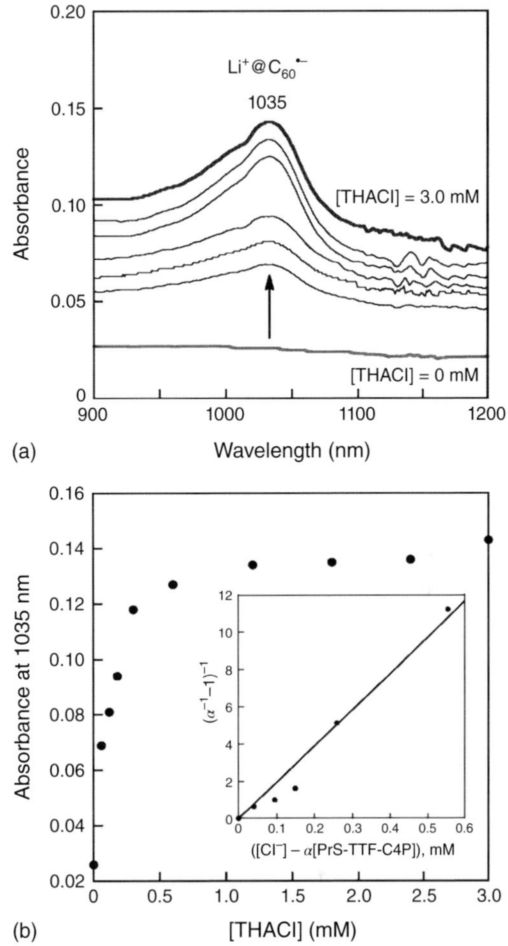

Figure 5.25 (a) Near-IR absorption spectral change in Cl^--promoted electron transfer from TTF-C4P (5.0×10^{-5} M) to $Li^+@C_{60}$ (5.0×10^{-5} M) in the presence of increasing concentrations of THACl in PhCN. (b) Plot of absorbance at 1035 nm vs. concentration of Cl^-. Inset: Plot used to determine the approximate binding constant for the interaction of THACl with TTF-C4P in PhCN at 298 K; $\alpha = (A - A_0)/(A_\infty - A_0)$. Source: Fukuzumi et al. 2011 [172]. Reproduced with permission of American Chemical Society; Yamada et al. 2014 [169]. Reproduced with permission of American Chemical Society.

rather than delocalized within a π-dimer radical cation complex formed by reaction with another TTF moiety, is observed [172]. This finding provides support for the proposed inclusion of $Li^+@C_{60}{}^{\cdot-}$ as a guest within the cavity provided by the cone conformation of TTF-C4P$^{\cdot+}$(Cl^-). Such a binding mode would preclude the formation of a π-dimer radial cation.

The observation of the signal at $g = 4.43$ (Figure 5.26b) provides support for the notion that two unpaired electrons interact to form a triplet state. From the zero-field splitting value ($D = 100$ G) in Figure 5.26c, the distance (r) between two electron spins was estimated using the relation $D = 27\,800/r^3$ to be 6.5 Å [172]. This distance is commensurate with the distance between the TTF-C4P$^{\cdot+}$(Cl^-) receptor and the bound $Li^+@C_{60}{}^{\cdot-}$ guest observed in the X-ray crystal structure of the radical ion pair, as shown in Figure 5.27 [172]. The lack of any additional charged species within the crystal lattice provides further support for the suggestion that an overall neutral complex is formed between a bowl-like TTF-C4P$^{\cdot+}$ radical species and a tightly encapsulated $Li^+@C_{60}{}^{\cdot-}$ guest.

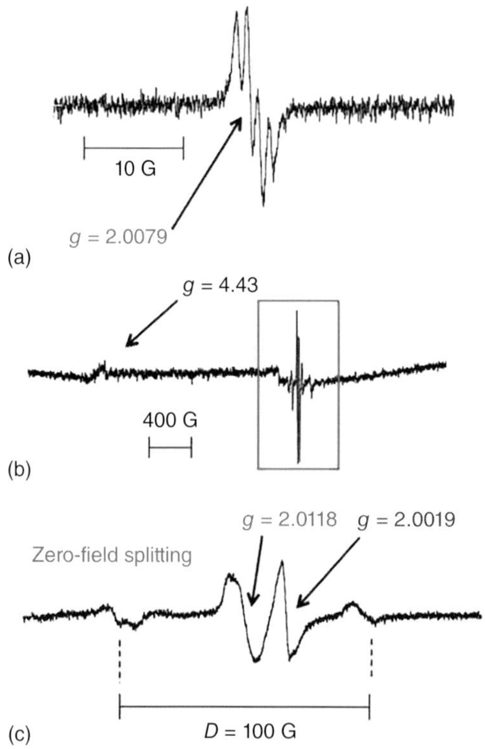

Figure 5.26 (a) EPR spectrum of the products of electron transfer from TTF-C4P (5.0×10^{-5} M) to $Li^+@C_{60}$ (5.0×10^{-5} M) in the presence of TEACl (3.0×10^{-4} M) in PhCN at 298 K. (b) EPR spectrum of the radical ion pair ($TTF\text{-}C4P^{\cdot+}/Li^+@C_{60}^{\cdot-}$) at 77 K. (c) Expanded view of the magnetic field region highlighted by the rectangular frame in (b). Source: Fukuzumi et al. 2011 [172]. Reproduced with permission of American Chemical Society.

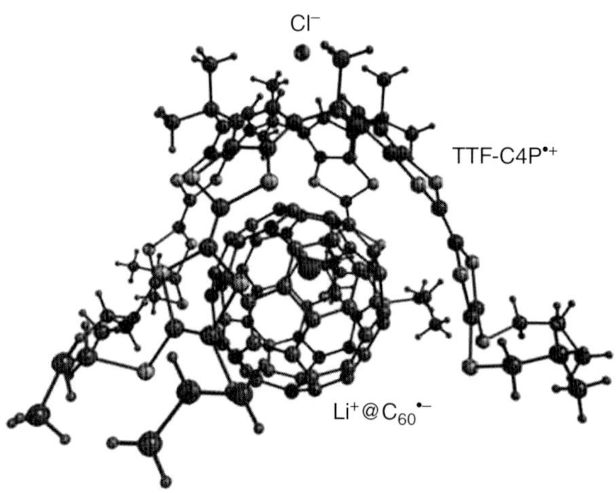

Figure 5.27 Single crystal X-ray structure of the product of electron transfer from TTF-C4P to $Li^+@C_{60}$ produced in the presence of THACl in PhCN. Note that the Li^+ cation is disordered over two positions but is seen in the difference map. Disordered solvent molecules have been removed for clarity. However, no other charged species are seen in the crystal lattice. Source: Fukuzumi et al. 2011 [172]. Reproduced with permission of American Chemical Society.

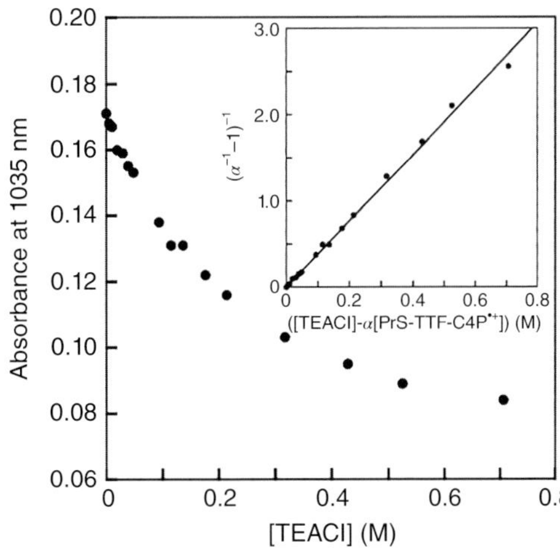

Figure 5.28 Change in absorbance at 1035 nm seen upon the addition of increasing concentration of TEACl to a preformed mixture of TTF-C4P, $Li^+@C_{60}$, and THACl in PhCN. Inset: Plot for determination of the binding constant for the interaction of TEACl with TTF-C4P; $\alpha = (A - A_0)/(A_\infty - A_0)$. Source: Fukuzumi et al. 2011 [172]. Reproduced with permission of American Chemical Society.

When tetraethylammonium chloride (TEACl) was added to the electron-transfer products produced from TTF-C4P and $Li^+@C_{60}$ in the presence of THACl in mixed PhCN/$CHCl_3$ solution, the absorbance due to $Li^+@C_{60}{}^{\bullet-}$ was found to decrease as a function of increasing TEACl concentration as shown in Figure 5.28 [172]. The TEA^+ binds effectively to the TTF-C4P cavity, thereby repelling $Li^+@C_{60}{}^{\bullet-}$ from the cavity; this results in back electron transfer from $Li^+@C_{60}{}^{\bullet-}$ to TTF-C4P$^{\bullet+}$ to produce the original $Li^+@C_{60}$ and the TEA^+-encapsulated form of TTF-C4P (Scheme 5.8) [172]. The effective binding constant of TEA^+ to TTF-C4P was determined to be 3.8 M^{-1}, as determined from the decrease in absorbance at 1035 nm as a function of [TEACl] (a linear plot to determine the binding constant is shown as an inset to Figure 5.28) [172].

5.4 Dendrimers

Because the morphology and the photochemical function of porphyrin dendrimers are similar to those of the light-harvesting units in photosynthesis, porphyrin dendrimers have attracted considerable interest as a unique way to assemble porphyrins, harvesting solar light in artificial photosynthesis [173–176]. Porphyrin dendrimers can also be used to combine light-harvesting units with reaction center units. For example, a zinc porphyrin dendrimer [D(ZnP)$_{16}$] [177] is combined with an electron acceptor, *N*-methyl-2-(4′-pyridyl)-3,4-fulleropyrrolidine (PyC_{60}), which has a pyridine binding site [175], to form a supramolecular complex as shown in Figure 5.29 [178–180]. The porphyrin dendrimers were synthesized by coupling of the porphyrin activated ester [5-amino-2-{5,10,15,20-tetrakis(3,5-di-*tert*-butylphenyl)}] -5-oxopentanoic acid 2,5-dioxopyrrolidin-1-yl ester with the appropriate first, second, or third generation polypropylenimine dendrimer [178].

Photoexcitation of the Soret band of D(ZnP)$_{16}$ at 438 nm in PhCN results in fluorescence at λ_{max} = 609 and 645 nm [180]. The fluorescence was efficiently quenched by addition of PyC_{60} to a PhCN solution of D(ZnP)$_{16}$ [180]. The

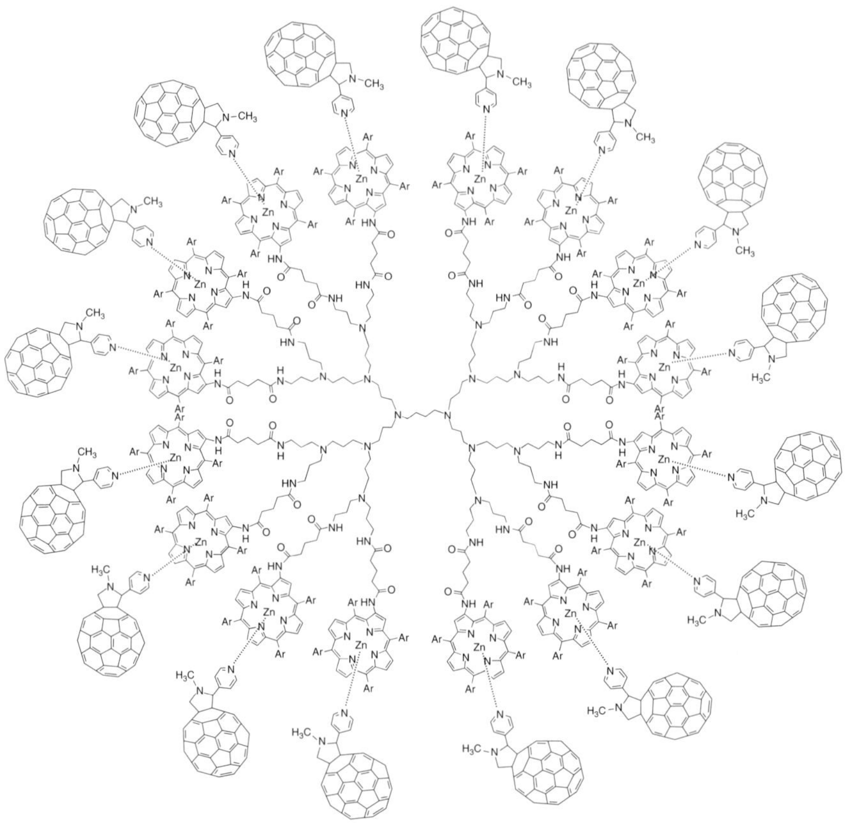

Figure 5.29 A supramolecular complex between a zinc porphyrin dendrimer [$D(ZnP)_{16}$] and PyC_{60}. Source: Fukuzumi et al. 2011 [180]. Reproduced with permission of Royal Society of Chemistry.

fluorescence intensity decreases to reach a constant value with increasing concentration of PyC_{60}. The formation constant (K) of 1 : 1 ZnP monomer unit–PyC_{60} complex was determined from the fluorescence quenching to be $5.0 \times 10^4\ M^{-1}$, which is significantly larger than the K value ($1.6 \times 10^4\ M^{-1}$) determined from the absorption spectral change [180]. The larger K value determined from the fluorescence quenching results from the excited energy migration between the porphyrin units, because the fluorescence of the unbounded ZnP moiety is quenched by PyC_{60} bound to a different ZnP moiety via the energy transfer. In contrast to the case of $D(ZnP)_{16}$, a ZnP monomer cannot form the supramolecular complex with PyC_{60} in a coordinating PhCN solution, when no fluorescence quenching occurred by PyC_{60} [180]. This indicates clearly dendrimer effects on the binding and excited energy migration.

The occurrence of photoinduced charge separation in the supramolecular complex was confirmed by nanosecond laser flash photolysis measurements of the $D(ZnP)_{16}$–PyC_{60} complex in PhCN as shown in Figure 5.30 [180]. The absorption band due to $PyC_{60}{}^{\cdot-}$ is clearly observed at 1000 nm together with that due

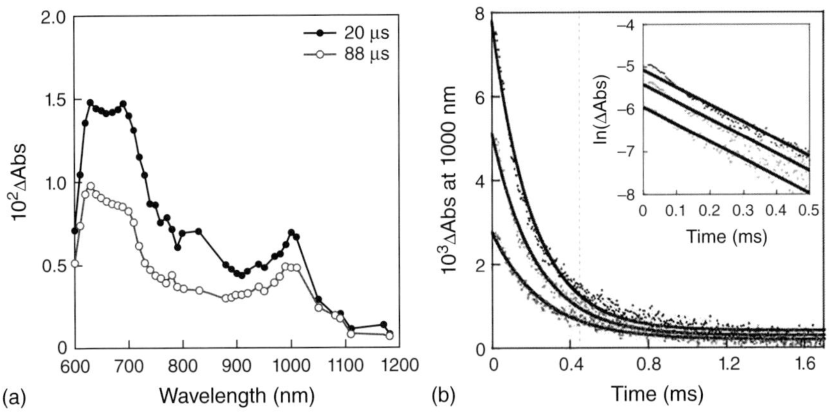

Figure 5.30 (a) Transient absorption spectra of $D(ZnP)_{16}$ (2.3×10^{-5} M) in the presence of PyC_{60} (5.2×10^{-5} M) in deaerated PhCN at 298 K taken at 20 μs (solid line with black circles) and 88 μs (solid line with white circles) after laser excitation at 561 nm, respectively. (b) Time profiles of the absorption at 1000 nm of $PyC_{60}^{\cdot-}$ with different laser intensities (3.0, 1.0, and 0.5 mJ/pulse) at 298 K. Inset: First-order plots. Source: Fukuzumi et al. 2011 [180]. Reproduced with permission of Royal Society of Chemistry.

to $ZnP^{\cdot+}$ at 630 nm after laser excitation at 561 nm where only the ZnP moiety is excited (Figure 5.30a) [180]. This indicates that the CS state of the supramolecular complex is formed via photoinduced electron transfer from the singlet excited state of the ZnP moiety to the PyC_{60} moiety. The quantum yield (Φ) of the CS state of the $D(ZnP)_{16}$–PyC_{60} complex is determined to be 25% [180]. The decay of the absorption band at 1000 nm due to $PyC_{60}^{\cdot-}$ obeyed clean first-order kinetics with the same slope irrespective of different laser power intensities (Figure 5.30b) [180]. Because intermolecular back electron transfer would obey second-order kinetics, the first-order decay of the CS state results from back electron transfer in the supramolecular complex. The lifetime of the CS state is determined as 0.25 ms at 298 K [180].

5.5 Supramolecular Solar Cells

Supramolecular assemblies of porphyrins and π-conjugated molecules, which undergo efficient photoinduced electron transfer as described above, provide excellent components to construct supramolecular solar cells. New approaches for the production of efficient and low-cost organic solar cells are desired because of increasing attention toward solar energy conversion to develop inexpensive renewable energy sources [181–183]. Typical examples of supramolecular solar cells are described in this section.

The first example is a supramolecular solar cell prepared using self-organization of π-complexes between porphyrin (donor) and fullerene (acceptor) units by clusterization with gold nanoparticles on nanostructured SnO_2 electrodes as shown in Figure 5.31a [184]. Porphyrin alkanethiolate monolayer-protected gold nanoparticles ($H_2PCnMPC$: n is the number of methylene groups in

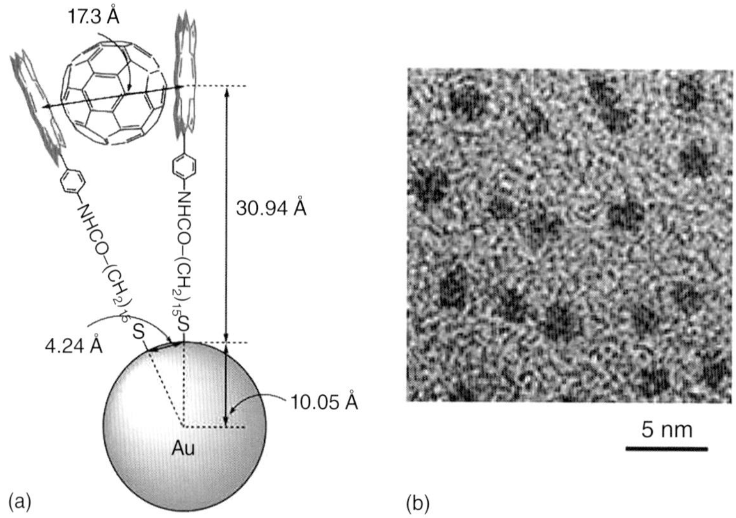

Figure 5.31 (a) Insertion of C_{60} between the porphyrin rings of H_2PC15MPC and (b) the TEM image of H_2PC15MPC. Source: Hasobe et al. 2003 [184]; Hasobe et al. 2005 [187].

the spacer) with spherical shape (Figure 5.31b) are prepared starting from porphyrin–alkanethiol [184–186]. Given the values of the elemental analysis of H_2PC11MPC (H: 4.88%; C: 44.77%; N: 3.10%), there are 57 porphyrin alkanethiolate chains on gold surface for H_2PC11MPC [184]. The distance between two porphyrins next to each other to accommodate C_{60} due to π–π interaction can be controlled by changing the number of methylene groups in the spacer and the best distance (17.3 Å) is obtained when $n = 15$ [187]. The diameter of the gold core in H_2PC15MPC is determined as 2.1 ± 0.4 nm (Figure 5.31b) [187]. These nanoparticles form π-complexes with C_{60} and they are clusterized in acetonitrile/toluene mixed solvent. The composite clusters were easily deposited onto the OTE/SnO_2 electrode by electrophoretic deposition method to prepare the OTE/SnO_2/(H_2PCnMPC+C_{60})$_m$ electrode as shown in Figure 5.32 [187]. Upon application of a DC electric field of 200 V between the OTE/SnO_2 and OTE electrodes (Figure 5.32), which were immersed together in a mixed acetonitrile/toluene (3/1, v/v) solution containing (H_2PCnMPC+C_{60})$_m$ clusters, the mixed clusters were readily deposited on the SnO_2 nanocrystallites.

The photoelectrochemical performance of the OTE/SnO_2/(H_2PCnMPC+C_{60})$_m$ electrode was examined in acetonitrile containing NaI (0.5 M) and I_2 (0.01 M) as redox electrolyte using a Pt gauge counter electrode [187]. The effect of the alkanethiolate chain length on the incident photon-to-photocurrent efficiency (IPCE) is shown in Figure 5.33a [187]. The action spectra indicate that higher IPCE and broader photoresponse are attained with the longer chain length of H_2PCnMPC [187]. The OTE/SnO_2/(H_2PC15MPC+C_{60})$_m$ ([H_2P] = 0.19 mM, [C_{60}] = 0.38 mM) electrode exhibits a maximum IPCE value (54%) and a very broad photoresponse (up to ~1000 nm), which extends to the near-IR region [187]. As mentioned above, a long methylene spacer of H_2PC15MPC provides the most appropriate space for fullerene molecules to be inserted between

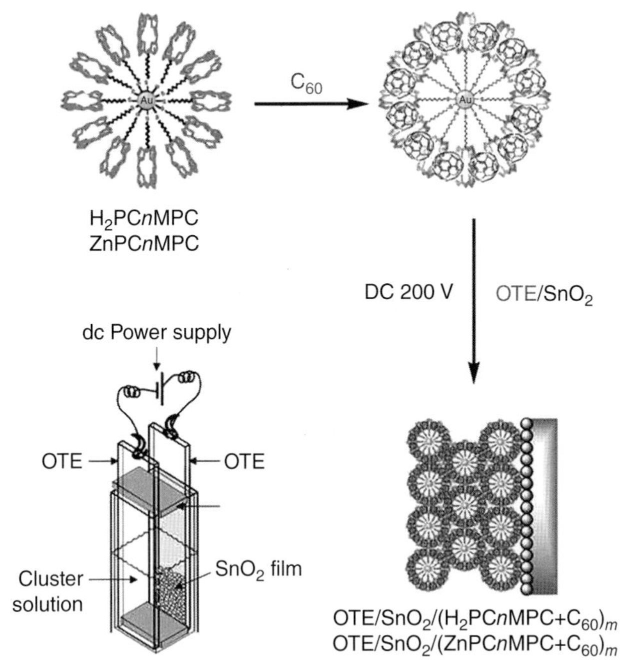

Figure 5.32 Illustration of the preparation of the $OTE/SnO_2/(H_2PC15MPC+C_{60})_m$ electrode. Source: Hasobe et al. 2005 [187]. Reproduced with permission of American Chemical Society.

neighboring two porphyrin rings due to π–π interaction as compared with the clusters with a shorter methylene spacer, leading to more efficient photocurrent generation [187].

The power conversion efficiency (η) of the photoelectrochemical cell of the $OTE/SnO_2/(H_2PC15MPC+C_{60})_m$ electrode was determined by varying the load resistance (Figure 5.33b) [187]. A drop in the photovoltage and an increase in the photocurrent are observed with decreasing load resistance. Power conversion efficiency η is determined by Eq. (5.2)

$$\eta = \mathrm{FF} \times I_{sc} \times V_{oc}/W_{in} \tag{5.2}$$

where the fill factor (*FF*) is defined as $FF = [IV]_{max}/I_{sc}\ V_{oc}$, where V_{oc} is the open circuit photovoltage and I_{sc} is the short circuit photocurrent [188]. The $OTE/SnO_2/(H_2PC15MPC+C_{60})_m$ system has a much larger fill factor (*FF*) of 0.43, open circuit voltage (V_{oc}) of 380 mV, short circuit current density (I_{sc}) of 1.0 mA cm^{-2}, and the overall power conversion efficiency (η) of 1.5% at input power (W_{in}) of 11.2 mW cm^{-2} as compared with the reference system [$OTE/SnO_2/(H_2P\text{-ref}+C_{60})_m$] [187].

A similar supramolecular solar cell was developed using porphyrin–peptide oligomers [porphyrin functionalized α-polypeptides: $P(H_2P)_n$ ($n = 1, 2, 4, 8, 16$)] and C_{60} assembled on a nanostructured SnO_2 electrode using an electrophoretic deposition method [189, 190]. Porphyrin oligomers with polypeptidic backbone (polylysine), which form π-complexes with carbon nanotubes (vide supra), provide the ideal space to form π-complex assemblies with C_{60} as shown

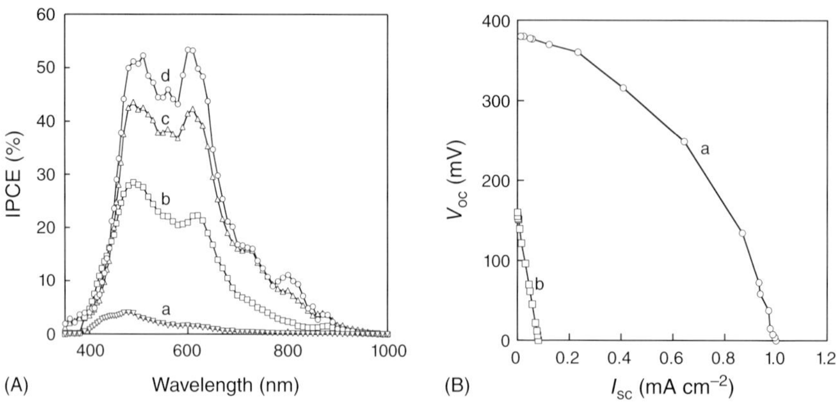

Figure 5.33 (A) Photocurrent action spectra of OTE/SnO_2/(H_2PCnMPC+C_{60})$_m$ electrode ([H_2P] = 0.19 mM; (a) $n = 5$, [C_{60}] = 0.31 mM; (b) $n = 11$, [C_{60}] = 0.31 mM; (c) $n = 15$, [C_{60}] = 0.31 mM; (d) $n = 15$, [C_{60}] = 0.38 mM); electrolyte: 0.5 M NaI and 0.01 M I_2 in acetonitrile. (B) Current–voltage characteristics of (a) OTE/SnO_2/(H_2PC15MPC+C_{60})$_m$ electrode and (b) OTE/SnO_2/(H_2P-ref+C_{60})$_m$ electrode prepared from cluster solution of ([H_2P] = 0.19 mM; [C_{60}] = 0.38 mM) under visible light illumination ($\lambda > 400$ nm); electrolyte: 0.5 M NaI and 0.01 M I_2 in acetonitrile; input power: 11.2 mW/cm^2. Source: Hasobe et al. 2005 [187]. Reproduced with permission of American Chemical Society.

schematically in Figure 5.34 [189, 190]. Remarkable enhancement in the photoelectrochemical performance as well as the broader photoresponse in the visible and near-infrared regions is seen with increasing number of porphyrin units in α-polypeptide structures [189, 190]. Formation of supramolecular clusters of porphyrins and fullerenes prepared in acetonitrile/toluene = 3/1 has been confirmed by transmission electron micrograph (TEM) and absorption spectra [189, 190]. The highly colored composite clusters of porphyrin–peptide oligomers and fullerenes have been assembled as three-dimensional arrays onto nanostructured SnO_2 films using an electrophoretic deposition method. A high power conversion efficiency (η) of ~1.6% and the maximum IPCE value (48%) were attained using composite clusters of free base and porphyrin–peptide hexadecamers [P(H_2P)$_{16}$] with fullerenes as shown in Figure 5.35 (A and B, respectively) [190]. The IPCE values increase with increasing number of porphyrins in a polypeptide unit in the (P(H_2P)$_n$+C_{60})$_m$ (n = 1, 2, 4, 8, 16) system. This suggests that efficient photoinduced electron transfer occurs from the singlet excited state of the porphyrin unit to C_{60} in the supramolecular complex with increasing number of porphyrins in the polypeptide unit. Femtosecond transient absorption and fluorescence measurements of porphyrin–fullerene composite films confirmed improved electron-transfer properties with increasing number of porphyrins in the polypeptide unit [190]. When P(H_2P)$_{16}$ was replaced by P(ZnP)$_{16}$, a larger IPCE value was obtained for (P(ZnP)$_{16}$+C_{60})$_m$ (56%) than that of (P(H_2P)$_{16}$+C_{60})$_m$ (48%) [190].

Photocurrent generation in this system (Scheme 5.9) is initiated by photoinduced electron transfer from the singlet excited state of free base porphyrin ($^1H_2P^*/H_2P^+ = -0.7$ V vs. NHE) or zinc porphyrin ($^1ZnP^*/ZnP^+ = -1.0$ V

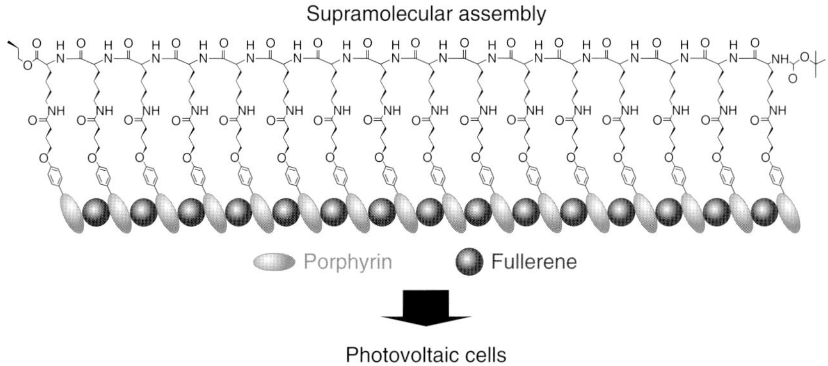

Figure 5.34 Supramolecular organization between porphyrins and fullerenes with a polypeptide structure. Source: Hasobe et al. 2007 [190]. Reproduced with permission of Royal Society of Chemistry.

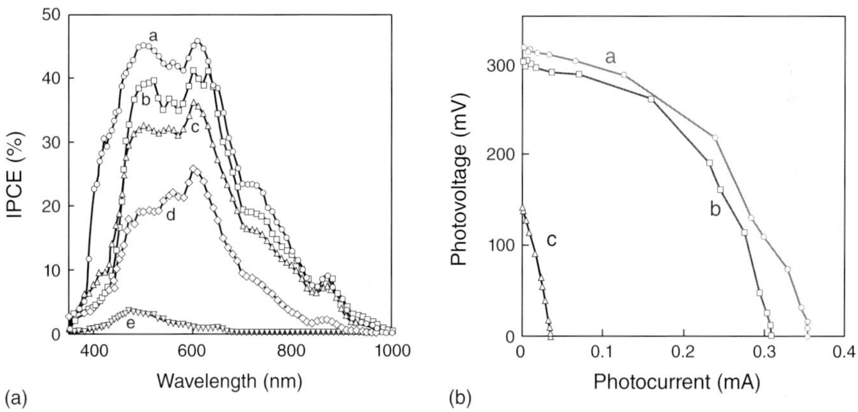

Figure 5.35 (A) The photocurrent action spectra (IPCE vs. wavelength) of (a) $(P(H_2P)_{16}+C_{60})_m$, (b) $(P(H_2P)_8+C_{60})_m$, (c) $(P(H_2P)_4+C_{60})_m$, (d) $P(H_2P)_2+C_{60})_m$, and (e) $(P(H_2P)_1+C_{60})_m$ modified OTE/SnO_2 electrodes. (B) Current–voltage characteristics of (a) OTE/SnO_2/$(P(H_2P)_{16}+C_{60})_m$, (b) OTE/SnO_2/$(P(H_2P)_8+C_{60})_m$, and (c) OTE/SnO_2/$(P(H_2P)_1+C_{60})_m$; $[P(H_2P)_{16}] = 0.012$ mM, $[P(H_2P)_8] = 0.024$ mM, $[P(H_2P)_1] = 0.19$ mM, $[C_{60}] = 0.31$ mM. Source: Hasobe et al. 2007 [190]. Reproduced with permission of Royal Society of Chemistry.

vs. NHE) in the porphyrin–peptide oligomer to C_{60} ($C_{60}/C_{60}^- = -0.2$ V vs. NHE) in the porphyrin–C_{60} complex rather than direct electron injection to the conduction band of SnO_2 (0 V vs. NHE) system [190]. The reduced C_{60} injects electrons into the SnO_2 nanocrystallites, whereas the oxidized porphyrin ($H_2P/H_2P^{\cdot+} = 1.2$ V or $ZnP/ZnP^+ = 1.0$ V vs. NHE) undergoes electron-transfer reduction with the iodide ($I_3^-/I^- = 0.5$ V vs. NHE) in the electrolyte system [190]. The driving force of electron transfer from $^1ZnP^*$ to C_{60} (0.8 eV) is significantly larger than that of electron transfer from $^1H_2P^*$ to C_{60} (0.5 eV) [190]. The larger IPCE value of $(P(ZnP)_{16}+C_{60})_m$ (56%) than that of $(P(H_2P)_{16}+C_{60})_m$ (48%) may be ascribed to the difference in the driving force of electron transfer [190]. It should be emphasized that the power conversion efficiency (η) of

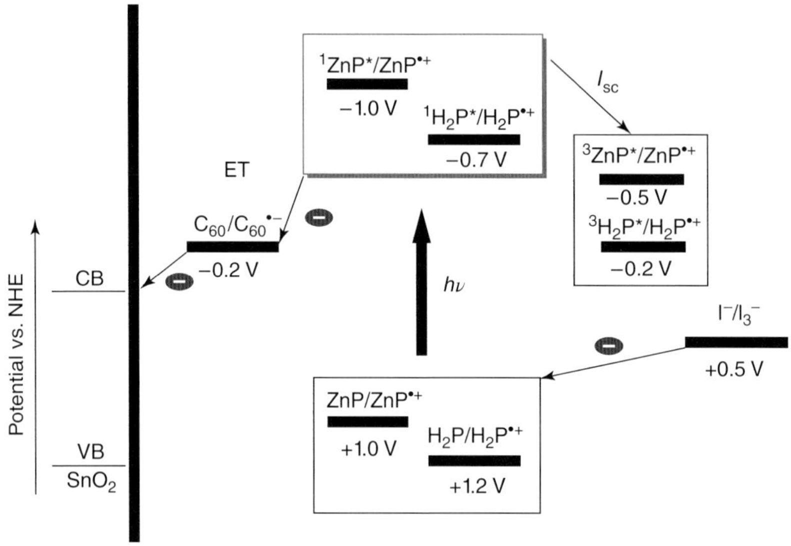

Scheme 5.9 Schematic illustration of photocurrent generation mechanism of $OTE/SnO_2/(P(H_2P)_{16}+C_{60})_m$ and $OTE/SnO_2/(Zn(H_2P)_{16}+C_{60})_m$. Source: Hasobe et al. 2007 [190]. Reproduced with permission of Royal Society of Chemistry.

$(P(H_2P)_{16}+C_{60})_m$ modified electrode reaches 1.6%, which is 40 times higher than the value (0.043%) of the porphyrin monomer $(P(H_2P)_1+C_{60})_m$ modified electrode [190]. Thus, organization approach between porphyrins and fullerenes with polypeptide structures is quite promising, and may make it possible to further improve the light energy conversion properties by using larger number of porphyrins in a polypeptide unit.

Supramolecular solar cells have also been constructed using molecular clusters of porphyrin dendrimer (D_nP_n in Figure 5.36) and C_{60} units assembled on SnO_2 electrodes [191]. Porphyrin dendrimers also form supramolecular complexes with C_{60} in toluene and they are clusterized in an acetonitrile/toluene mixed solvent [191]. The $OTE/SnO_2/(D_4P_4+C_{60})_m$ system exhibits the maximum IPCE value of 15% as well as the broad photoresponse, which extends well into the infrared (up to 1000 nm). However, the IPCE value decreases with increasing number of dendrimer generation: the $OTE/SnO_2/(D_{16}P_{16}+C_{60})_m$ system shows an even smaller IPCE value as compared with the reference system $[OTE/SnO_2/(H_2P\text{-ref}+C_{60})_m]$ [191]. In the case of $D_{16}P_{16}$, the more crowded dendrimer structure may not be suitable for efficient π-complex formation of the porphyrin moieties with C_{60} due to the lack of space to accommodate C_{60} insertion between two adjacent porphyrin rings [191]. This indicates that the distance between two adjacent porphyrin rings is an essential factor in the construction of efficient light energy conversion systems using supramolecular multi-porphyrin nanomaterials. By mixing PhCN solutions of the supramolecular complexes of $MTPPS^{4-}$ and $Li^+@C_{60}$ with MeCN, nanoclusters were produced and they were deposited on the optically transparent electrode (OTE) of nanostructured SnO_2 (OTE/SnO_2) by application of dc electric field (~100 V cm^{-1}) to construct

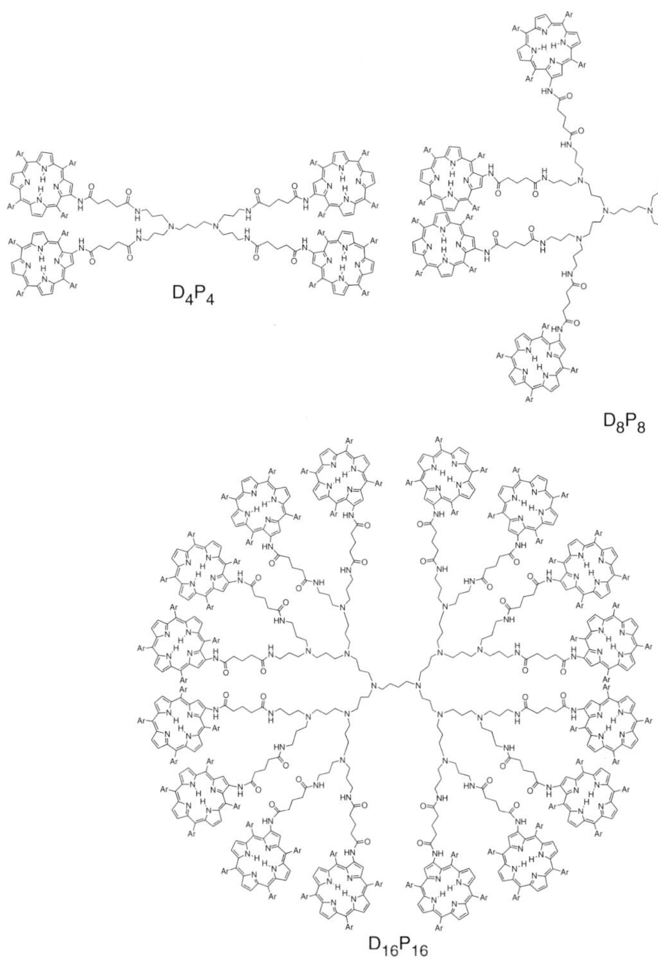

Figure 5.36 Porphyrin dendrimers employed for construction of supramolecular solar cells composed of multi-porphyrin/C_{60} supramolecular assembies. Source: Hasobe et al. 2004 [191]. Reproduced with permission of American Chemical Society.

photovoltaic cells [192]. The $(MTPPS^{4-}/Li^+@C_{60})_n$ films are composed of closely packed $Li^+@C_{60}$ clusters of about 80 nm size, which renders a nanoporous morphology to the film as shown in the TEM images in Figure 5.37 [192].

The photoelectrochemical measurements of a robust thin film of $OTE/SnO_2/(MTPPS^{4-}/Li^+@C_{60})_n$ were performed using a standard two-electrode system consisting of a working electrode and a Pt wire gauze electrode in air-saturated acetonitrile (MeCN) containing 0.5 M LiI and 0.01 M I_2 (Figure 5.38) [192]. The IPCE (incident photon-to-photocurrent efficiency) values were determined by normalizing the photocurrent values for incident light energy and intensity and using Eq. (5.2). The IPCE value of $OTE/SnO_2/(ZnTPPS^{4-}/Li^+@C_{60})_n$ is much higher than the sum of the two individual IPCE values of the individual systems, $OTE/SnO_2/(ZnTPPS^{4-})_n$ and $OTE/SnO_2/(Li^+@C_{60})_n$

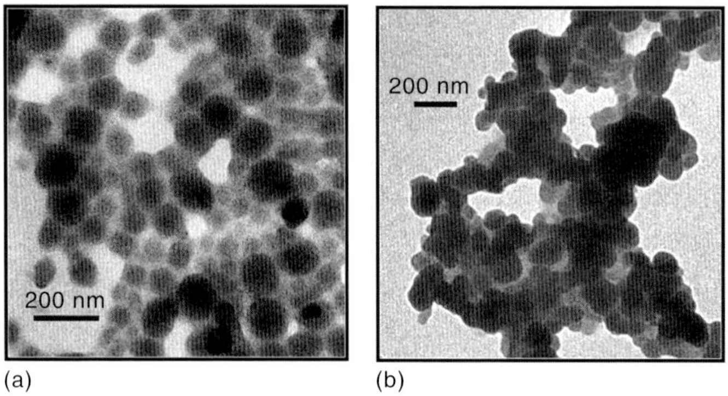

Figure 5.37 TEM images of (a) $Li^+@C_{60}/ZnTPPS^{4-}$ and (b) $Li^+@C_{60}/H_2TPPS^{4-}$ nanoclusters. Source: Ohkubo et al. 2013 [192]. https://pubs.rsc.org/en/content/articlelanding/2013/CC/c3cc41187g#!divAbstract and https://creativecommons.org/licenses/by/3.0/. Licensed under CCBY 3.0.

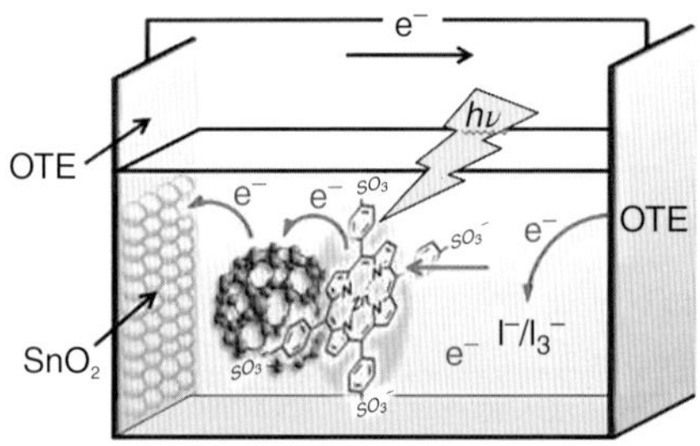

Figure 5.38 Schematic image of photoelectrochemical cell of $OTE/SnO_2/MTPPS^{4-}/Li^+@C_{60}$ and electron-transfer pathways to generate photocurrent. Source: Kamat et al. 2000 [188]. Reproduced with permission of John Wiley & Sons.

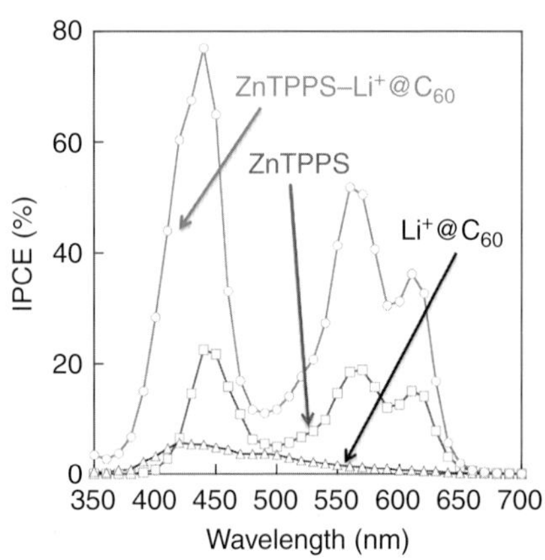

Figure 5.39 Photocurrent action spectra of $OTE/SnO_2/(ZnTPPS^{4-}/Li^+@C_{60})_n$, $OTE/SnO_2/(ZnTPPS^{4-})_n$, and $OTE/SnO_2/(Li^+@C_{60})_n$. Electrolyte: 0.5 M LiI and 0.01 M I_2 in MeCN/PhCN (3 : 1, v/v). Source: Ohkubo et al. 2013. https://pubs.rsc.org/en/content/articlelanding/2013/CC/c3cc41187g#!divAbstract and https://creativecommons.org/licenses/by/3.0/. Licensed under CCBY 3.0.

in the visible region (Figure 5.39) [192]. The maximum IPCE value of OTE/SnO_2/($ZnTPPS^{4-}$/Li^+@C_{60})$_n$ was 77% at 450 nm [192]. Such a high IPCE value indicates that photocurrent generation is initiated via photoinduced electron transfer from $ZnTPPS^{4-}$ to Li^+@C_{60}, followed by charge transport to the collective surface of OTE/SnO_2 electrode (Figure 5.38) [192]. When $ZnTPPS^{4-}$ was replaced by H_2TPPS^{4-}, a significantly low IPCE value was observed as 7% at 440 nm probably because of the self-aggregation of H_2TPPS^{4-} without binding with Li^+@C_{60} [192].

The power conversion efficiency (η) of the OTE/SnO_2/($ZnTPPS^{4-}$/Li^+@C_{60})$_n$ electrode was calculated using Eq. (5.2). The OTE/SnO_2/($ZnTPPS^{4-}$/Li^+@C_{60})$_n$ electrode has an overall power conversion efficiency (η) of 2.1% at an input power (W_{in}) of 28 mW cm^{-2}, whereas $FF = 0.37$, $V_{oc} = 460$ mV, and $I_{sc} = 3.4$ mA cm^{-2} in the OTE/SnO_2/($ZnTPPS^{4-}$/Li^+@C_{60})$_n$ [193]. The η value is 2 orders of magnitude greater than that of the previously reported simple porphyrin and C_{60} composite system (~0.03%) [193]. Such a significant enhancement of the η value indicates that the strong ordering in the clusters and the efficient charge separation in ($ZnTPPS^{4-}$/Li^+@C_{60})$_n$ improved the light energy conversion properties.

6

Effects of Metal Ions on Photoinduced Electron Transfer

Long-lived charge separation (CS) states have been successively achieved by minimizing the reorganization energy of electron transfer in the Marcus inverted region (vide supra). In principle, a totally opposite approach to attaining long-lived CS states is possible, that is, the use of a component with a large reorganization energy, which results in slow back electron transfer in the Marcus normal region ($-\Delta G_{ET} < \lambda$). In such a case, however, the rate of forward electron transfer with a much smaller driving force becomes much smaller than the back electron transfer rate, and thus the use of a component with a large organization energy has never been employed to design the artificial photosynthetic reaction center. If one can design a system in which the forward electron transfer has a small reorganization energy whereas the reorganization energy of the back electron transfer becomes much larger than that of the forward electron transfer in the Marcus normal region, long-lived charge separated states would be attained efficiently. In the Marcus normal region, the smaller the driving force, the slower is the electron transfer rate (Eq. (2.1)). Thus, a decrease in the driving force of back electron transfer is essential to attain the long-lived charge-separated states in the Marcus normal region. The driving force of electron transfer has been shown to be finely controlled by complexation of radical anions, produced in the electron transfer, with metal ions that act as Lewis acids, in a variety of intermolecular and intramolecular electron-transfer systems [194–201]. Quantitative measurements to determine the Lewis acidity of a variety of metal ions have been well established in relation with the promoting effects of metal ions on the electron-transfer reactions as discussed in detail later [201, 202].

Significant effects of metal ions on photodynamics of photoinduced electron transfer in porphyrin-containing donor–acceptor ensembles, a zinc porphyrin–naphthalenediimide (ZnP–NIm) dyad has been reported as shown in Scheme 6.1 [203]. The photoexcitation of ZnP–NIm results in the formation of the singlet excited state (^{1}ZnP*–NIm: 2.12 eV) in which electron transfer from ^{1}ZnP* to NIm occurs to give the CS state (ZnP$^{\cdot+}$–NIm$^{\cdot-}$) with the rate constant $k_{CS} = 3.4 \times 10^9$ s^{-1} in competition with the decay to the ground state with $k_0 = 4.0 \times 10^8$ s^{-1} [203]. The back electron transfer from NIm$^{\cdot-}$ to ZnP$^{\cdot+}$ (charge recombination [CR]) occurs with $k_{CR} = 7.7 \times 10^5$ s^{-1} [203]. In the presence of metal ions (M^{n+}), the CS process takes place mainly from ^{1}ZnP*–NIm rather than from ^{1}ZnP*–NIm/M^{n+} complex due to very weak binding of M^{n+} with

Electron Transfer: Mechanisms and Applications, First Edition. Shunichi Fukuzumi.

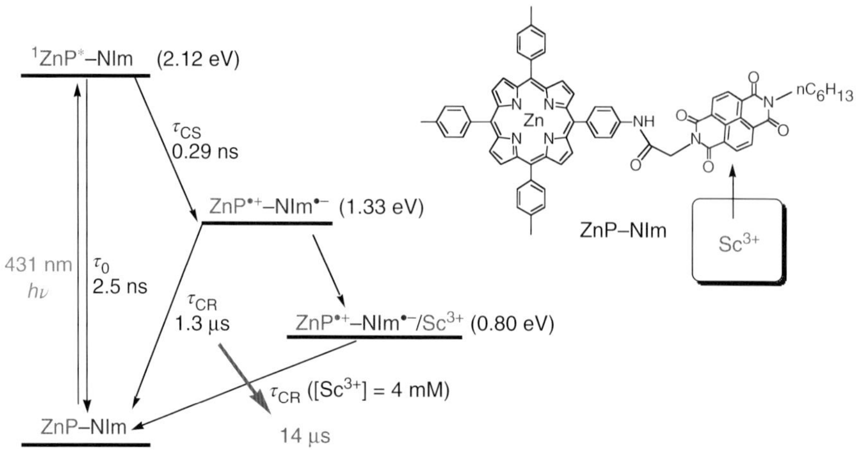

Scheme 6.1 Energy diagram and photodynamics of a zinc porphyrin–naphthalenediimide (ZnP–NIm) dyad in the absence and presence of a metal ion. Source: Okamoto et al. 2004 [203]. Reproduced with permission of John Wiley & Sons.

neutral species of NIm [203]. In such a case, the k_{CS} values in the presence of M^{n+} are determined by the $-\Delta G_{CS}$ value in the absence of M^{n+} rather than the $-\Delta G_{CS}$ values in the presence of M^{n+}, although the $-\Delta G_{CS}$ values increase with increasing $[M^{n+}]$ [203]. In contrast, the k'_{CR} values in the presence of M^{n+} are determined by the $-\Delta G_{CR}$ values in the presence of M^{n+}, since the CR process occurs mainly from the $ZnP^{\cdot+}–NIm^{\cdot-}/M^{n+}$ complex rather than from $ZnP^{\cdot+}–NIm^{\cdot-}$ [203]. The back electron transfer from $NIm^{\cdot-}$ to $ZnP^{\cdot+}$ (CR) in the presence of 1.0×10^{-3} M Sc^{3+} occurs with $k'_{CR} = 6.9 \times 10^4\ s^{-1}$ [203]. The decrease in the CR driving force together with an increase in the λ value in the presence of M^{n+} results in elongation of the CS lifetime [203]. The effect of metal ion is the largest for Sc^{3+} ion, which is the strongest Lewis acid among metal ions [201, 202].

Remarkable seven million times elongation of the CS lifetime has been attained by complex formation of yttrium triflate [$Y(OTf)_3$] with the CS state in photoinduced electron transfer of a ferrocene–anthraquinone dyad (**Fc–AQ**) involving a rigid amide spacer (Scheme 6.2) [204]. Photoexcitation of the AQ moiety in **Fc–AQ** in deaerated PhCN with femtosecond (150 fs width) laser light results in the appearance of the absorption bands around 420 and 600 nm at 500 fs [204]. The transient absorption bands around 420 and 600 nm are assigned to $AQ^{\cdot-}$ by comparison with the absorption spectrum of $AQ^{\cdot-}$ produced by the chemical reduction of AQ with naphthalene radical anion [204]. The decay process obeys first-order kinetics with a lifetime of 12 ps [204]. This indicates that electron transfer from Fc to the singlet excited state of AQ occurs efficiently to produce the CS state (**Fc^+–$AQ^{\cdot-}$**) within 500 fs and decays via back electron transfer to the ground state with a lifetime of 12 ps (Scheme 6.2a) [204].

In the presence of $Y(OTf)_3$ (1.0×10^{-2} M) as well, photoexcitation of the **Fc–AQ** system using a femtosecond laser results in efficient electron transfer from Fc to AQ within 500 fs [204]. However, the transient absorption band observed at 700 nm in the presence of $Y(OTf)_3$ is significantly red-shifted as

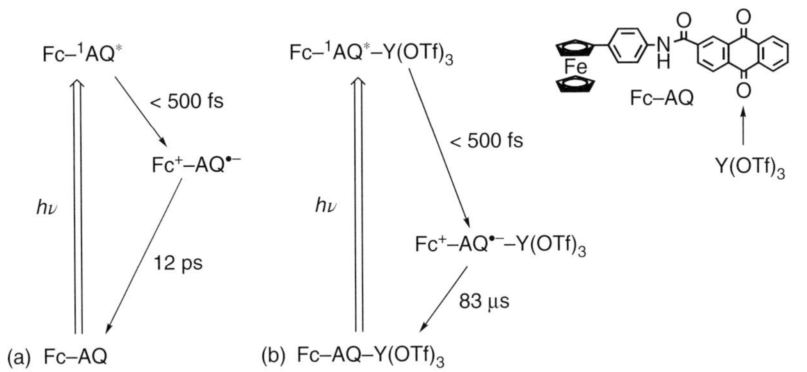

Scheme 6.2 Photodynamics of a ferrocene–anthraquinone dyad (Fc–AQ) (a) in the absence and (b) in the presence of $Y(OTf)_3$. Source: Okamoto et al. 2004 [204]. Reproduced with permission of American Chemical Society.

compared with that observed at 600 nm in the absence of $Y(OTf)_3$ [204]. The decay of absorbance at 420 and 700 nm due to the Fc^+–$AQ^{\cdot-}$–$Y(OTf)_3$ complex obeys first-order kinetics to afford an identical CS lifetime, which is determined as 83 μs [204]. The CS lifetime is seven million times longer than the lifetime in the absence of $Y(OTf)_3$ (12 ps) [204]. The strong binding of $Y(OTf)_3$ with $AQ^{\cdot-}$ results in a substantial deceleration of the back electron transfer from $AQ^{\cdot-}$ to Fc^+, leading to a remarkable elongation of the CS lifetime, whereas the forward photoinduced electron transfer in both the absence and the presence of $Y(OTf)_3$ takes place within 500 fs (Scheme 6.2b) [204].

Such a remarkable effect of Y^{3+} on controlling the electron-transfer processes of quinones has enabled us to develop a highly Y^{3+}-selective fluorescence sensor using a zinc porphyrin–CONH–quinone dyad (ZnP–CONH–Q) as shown in Figure 6.1 [205]. ZnP–NHCO–Q linked dyad and the reference compound (ZnP–CONH-ref) are also employed for comparison. The fluorescence intensity of ZnP–CONH–Q as well as ZnP–NHCO–Q is significantly quenched as compared with that of ZnP–CONH-ref in deaerated PhCN via photoinduced electron transfer from the singlet excited state of the ZnP moiety to the Q moiety [205].

Addition of a small concentration of $Y(OTf)_3$ (10^{-6} M – 10^{-3} M) to a deaerated PhCN solution of ZnP–CONH–Q results in a remarkable enhancement of the ZnP fluorescence intensity. From the fluorescence titration of Y^{3+} in ZnP–CONH–Q, the binding constant (K) between Y^{3+} and ZnP–CONH–Q is estimated as 3900 M^{-1} in deaerated PhCN [205]. The fluorescence intensities of ZnP–CONH–Q (3 mM) and ZnP–NHCO–Q (3 mM) in the presence of 400 mM concentration of various metal ions in reference to those in the absence of metal ion are determined in deaerated PhCN, as shown in Figure 6.1 [205]. A remarkable enhancement of the fluorescence intensity is observed exclusively in the case of ZnP–CONH–Q in the presence of Y^{3+}, whereas ZnP–NHCO–Q exhibits only moderate enhancement of the fluorescence intensity in the presence of metal ions. Such a significant difference in the fluorescent intensity between ZnP–CONH–Q and ZnP–NHCO–Q in the presence of Y^{3+} indicates that the binding sites of ZnP–CONH–Q to Y^{3+} are more suitable

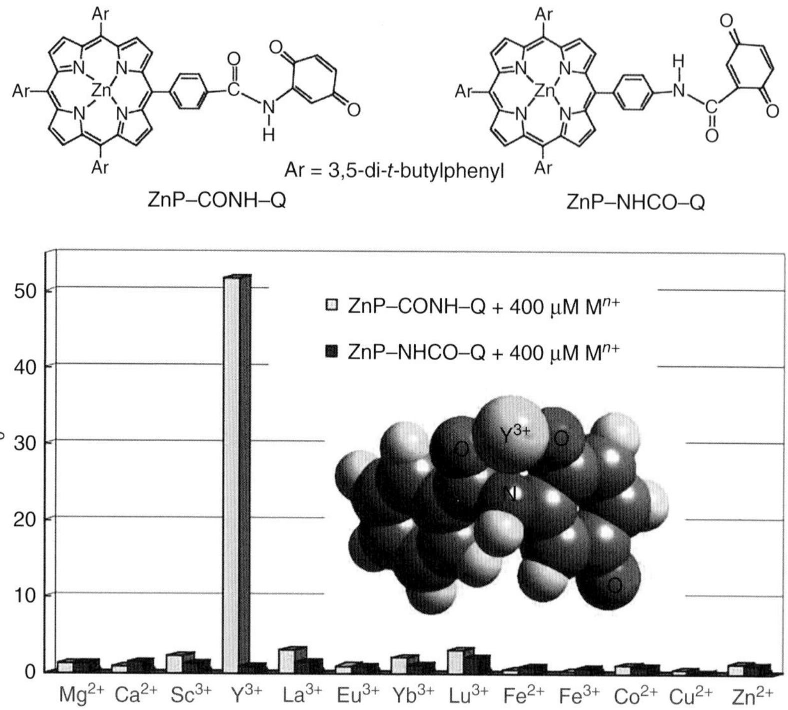

Figure 6.1 Fluorescence responses (I/I_0 at 610 nm) of **ZnP–CONH–Q** and **ZnP–NHCO–Q** (3 μM) in the presence of 400 μM metal ions excited at 560 nm in reference to those in the absence of metal ion in deaerated PhCN; $CF_3OSO_3^-$ salt (Ca^{2+}, Sc^{3+}, Y^{3+}, La^{3+}, Eu^{3+}, Yb^{3+}, Lu^{3+}, Zn^{2+}), ClO_4^- salt (Mg^{2+}, Fe^{2+}, Fe^{3+}, Co^{2+}, Cu^{2+}) [205]. The optimized structure of **Ph–CONH–Q/Y^{3+}** complex is obtained by ADF calculation with II (large) basis set. Source: Okamoto et al. 2004 [205]. Reproduced with permission of American Chemical Society.

than those of ZnP–NHCO–Q. This is supported by the optimized structure of Ph–CONH–Q/Y^{3+} complex evaluated by Amsterdam density function (ADF) calculation with II (large) basis set as shown in the inset of Figure 6.1, which demonstrates the strong binding of Y^{3+} with two carbonyl oxygen of ZnP–CONH–Q [205]. The Y^{3+}-selective enhancement of the fluorescence of ZnP–CONH–Q results from the strong binding of Y^{3+} to the quinone moiety, which causes a significant increase in the λ value to retard the photoinduced electron transfer despite the larger driving force of electron transfer. Thus, the fluorescent property of an electron donor–acceptor dyad can be finely modulated by coordination of metal ions. This has led to a rational design to develop fluorescence sensors [206].

7

Photoredox Catalysis

7.1 Photocatalytic Oxygenation

As described earlier, photoexcitation of Acr^+–Mes results in the formation of the electron transfer (ET) state ($Acr^{\cdot}$–$Mes^{\cdot+}$), which has both high oxidizing and reducing abilities with a sufficiently long lifetime for the occurrence of ET reactions with electron donors and acceptors in competition with the back electron transfer (BET). In such a case, Acr^+–Mes can be utilized as an efficient photocatalyst for radical coupling reactions between radical cations and radical anions, which are produced by the ET oxidation and ET reduction of external electron donors and acceptors, respectively (Scheme 7.1).

Visible light irradiation ($\lambda > 430$ nm) of the absorption band of Acr^+–Mes (1.0×10^{-3} M) in an O_2-saturated acetonitrile (MeCN) solution containing anthracene derivatives (1.0×10^{-2} M) results in the formation of the oxygenation products, i.e. epidioxyanthracenes (Scheme 7.2) [207]. The photocatalytic oxygenation of Me_2An with O_2 in a preparative scale (100 mg, 5.0×10^{-4} mol) with Acr^+–Mes (8.2 mg, 2.0×10^{-5} mol) resulted in isolation of dimethylepidioxyanthracene (Me_2An–O_2) in 80% yield [207]. The addition of Me_2An to MeCN solution of Acr^+–Mes and the laser photoexcitation result in the formation of Me_2An radical cation ($Me_2An^{\cdot+}$: λ_{max} = 660 nm) [208] with a concomitant decrease in the absorption band due to the $Mes^{\cdot+}$ moiety.

The rate constant (k_{et}) of electron transfer from Me_2An to the $Mes^{\cdot+}$ moiety of $Acr^{\cdot}$–$Mes^{\cdot+}$ is determined to be 1.4×10^{10} M^{-1} s^{-1}, which is close to be the diffusion-limited value as expected from the exergonic electron transfer [207]. It is important to note that the absorption band due to the $Acr^{\cdot}$ moiety remains virtually the same in the absence of O_2. In the presence of O_2, the absorption band due to $Acr^{\cdot}$ moiety decays by electron transfer from the $Acr^{\cdot}$ moiety to O_2. The formation of $O_2^{\cdot-}$ was confirmed by the EPR spectrum measured at 233 K with the typical anisotropic g values due to $O_2^{\cdot-}$ ($g_{\parallel}$ = 2.1050 and $g_{\perp}$ = 2.0032) [207]. The rate constant of electron transfer from the $Acr^{\cdot}$ moiety to O_2 (k'_{et}) was determined to be 6.8×10^8 M^{-1} s^{-1} [207]. The absorbance at 660 nm due to $Me_2An^{\cdot+}$ in the presence of O_2 decays obeying second-order kinetics by the bimolecular reaction between $Me_2An^{\cdot+}$ and $O_2^{\cdot-}$ [207]. The second-order rate constant was determined as 1.7×10^{10} M^{-1} s^{-1}, which is close to the diffusion-limited value in MeCN [207]. The quantum yields of the formation of anthracene radical cations

Electron Transfer: Mechanisms and Applications, First Edition. Shunichi Fukuzumi.

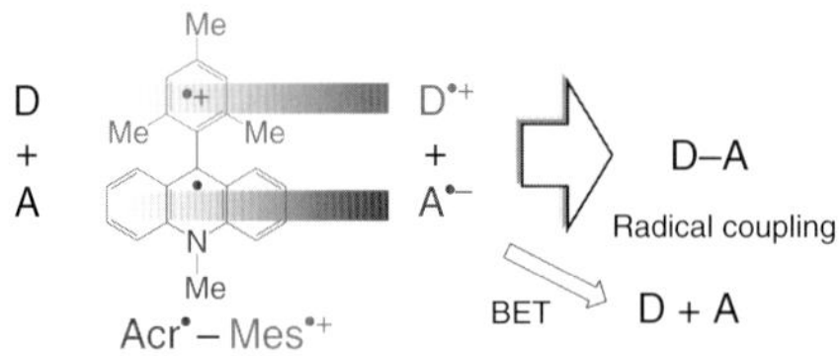

Scheme 7.1 Photocatalytic reaction of an electron donor (D) and an acceptor (A) with the ET state of Acr^+–Mes to form D–A via radical coupling between $D^{\cdot+}$ and $A^{\cdot+}$.

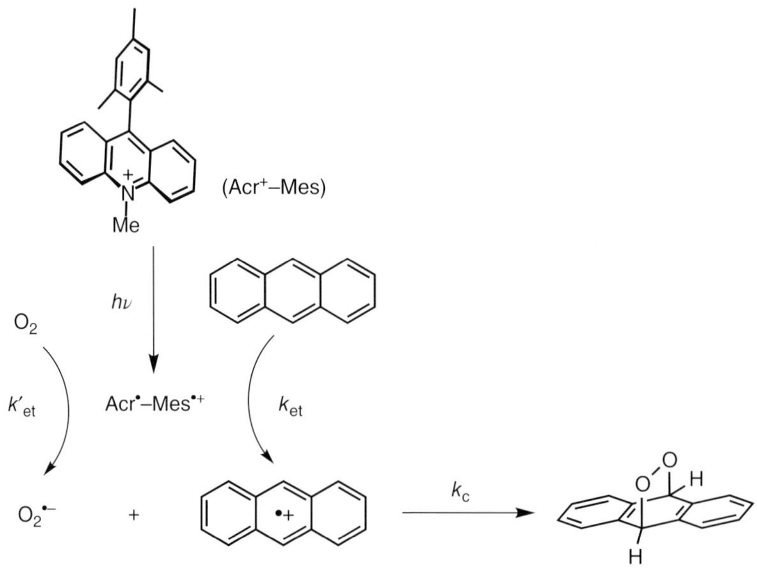

Scheme 7.2 Photocatalytic oxygenation of anthracene with O_2 using Acr^+–Mes. Source: Kotani et al. 2004 [207]. Reproduced with permission of American Chemical Society.

in the presence of O_2 are significantly larger than those of the epidioxyanthracenes formation [207]. This indicates that the back electron transfer from $O_2^{\cdot-}$ to anthracene radical cations to regenerate the reactant pair is also involved in the second-order decay of anthracene radical cations in addition to the radical coupling between anthracene radical cations and $O_2^{\cdot-}$ to afford the corresponding epidioxyanthracenes (Scheme 7.2) [207]. It was confirmed that An–O_2 is formed exclusively by the radical coupling between anthracene radical cation and $O_2^{\cdot-}$ (k_c) rather than the reaction of anthracene and $^1O_2$ [209, 210]. The [4+2] cycloaddition of O_2 to anthracene is known to occur also by the reaction of anthracene with singlet oxygen ($^1O_2$) [209, 210]. In order to evaluate the contribution of the singlet oxygen pathway, the rate constant of the reaction of $^1O_2$ with DMA was determined by emission decay rates of $^1O_2$ ($\lambda_{em} = 1270$ nm) [211, 212] in the presence of various concentrations of DMA to be $2.4 \times 10^5\ M^{-1}\ s^{-1}$ [207]. This value is much smaller than the second-order rate constant (k_c) of the radical coupling between $DMA^{\cdot+}$ and $O_2^{\cdot-}$ ($1.7 \times 10^{10}\ M^{-1}\ s^{-1}$) [207]. It was confirmed that no singlet oxygen emission was observed during the photocatalytic

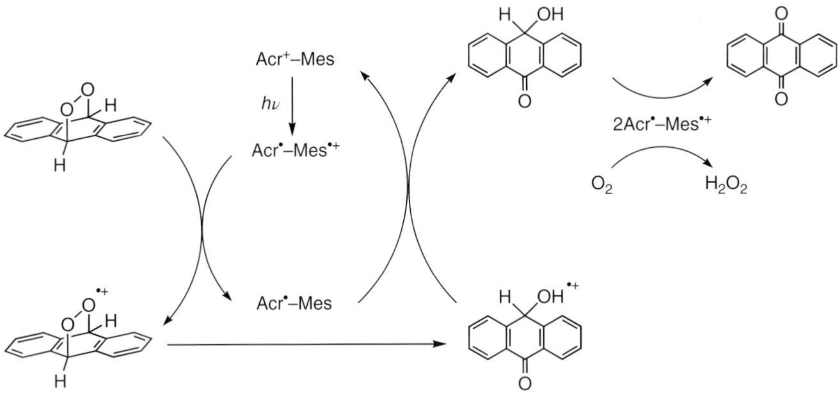

Scheme 7.3 Photocatalytic oxygenation of AnO_2 to anthraquinone with $Acr^{+\cdot}$–Mes. Source: Kotani et al. 2004 [207]. Reproduced with permission of American Chemical Society.

oxygenation of DMA with Acr^+–Mes in an O_2-saturated CD_3CN [207]. Thus, the [4 + 2] cycloaddition of O_2 to anthracene occurs exclusively by the radical coupling between anthracene radical cation and $O_2^{\cdot-}$ rather than the reaction of anthracene and $^1O_2$, although both pathways yield the same product [207].

In the case of anthracene, An–O_2 is converted to 10-hydroxyanthrone, which is further oxidized to yield the final six-electron oxidation product, i.e. anthraquinone, accompanied by generation of H_2O_2 with the further photoirradiation ($\lambda > 430$ nm) of an O_2-saturated CD_3CN solution of An–O_2 and Acr^+–Mes [207]. When the reaction is started from the isolated An–O_2, no photochemical reaction has occurred without Acr^+–Mes or no thermal reaction has taken place with Acr^+–Mes [207]. Under photoirradiation in the presence of Acr^+–Mes, electron transfer from An–O_2 to the $Mes^{\cdot+}$ moiety of $Acr^{\cdot}$–$Mes^{\cdot+}$ results in the O—O bond cleavage of An–$O_2^{\cdot+}$, followed by facile intramolecular hydrogen transfer to produce 10-hydroxyanthrone radical cation as shown in Scheme 7.3 [207]. The back electron transfer from the $Acr^{\cdot}$ moiety to 10-hydroxyanthrone radical cation affords 10-hydroxyanthrone, accompanied by regeneration of Acr^+–Mes. The electron-transfer oxidation of 10-hydroxyanthrone by the $Mes^{\cdot+}$ moiety of $Acr^{\cdot}$–$Mes^{\cdot+}$ results in further two-electron oxidation to yield anthraquinone by releasing two protons, whereas the two-electron reduction of O_2 by the $Ac^{\cdot}$ moiety of $Acr^{\cdot}$–$Mes^{\cdot+}$ with two protons yields H_2O_2 (Scheme 7.3) [207].

The radical coupling reaction between anthracene radical cation and $O_2^{\cdot-}$ to produce An–O_2 in Scheme 7.2 can be expanded to the dioxetane formation from olefins. Thus, the use of Acr^+–Mes as an electron-transfer photocatalyst in the presence of O_2 provides a convenient methodology to produce radical cations of electron donors of anthracenes and olefins as well as $O_2^{\cdot-}$, which can combine together to yield the oxygenated products selectively in a preparative scale [213, 214]. In general, the most common preparation of 1,2-dioxetanes is through the formal [2 + 2] cycloaddition of singlet oxygen ($^1O_2$) to electron-rich alkenes [215]. If alkenes are too electron poor to react with $^1O_2$, however, no oxygenated products were obtained. For example, it was reported that no products were formed in an oxygen-saturated MeCN solution of TPE (tetraphenylethylene) in the presence of $^1O_2$ sensitizers such as [60]fullerene and porphyrin derivative under photoirradiation [216]. When Acr^+–Mes is used as a photocatalyst, however, the photocatalytic oxygenation of TPE with O_2 occurs efficiently with Acr^+–Mes via the radical coupling between TPE radical cation ($TPE^{\cdot+}$) and $O_2^{\cdot-}$, both of which were produced by ET reactions of TPE and O_2 with the photogenerated ET state of Acr^+–Mes ($Acr^\cdot$–$Mes^{\cdot+}$), leading to successful isolation of the corresponding 1,2-dioxetane for the first time (left-hand side in Scheme 7.4) [213]. The one-electron oxidation potential of TPE (E_{ox} = 1.56 V vs. saturated calomel electrode [SCE]) is less positive than the value of one-electron reduction potential of $Mes^{\cdot+}$ (E_{red} = 1.88 V vs. SCE) [213]. Thus, electron-transfer oxidation of TPE with the $Mes^{\cdot+}$ moiety of $Acr^\cdot$–$Mes^{\cdot+}$ is thermodynamically favorable, resulting in the formation of $TPE^{\cdot+}$ and $Acr^\cdot$–Mes. The second-order rate constant (k_{et}) of electron transfer from TPE to the $Mes^{\cdot+}$ moiety of $Acr^\cdot$–$Mes^{\cdot+}$ in Scheme 7.4 was determined to be $2.5 \times 10^9\ M^{-1}\ s^{-1}$ in $CHCl_3$ by laser flash photolysis measurements [213]. This value is close to the diffusion-limited value as expected from the exergonic electron transfer. On the other hand, the electron-transfer reduction of O_2 with the $Acr^\cdot$ moiety of $Acr^\cdot$–$Mes^{\cdot+}$ also efficiently occurs with the second-order rate constant of electron transfer (k'_{et}) that was determined to be $3.8 \times 10^8\ M^{-1}\ s^{-1}$. The 1,2-dioxetane was isolated by column chromatography. The isolated yield was 27% after four-hour photoirradiation [213]. The quantum yield of 1,2-dioxetane increases with an increase in concentration of TPE to approach a limiting value as 0.17 and 0.022 in $CHCl_3$ and MeCN, respectively [214].

The dioxetane is further oxidized by $Acr^\cdot$–$Mes^{\cdot+}$ to produce the dioxetane radical cation, which undergoes the O—O bond homolysis to produce benzophenone and the radical cation (right-hand side in Scheme 7.4) [214]. The benzophenone radical cation is reduced by $Acr^\cdot$–Mes to produce another benzophenone, accompanied by regeneration of Acr^+–Mes (right-hand side in Scheme 7.4) [214].

Acr^+–Mes also acts as an efficient photoredox catalyst for the cis–trans isomerization of stilbene via the radical cation (Scheme 7.5) [214]. It is known that cis–trans isomerization occurs rapidly in the stilbene radical cation [217]. The steady-state cis–trans ratio of stilbene has been reported to be 98.8 : 1 [217]. The observed yield of *trans*-stilbene from *cis*-stilbene was 96% after 60 minutes photoirradiation, when the total consumption of *cis*- and *trans*-stilbene by the photocatalytic oxidation by O_2 was still 4% [214].

Photocatalytic oxygenation of *p*-xylene with O_2 also occurs under visible light irradiation of $[Acr^+–Mes]ClO_4^-$ (λ_{max} = 430 nm) in oxygen-saturated MeCN containing *p*-xylene (4.0 mM) to yield the oxygenated product, *p*-tolualdehyde (34%), *p*-methylbenzyl alcohol (10%), and the reduced product of O_2, H_2O_2 (30%) [218]. The photocatalytic reactivity was enhanced by the presence of H_2O (0.9 M) and sulfuric acid (1.0 mM) to yield *p*-tolualdehyde (75%), *p*-methylbenzyl alcohol (15%), and H_2O_2 (74%) with a high quantum yield (0.25) [218]. The 100% yield of *p*-tolualdehyde and H_2O_2 with a higher quantum yield (0.37) was achieved by using 9-mesityl-2,7,10-trimethylacridinium ion (Me_2Acr^+–Mes), where the hydrogens at 2 and 7 positions of the acridinium ring are replaced by the methyl groups [218]. The E_{red} value of Me_2Acr^+–Mes (−0.67 V vs. SCE) is

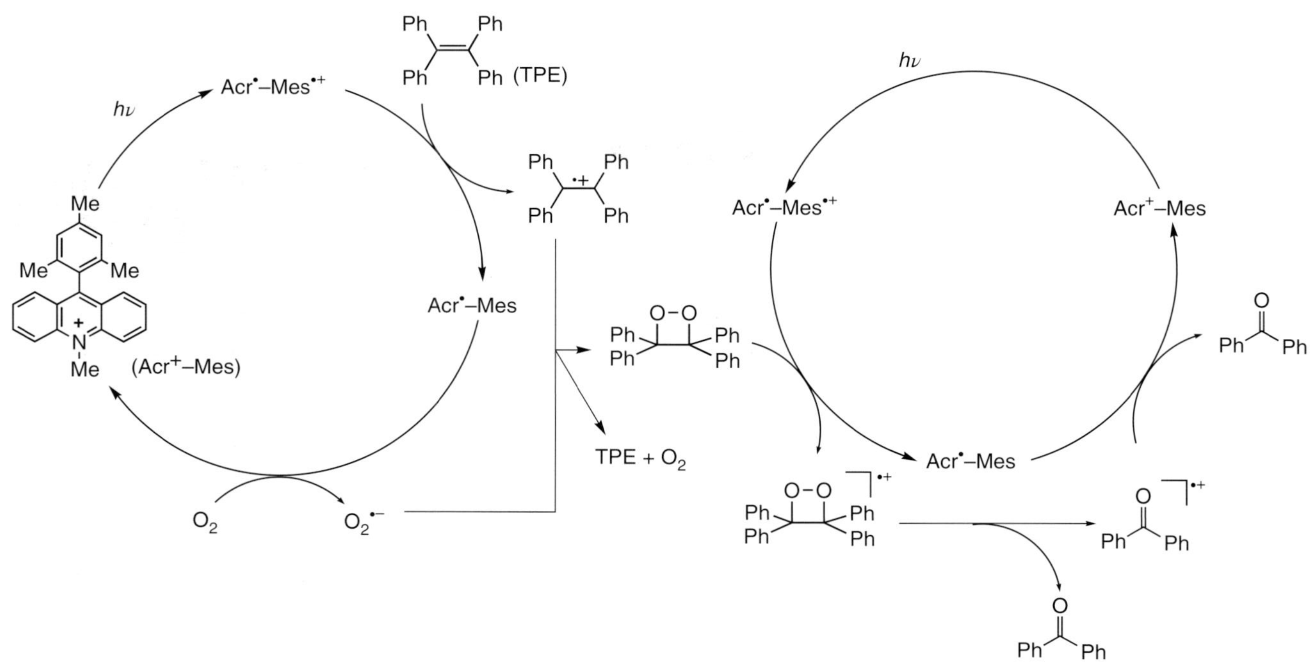

Scheme 7.4 Photocatalytic [2 + 2] cycloaddition of O_2 to tetraphenylethylene (TPE) via photoinduced electron transfer. Source: Ohkubo et al. 2005 [213] and Ohkubo et al. 2006 [214].

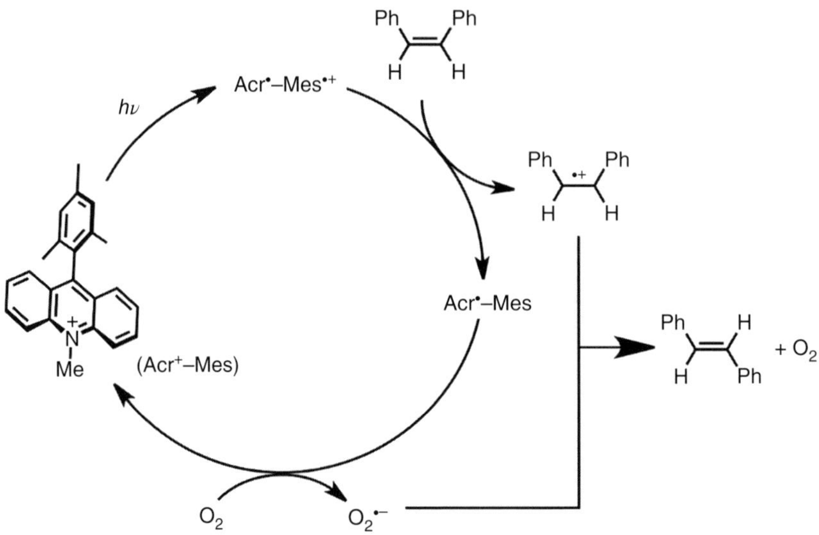

Scheme 7.5 Photocatalytic cis–trans isomerization of stilbene with Acr^+–Mes. Source: Ohkubo et al. 2006 [214]. Reproduced with permission of Elsevier.

by 0.1 eV more negative than that of Acr^+–Mes (−0.57 V), indicating that the $Me_2Acr^·$ moiety acts as a stronger electron donor [218]. The rate constants of the electron-transfer reduction of O_2 were determined from the quenching of the transient absorption due to the ET state by O_2 to be $6.8 \times 10^8\ M^{-1}\ s^{-1}$ for $Acr^·$–$Mes^{·+}$ and $2.0 \times 10^{10}\ M^{-1}\ s^{-1}$ for $Me_2Acr^·$–$Mes^{·+}$ in MeCN at 298 K [218]. Thus, the reducing ability of $Me_2Acr^·$–$Mes^{·+}$ was significantly enhanced by the electron-donating methyl substitution of the acridinium ring of Acr^+–Mes. This may be the reason why the 100% yield of tolualdehyde and H_2O_2 with a higher quantum yield (0.37) was achieved by using Me_2Acr^+–Mes (vide infra). No further oxygenated product, *p*-toluic acid or *p*-phthaladehyde, was produced during the photocatalytic reaction.

The photocatalytic oxygenation also occurred when using durene and mesitylene as substrates [218]. The E_{ox} values of toluene derivatives are lower than the one-electron reduction potential (E_{red}) of the ET state of R_2Acr^+–Mes ($R_2Acr^·$–$Mes^{·+}$; R = H and Me: 2.06 V vs. SCE in MeCN) [218]. Thus, electron transfer from toluene derivatives such as *p*-xylene to the $Mes^{·+}$ moiety of $R_2Acr^·$–$Mes^{·+}$ is energetically feasible, whereas electron transfer from toluene (E_{ox} = 2.20 V) [57] to the $Mes^{·+}$ moiety is energetically unfavorable when no photocatalytic oxidation of toluene by O_2 occurred with Acr^+–Mes under the same experimental conditions [218]. The E_{ox} values of the oxygenated products of the corresponding benzaldehydes are also higher than the E_{red} value of $R_2Acr^·$–$Mes^{·+}$ [218]. This is the reason why the selective oxygenation of *p*-xylene to *p*-tolualdehyde was achieved without further oxygenation of *p*-tolualdehyde.

The photocatalytic reaction is also initiated by electron transfer from *p*-xylene to the $Mes^{·+}$ moiety of $R_2Acr^·$–$Mes^{·+}$ to produce *p*-xylene radical cation, which undergoes fast deprotonation to afford the deprotonated radical [218]. This is followed by rapid O_2 addition to afford the peroxyl radical [218]. The

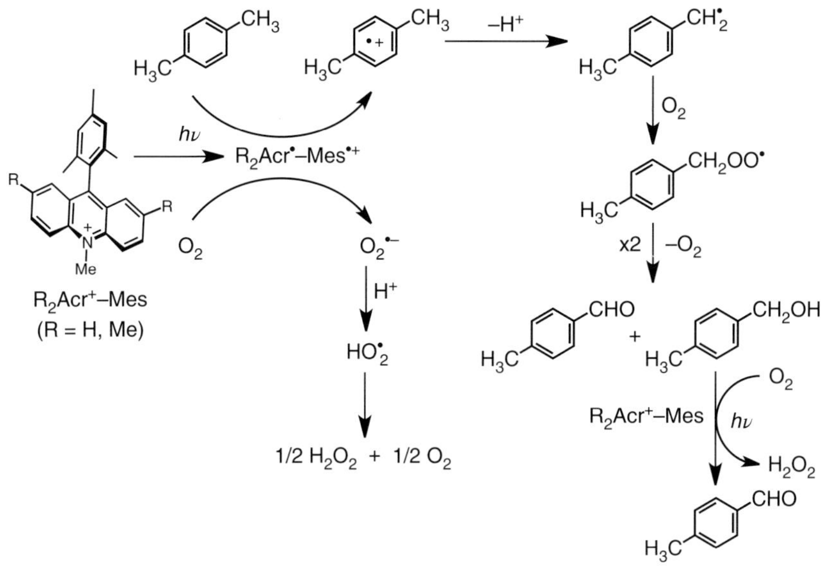

Scheme 7.6 Reaction scheme of photocatalytic oxygenation of *p*-xylene and formation of H_2O_2 catalyzed by R_2Acr^+–Mes. Source: Ohkubo et al. 2010 [218]. Reproduced with permission of Royal Society of Chemistry.

disproportionation of the peroxyl radical affords *p*-tolualdehyde, *p*-methylbenzyl alcohol, and O_2 [218]. *p*-Methylbenzyl alcohol is readily oxygenated to yield *p*-tolualdehyde with $Acr^{\cdot}$–$Mes^{\cdot+}$ [218]. On the other hand, $O_2^{\cdot-}$ undergoes disproportionation with proton to yield H_2O_2 and O_2 (Scheme 7.6) [218]. The radical intermediates involved in Scheme 7.6 were detected by electron paramagnetic resonance (EPR) ($g_{\parallel} = 2.101$, $g_{\perp} = 2.009$ for $O_2^{\cdot-}$ and $g_{\parallel} = 2.033$, $g_{\perp} = 2.006$ for *p*-methylbenzylperoxyl radical) in frozen MeCN [218].

Addition of aqueous sulfuric acid enhanced the deprotonation of *p*-xylene radical cation and the disproportionation process of $O_2^{\cdot-}$, respectively, leading to a remarkable enhancement of photocatalytic reactivity as mentioned above [218]. The photocatalytic reactivity and stability of Acr^+–Mes was further improved by incorporating Acr^+–Mes into mesoporous silica–alumina with a copper complex $[(tmpa)Cu^{II}]^{2+}$ (tmpa = tris(2-pyridylmethyl)amine) for the selective oxygenation of *p*-xylene by O_2 to produce *p*-tolualdehyde [69], because the $[(tmpa)Cu^{II}]^{2+}$ complex acts as an efficient catalyst for the O_2 reduction [219–221].

Methyl-substituted naphthalenes were also oxygenized with O_2 using Acr^+–Mes as a photoredox catalyst [222, 223]. It should be noted that 2-methylnaphthalene does not react with $^1O_2$ to produce any oxygenated products [222]. This underscores the utility of Acr^+–Mes in photocatalytic oxygenation of substrates as compared with $^1O_2$ photosensitizers. Photocatalytic oxidation of triphenylphosphine (Ph_3P) and benzylamine ($PhCH_2NH_2$) with O_2 also occurs efficiently using Acr^+–Mes as an photoredox catalyst to yield $Ph_3P{=}O$ and $PhCH_2N{=}CHPh$, respectively [224].

The scope of the use of Acr^+–Mes as an efficient electron-transfer photocatalyst for oxidation of organic compounds with O_2 has been shown by

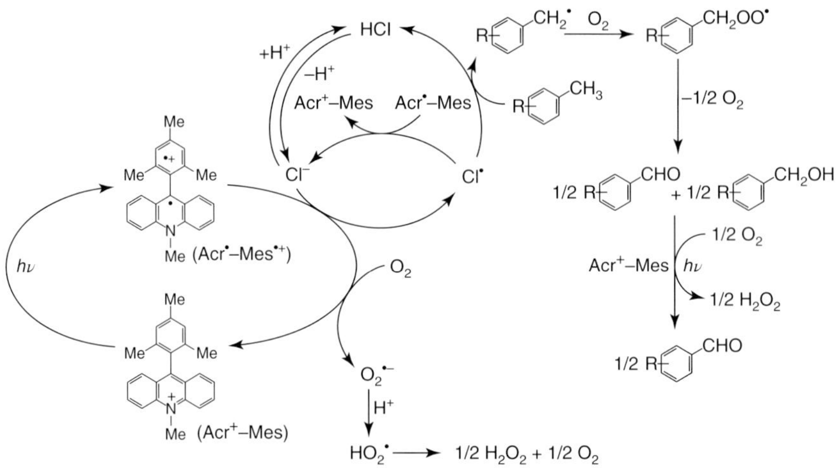

Scheme 7.7 Photocatalytic mechanism of oxygenation of toluene derivatives (R-$C_6H_4CH_3$: R = H, CN, NO_2, Br, Cl, CH_3). Source: Ohkubo et al. 2013 [225]. Reproduced with permission of Springer Nature.

examining photocatalytic oxygenation of toluene and cyclohexane with O_2 and hydrochloric acid (Schemes 7.7 and 7.8, respectively) [225, 226]. These reactions are initiated by the electron transfer oxidation of Cl^- with electron-transfer state of Acr^+–Mes to form $Cl^\cdot$, which cleaves the C—H bond of the substrates [225, 226]. The carbon-centered radicals thus produced are readily trapped by O_2 to produce the peroxoyl radicals, which disproportionate to yield the alcohol, aldehyde (ketone), and O_2 [225, 226]. The second-order rate constant of electron transfer from Cl^- to $Acr^\cdot$–$Mes^{\cdot+}$ was determined to be $9.5 \times 10^8\ M^{-1}\ s^{-1}$. Because $Cl^\cdot$ can abstract a hydrogen atom from methane [227], Acr^+–Mes may act as an efficient photocatalyst for the photocatalytic oxidation of methane with HCl.

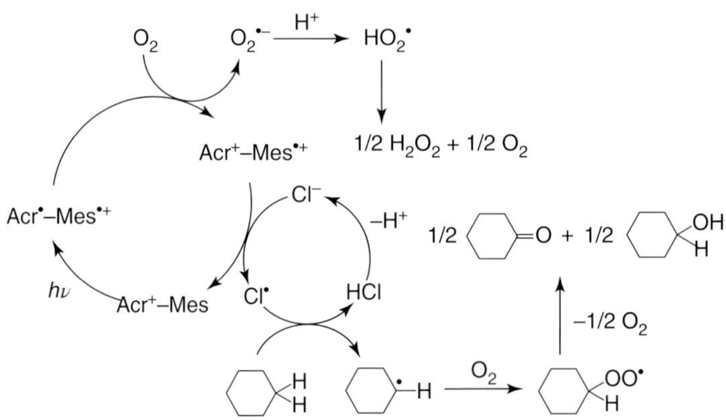

Scheme 7.8 Photocatalytic mechanism of oxygenation of cyclohexane with Acr^+–Mes and HCl. Source: Ohkubo et al. 2011 [226]. Reproduced with permission of Royal Society of Chemistry.

7.2 Photocatalytic Oxibromination

Bromination of aromatic compounds has been one of the most important and fundamental reactions in organic synthesis, providing key precursors for various transformations such as Grignard reactions and Suzuki–Miyaura coupling [228, 229]. Electrophilic bromination in nature mainly occurs by oxidative bromination through the catalyzed oxidation of the halide ion to form a brominating reagent, whereas bromination is usually carried out with hazardous, toxic, and corrosive molecular bromine, which is better avoided from the ecological point of view [230]. The best candidate for oxidants would be oxygen since hydrogen peroxide or water would be the only side product [231]. In this context, Acr^+–Mes was reported to act as an efficient organic photocatalyst for the oxidative bromination of aromatic hydrocarbons by O_2 with hydrogen bromide to produce the monobrominated products selectively [232]. Both the product yield and selectivity for the bromination of 1,3,5-trimethoxybenzene (TMB) were 100% with a quantum yield of 4.8% [232]. The photocatalytic turnover number was 900 based on the initial concentration of Acr^+–Mes [232]. When methoxy-substituted aromatic compounds were replaced by toluene derivatives, the consumption of substrate occurred efficiently under the same experimental conditions [232]. However, the yield of the brominated product and selectivity were significantly lower as compared with methoxy-substituted benzenes, because the photobromination competes with photooxygenation with oxygen to yield the corresponding aromatic aldehyde (Scheme 7.6) [232].

The photocatalytic reaction is also initiated by intramolecular photoinduced electron transfer from the Mes moiety to the singlet excited state of the Acr^+ moiety of Acr^+–Mes to generate the ET state ($Acr^\cdot$–$Mes^{\cdot+}$) as shown in Scheme 7.9, where the $Mes^{\cdot+}$ moiety can oxidize TMB to produce $TMB^{\cdot+}$, whereas the $Acr^\cdot$ moiety can reduce O_2 with proton to $HO_2\cdot$ [232]. The $TMB^{\cdot+}$ reacts with Br^- to form the Br-adduct radical, which undergoes dehydrogenation with $HO_2\cdot$ to afford the corresponding monobrominated product and hydrogen peroxide. Hydrogen peroxide further reacts with HBr and the substrate to produce another monobrominated product and H_2O [232]. The selectivity of monobromination resulted from the lower reactivity of the brominated benzene radical cations with Br^- [232]. Although the substrates that can be brominated are limited by their one-electron oxidation potentials, which should be less positive than the E_{ox} value of Acr^+–Mes (2.06 V vs. SCE), this limitation is compensated for by the high selectivity for the bromination to avoid over-bromination [232]. When HBr was replaced by HCl, photocatalytic chlorination of aromatic substrates with Acr^+–Mes also occurred under otherwise same experimental conditions [232].

7.3 Carbon—Carbon Bond Formation

When the ET oxidation of an electron donor by the $Mes^{\cdot+}$ moiety of $Acr^\cdot$–$Mes^{\cdot+}$ is coupled with deprotonation of the resulting radical cation and the dissociative ET reduction of a halogenated compound by the $Acr^\cdot$ moiety of $Acr^\cdot$–$Mes^{\cdot+}$, the

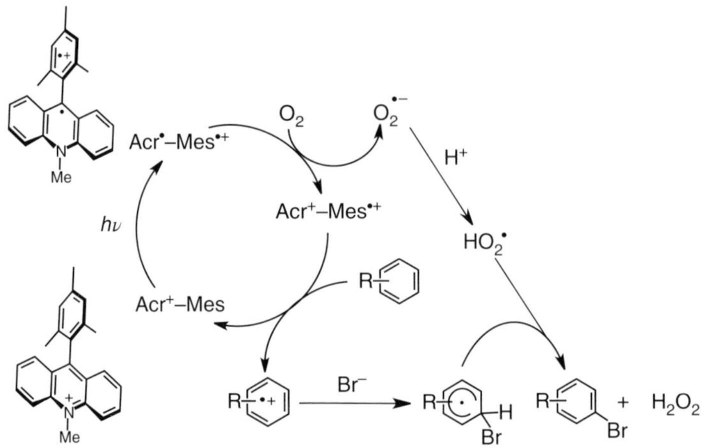

Scheme 7.9 Photocatalytic mechanism of bromination of aromatic compounds with HBr and O_2 using Acr$^+$–Mes as an organic photocatalyst. Source: Yamada et al. 2012 [221]. Reproduced with permission of Royal Society of Chemistry.

homocoupling product can be obtained. For example, photocatalytic coupling of 9,10-dimethylanthracene (DMA) in chloroform occurs efficiently via the ET oxidation of DMA by the Mes$^{\cdot+}$ moiety of Acr$^{\cdot}$–Mes$^{\cdot+}$, followed by deprotonation from the methyl group of DMA radical cation and the radical coupling reaction between the resulting anthracenylmethyl radicals to produce lepidopterene (5,6,11,12-tetrahydro-4b,12-[1′,2′],6,10b[1″,2″]-dibenzenochrysene) together with 1,2-bis(9-anthracenyl)ethane as shown in Scheme 7.10 [233]. The Acr$^{\cdot}$ moiety of Acr$^{\cdot}$–Mes, which is produced by electron transfer from DMA to Acr$^{\cdot}$–Mes$^{\cdot+}$, is oxidized by dissociative electron transfer to $CHCl_3$ to produce $CHCl_2^{\cdot}$ and Cl^-. The $CHCl_2^{\cdot}$ radicals dimerize to yield 1,1,2,2-tetrachloroethane ($CHCl_2CHCl_2$) [233]. The deprotonation from the methyl group of DMA$^{\cdot+}$ is the key step for the formation of lepidopterene in Scheme 7.10 [233]. In the case of unsubstituted anthracene, there is no methyl group to be deprotonated in the radical cation. In the case of 9,10-diethylanthracene, the deprotonation from the ethyl group in the radical cation may be too slow to compete with the back electron transfer [233]. Thus, in neither case photodimerization has occurred.

If one chooses a substrate that has a smaller HOMO–LUMO gap than the ET state energy of Acr$^+$–Mes (2.73 eV), both the radical cation and radical anion of the substrate can be formed by the ET oxidation and reduction with Acr$^{\cdot}$–Mes$^{\cdot+}$, respectively. Fullerenes are known to have relatively small HOMO–LUMO gap and are also suitable for efficient ET oxidation and reduction because of the small ET reorganization energy [31–33]. The free energy change of electron transfer from C_{60} (E_{ox} = 1.73 V vs. SCE) [234] to the Mes$^{\cdot+}$ moiety of Acr$^{\cdot}$–Mes$^{\cdot+}$ (E_{ox} = 1.88 V) [60] and that of electron transfer from the Acr$^{\cdot}$ moiety (E_{red} = −0.49 V) [60] to C_{60} (E_{red} = −0.43 V) [234] are both negative and thereby both the ET oxidation and reduction of C_{60} with Acr$^{\cdot}$–Mes$^{\cdot+}$ are energetically feasible. Thus, C_{60} acts as both an electron donor and acceptor in ET reactions of the ET state (Acr$^{\cdot}$–Mes$^{\cdot+}$) with C_{60} to produce $C_{60}^{\cdot+}$ and $C_{60}^{\cdot-}$ at the same

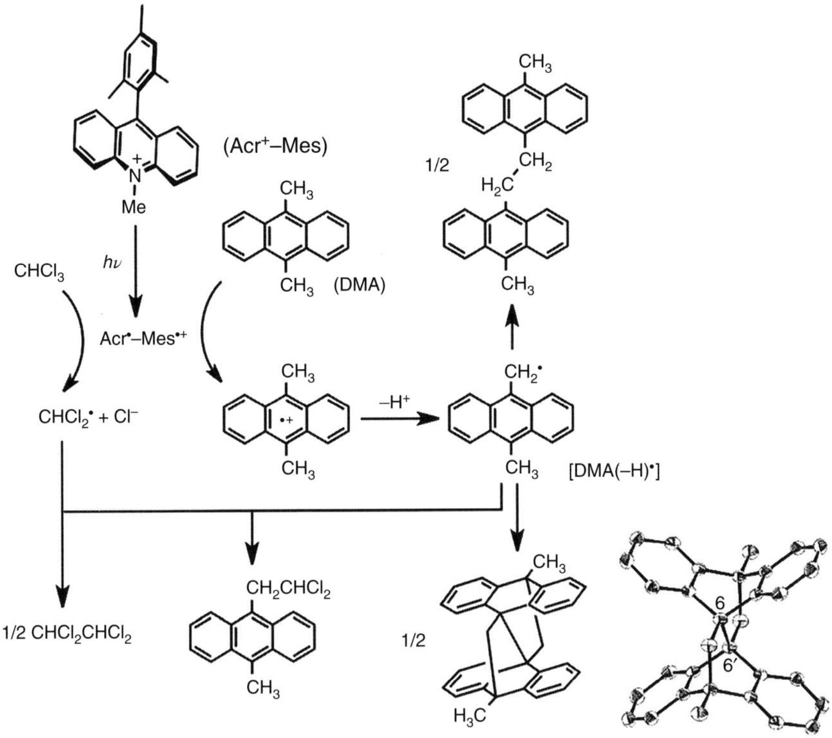

Scheme 7.10 Photocatalytic dimerization of 9,10-dimethylanthracene with Acr+–Mes in $CHCl_3$. Source: Ohkubo et al. 2006 [233]. Reproduced with permission of American Chemical Society.

time. The formation of both $C_{60}^{\cdot+}$ and $C_{60}^{\cdot-}$ is confirmed by a nanosecond laser excitation at 430 nm of a deaerated PhCN solution containing Acr$^+$–Mes and C_{60} as shown in Figure 7.1a, where the transient absorption bands due to $C_{60}^{\cdot+}$ (960 nm) [235] and $C_{60}^{\cdot-}$ (1080 nm) are observed [236]. The transient absorption bands due to $C_{60}^{\cdot+}$ and $C_{60}^{\cdot-}$ disappear, and this coincides with the appearance of a new absorption band at 740 nm due to $^3C_{60}^*$ (Figure 7.1b) [236]. The formation rate constant of $^3C_{60}^*$ generated by the charge recombination between $C_{60}^{\cdot+}$ and $C_{60}^{\cdot-}$ was determined as $5.2 \times 10^5\ s^{-1}$ [236]. The decay at 740 nm obeys first-order kinetics and the decay rate constant of $^3C_{60}^*$ is determined as $1.0 \times 10^5\ s^{-1}$ [236].

The [2 + 2] cycloaddition occurs efficiently between $C_{60}^{\cdot+}$ and $C_{60}^{\cdot-}$ in a nonpolar solvent (toluene/MeCN) to afford $^3C_{120}^*$ (Scheme 7.11) [236], because the driving force of charge recombination (2.16 eV) is larger than the triplet excited state energy of C_{120} (ca. 1.5 eV) [237]. The slower charge recombination in a nonpolar solvent (Tol/MeCN) than that in PhCN leads to the efficient formation of C_{120}. Further oligomerization may occur by the same process (Scheme 7.11) [236]. The quantum yield of dimer formation (Φ) under photoirradiation with monochromatized light (λ = 430 nm) was determined from an increase in absorbance at 700 nm due to C_{60} adducts to be 0.75 [236]. It should be emphasized that there was no photochemical reaction of C_{60} without Acr$^+$–Mes under

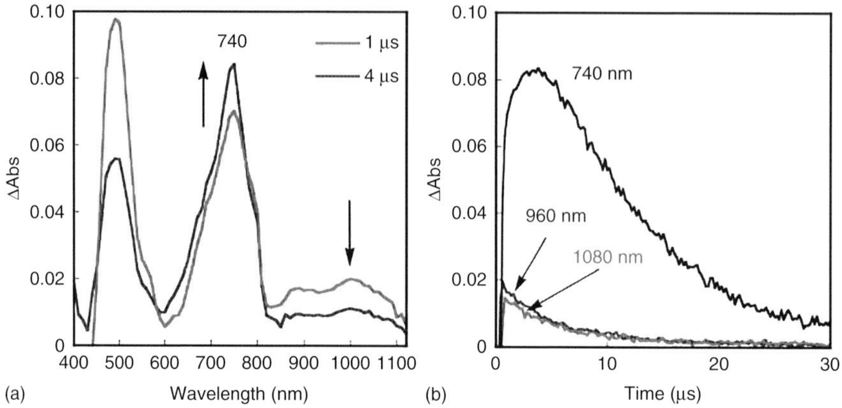

Figure 7.1 (a) Transient absorption spectra in PhCN observed in photoinduced ET oxidation of C_{60} (1.0×10^{-4} M) with Acr^+–$MesClO_4^-$ (1.0×10^{-4} M) taken 1.0 and 4.0 μs after laser excitation at 430 nm at 298 K. (b) Time profiles at 740, 960, and 1080 nm in PhCN. Source: Ohkubo et al. 2007 [236]. Reproduced with permission of Royal Society of Chemistry.

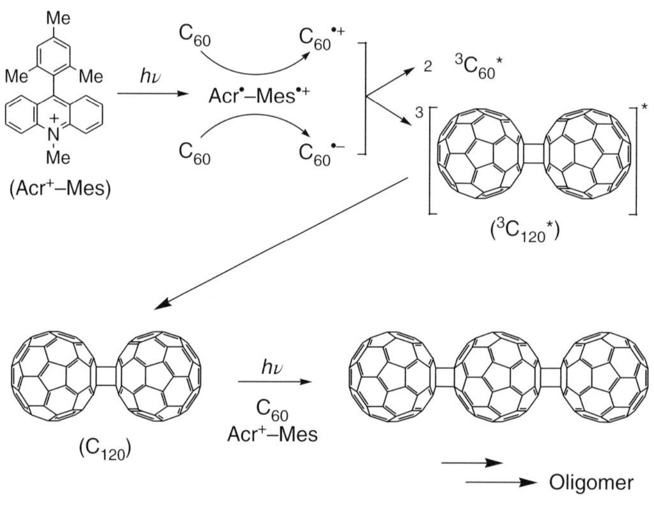

Scheme 7.11 Photocatalytic oligomerization of C_{60} with Acr^+–Mes. Source: Ohkubo et al. 2007 [236]. Reproduced with permission of Royal Society of Chemistry.

otherwise same experimental conditions [236]. The photocatalytic cycloaddition via formation of both radical cation and radical anion of a substrate with Acr^+–Mes provides a new synthetic pathway that could not be exploited using a classical concerted pathway.

The ET state of Acr^+–Mes is also useful for the detection of the transient absorption spectra of radical cations of electron donors, which can be readily produced by ET from electron donors to the $Mes^{\cdot+}$ moiety. The transient absorption band at 460 nm due to the radical cation of linoleic acid was detected by photoexcitation of Acr^+–Mes in the presence of linoleic acid in MeCN [238].

The deprotonation of the radical cation of linoleic acid was determined from the first-order decay of the absorbance at 460 nm as $1.1 \times 10^4\ s^{-1}$ [238].

7.4 DNA Cleavage

The oxidizing ability of the $Mes^{\cdot+}$ moiety of $Acr^{\cdot}–Mes^{\cdot+}$ is strong enough to produce radical cations of all types of DNA bases as well as DNA by the one-electron oxidation [239]. Radical cations of GMP (guanosine 5′-monophosphate), AMP (adenosine 5′-monophosphate), CMP (cytidine 5′-monophosphate), and TMP (thymidine 5′-monophosphate) have been detected as the transient absorption spectra in the laser flash photolysis measurements in photoinduced ET oxidation of DNA bases with $Acr^{\cdot}–Mes^{\cdot+}$ in an aqueous solution [239]. The absorption maxima of the radical cations of CMP, TMP, and AMP in the range from 450 to 550 nm are similar to that of $GMP^{\cdot+}$ [240], although the absorption intensity varies depending on DNA bases [239]. The k_{et} values of electron transfer at pH 7.0 from CMP, TMP, and AMP are also determined to be 2.2×10^5, 1.2×10^5, and $7.6 \times 10^5\ M^{-1}\ s^{-1}$, respectively [239]. Calf thymus DNA was also efficiently oxidized by the $Mes^{\cdot+}$ moiety of $Acr^{\cdot}–Mes^{\cdot+}$ [239]. It should be noted that there is no intercalation of a bulky Acr^+–Mes with double-stranded DNA because the dihedral angle between Acr^+ and Mes moieties of Acr^+–Mes is perpendicular [60]. The largest k_{et} value of the ET oxidation of GMP together with the lowest oxidation potential of GMP among DNA bases indicates that guanine is eventually oxidized in ET from DNA to the $Mes^{\cdot+}$ moiety of $Acr^{\cdot}–Mes^{\cdot+}$, leading to the efficient DNA cleavage (Scheme 7.12) [239]. The reactivity of DNA cleavage after five-minute photoirradiation of pBR 322 (one of the first widely used *E. coli* cloning vectors) with the monochromatized light (λ = 360 nm) in the presence of Acr^+–Mes is compared with those in the presence of 9-substitued acridinium ions without an electron donor moiety ($AcrR^+$, R=H, ^{i}Pr and Ph) as shown in Figure 7.2 [239]. The reactivity of DNA cleavage increases in the following order: $AcrPh^+ < AcrH^+$, $Acr^iPr^+ \ll Acr^+$–Mes [239]. The low reactivity of $AcrPh^+$ results from the short fluorescence lifetime (1.3 ns) of $^1AcrPh^{+*}$ (* denotes the excited state) as compared with those of $^1Acr^iPr^{+*}$ (26 ns) and $^1AcrH^{+*}$ (31 ns) [239]. The highest reactivity of Acr^+–Mes may result from the extremely long-lived ET state [60]. It is interesting to note that the DNA cleavage activity with Acr^+–Mes in the absence of O_2 is much higher than that in the presence of O_2 at pH 5.0 and 7.0 as shown in Figure 7.2.

7.5 Anti-Markovnikov Hydroetherification

Nicewicz and coworker utilized the high oxidizing ability of the ET state of Acr^+–Mes for the anti-Markovnikov hydroetherification of alkenols with 2-phenylmalononitrile as a redox-cycling source of a hydrogen atom, with complete regioselectivity without any trace of the undesired Markovnikov regioisomer [241, 242]. The utility of Acr^+–Mes as an organic photoredox catalyst is underscored when compared directly with the frequently employed

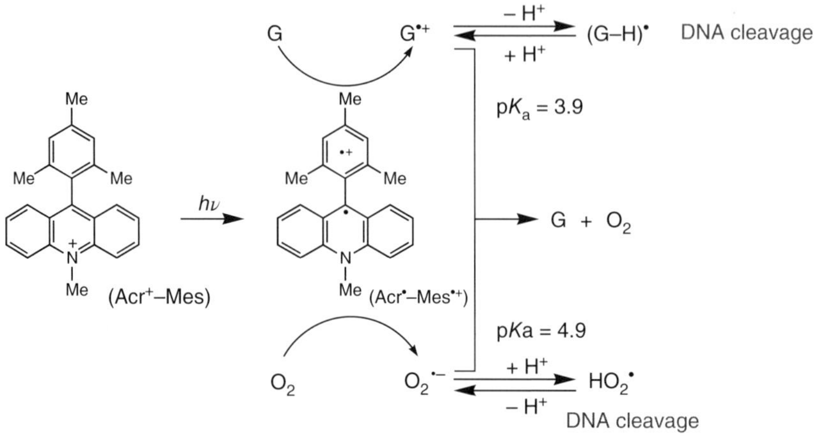

Scheme 7.12 Photocatalytic DNA cleavage with Acr^+–Mes. Source: Ohkubo et al. 2006 [239]. Reproduced with permission of Royal Society of Chemistry.

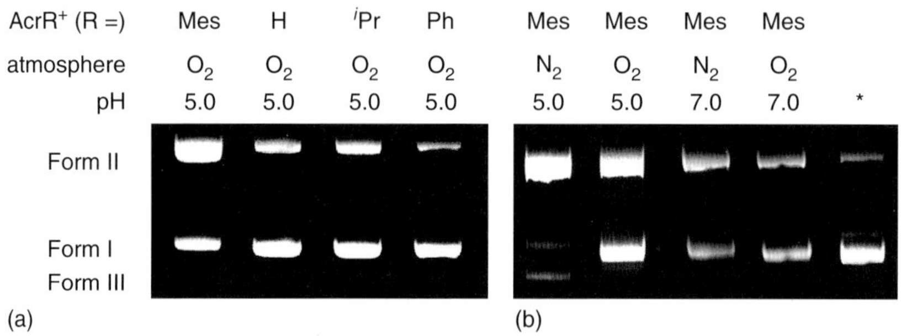

Figure 7.2 Agarose gel electrophoresis of photoinduced cleavage of supercoiled pBR322 DNA (0.051 μg μl^{-1}) with (a) various 9-substituted acridinium ions ($AcrR^+$: 1.0×10^{-4} M) in an O_2-saturated buffer solution (10 mM CH_3COOH/KOH, pH 5.0) after five minutes and (b) Acr^+–Mes (1.0×10^{-4} M) in an N_2- or O_2-saturated buffer solution after four minutes photoirradiation of monochromatized light (λ = 360 nm). Asterisk denotes the control experiment: pBR322 DNA in the presence of Acr^+–Mes before photoirradiation. (b) The retarding effect of O_2 may result from more efficient back ET from $O_2^{\cdot-}$ to DNA radical cation as compared with that from the Acr$^\cdot$ moiety to DNA radical cation before oxidizing DNA (Scheme 7.12). Source: Ohkubo et al. 2006 [239]. Reproduced with permission of Royal Society of Chemistry.

$[Ru(bpy)_3]^{2+}$, which failed to give any of the desired products [241]. The high oxidizing ability of the ET state of Acr^+–Mes allowed for greater latitude in potential substrates with alkenes possessing the one-electron oxidation potentials ranging up to +2.0 V vs. SCE [241].

The photocatalytic cycle is shown in Scheme 7.13 [241]. The ET state of Acr^+–Mes oxidizes the alkenol via the electron transfer from the alkenol to the $Mes^{\cdot+}$ moiety of the ET state to produce the corresponding radical cation, which is cyclized followed by hydrogen atom transfer from 2-phenylmalononitrile. The resulting radical could serve as an oxidant for the Acr$^\cdot$ radical to produce

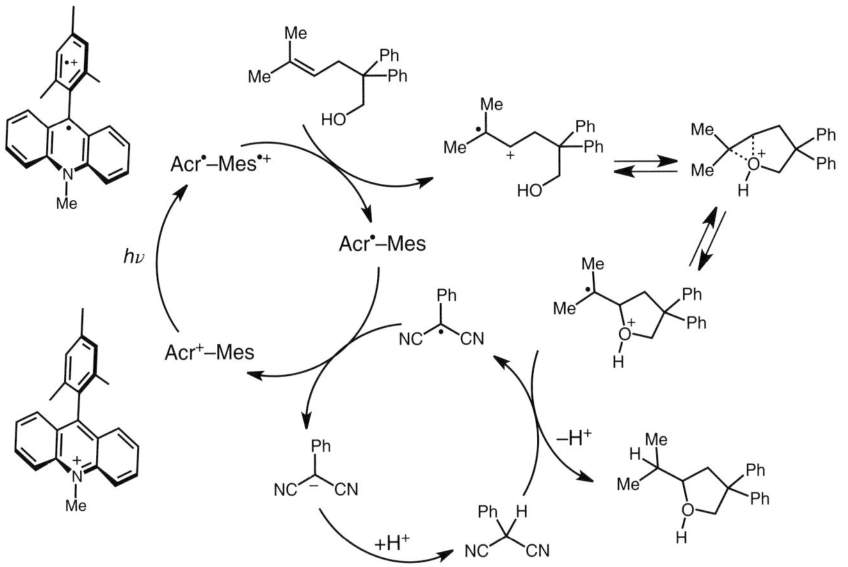

Scheme 7.13 Photocatalytic cycle for intramolecular anti-Markovnikov hydroetherification of alkenols. Source: Hamilton and Nicewicz 2012 [241]. Reproduced with permission of American Chemical Society.

the carbanion, regenerating the ground state Acr$^+$–Mes. Proton transfer from the cyclized cation to the carbanion regenerates the hydrogen-atom donor (2-phenylmalononitrile) and yields the desired product (Scheme 7.13) [241]. The scope of the intramolecular anti-Markovnikov hydroalkoxylation of alkenols was shown by using both electron-rich (4-(MeO)C_6H_4, 80% yield) and electron-deficient (4-ClC_6H_4, 60% yield) compounds, which provided good yields of the desired 5-exo adducts [241]. The anti-Markovnikov hydroetherification of alkenols shows sharp contrast to Brønsted acid assisted Markovnikov hydroetherification [241].

The same strategy used for intramolecular hydroetherification of alkenols where the radical cations gave rise to anti-Markovnikov reactivity in Scheme 7.13 has also been applied for intramolecular anti-Markovnikov hydroamination of unsaturated amines in which thiophenol was used as a hydrogen-atom donor [243]. The photocatalytic system is effective for a range of cyclization modes to give important nitrogen-containing heterocycles [243].

7.6 Photocatalytic Cycloaddition

The intramolecular anti-Markovnikov hydroetherification of alkenols in Scheme 7.13 has been extended to intermolecular cycloaddition of *trans*-*β*-methylstyrene and allyl alcohol in Scheme 7.14 [244]. The *β*-methylstyrene radical cation produced by electron transfer from *β*-methylstyrene to the ET state of Acr$^+$–Mes reacts with allyl alcohol to produce the adduct radical cation,

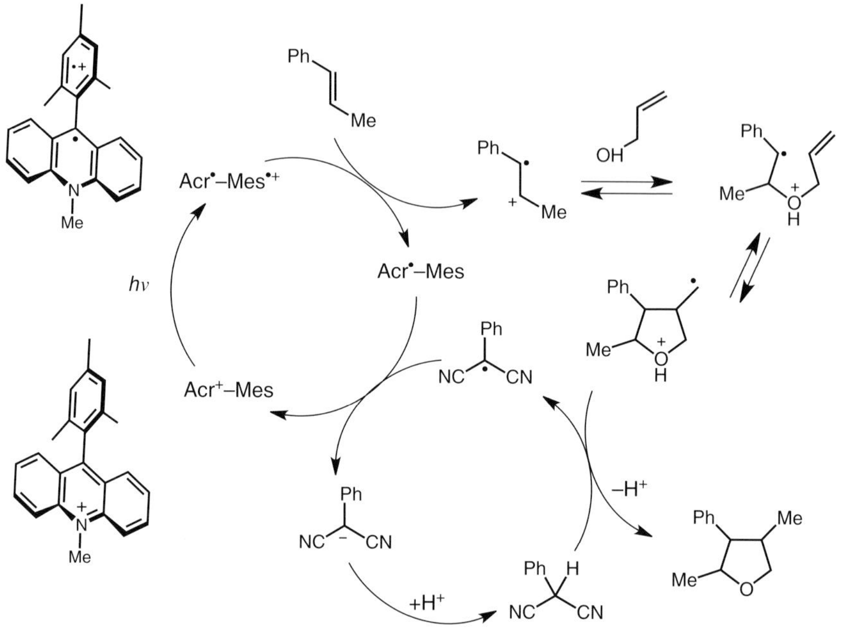

Scheme 7.14 Photocatalytic cycle of intermolecular cycloaddition between *β*-methylstyrene and allyl alcohol. Source: Grandjean and Nicewicz 2013 [244]. Reproduced with permission of John Wiley & Sons.

which undergoes a 5-exo radical cyclization with the pendant alkene [244]. Hydrogen-atom abstraction from 2-phenylmalononitrile and the loss of a proton yield the tetrahydrofuran adduct (63% yield) [244]. The phenyl malononitrile anion is neutralized by the generated proton to regenerate the hydrogen-atom donor (2-phenylmalonitrile) [244].

Employing *cis-β*-methylstyrene gave an identical mixture of diastereomers as *trans-β*-methylstyrene (80% yield), demonstrating the loss of alkene geometry upon the one-electron oxidation [244]. 4-Chloro-*β*-methylstyrene gave the corresponding tetrahydrofuran adduct in good yield (70% yield), whereas 4-methoxy-*β*-methylstyrene was not reactive under these conditions, probably due to the stability of the resultant radical cation intermediate [244]. Cyclic alkene substrates such as indene and 1-phenylcyclohexene also afforded good yields of the corresponding cyclic ether adducts [244]. Aliphatic trisubstituted alkenes with higher oxidation potentials, such as 2-methylbut-2-ene, also afforded highly substituted cyclic ethers [244]. Thus, Acr^+–Mes is used as an effective organic photoredox catalyst to synthesize highly substituted tetrahydrofurans from readily available allylic alcohols and alkenes [244].

The photocatalytic cycle in intermolecular cycloaddition between *β*-methylstyrene and allyl alcohol in Scheme 7.14 has also been applied to anti-Markovnikov hydroacetoxylation of styrenes, trisubstituted aliphatic alkenes, and enamides, with a variety of carboxylic acids to afford the anti-Markovnikov addition adducts exclusively (Scheme 7.15) [245]. Electron-transfer oxidation of the alkene by the $Mes^{\cdot+}$ moiety of the ET state of Acr^+–Mes

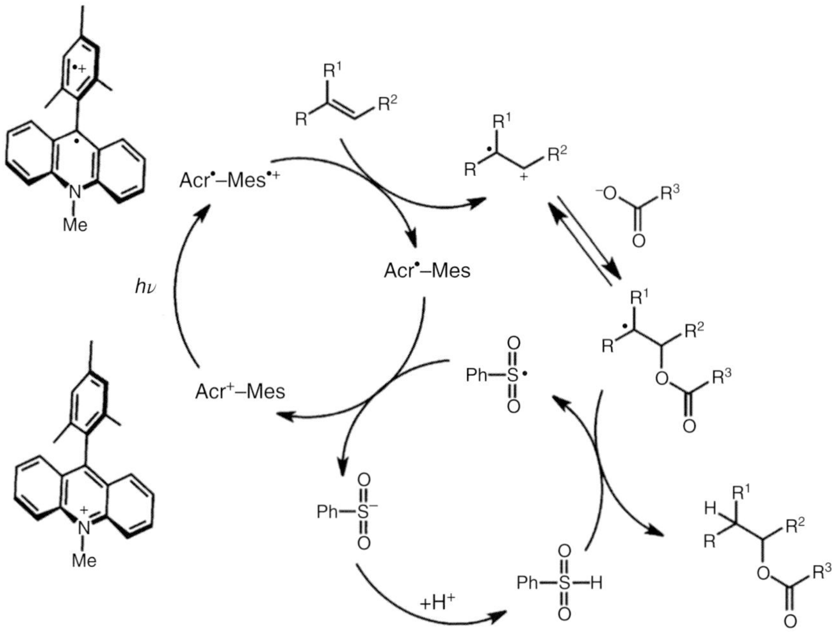

Scheme 7.15 Photocatalytic cycle of anti-Markovnikov alkene hydroacetoxylation. Source: Perkowski and Nicewicz 2013 [245]. Reproduced with permission of American Chemical Society.

results in the formation of the alkene cation radical to which the carboxylate nucleophile is added to the less substituted position of the cation radical to produce the adduct radical [245]. A rapid acid–base equilibrium with the excess carboxylic acid generates small quantities of benzenesulfinic acid, which acts as the active hydrogen-atom donor [245]. Hydrogen-atom transfer from benzenesulfinic acid yields the anticipated anti-Markovnikov adduct [245]. The hydrogen-atom transfer step is found to be rate determining because of large deuterium kinetic isotope effect [245]. The resultant benzenesulfinyl radical oxidizes the Acr$^{\cdot}$ moiety to regenerate Acr$^+$–Mes and benzenesulfinate [245].

7.7 Photocatalytic Hydrotrifluoromethylation

The development of new methodologies for highly efficient and selective incorporation of a CF_3 group into diverse skeletons has merited significant interest of synthetic chemists [246], because the CF_3 group is a useful structural motif in many biologically active molecules as well as materials [247]. There have been several reports on photocatalytic trifluoromethylation using metal complexes as photosensitizers [248–251]. Nicewicz and coworkers reported metal-free hydrotrifluoromethylation of alkenes using Acr$^+$–Mes as an efficient organic photoredox catalyst as shown in Scheme 7.16 [252]. The electron-transfer oxidation of sodium trifluoromethanesulfinate (CF_3SO_2Na, Langlois reagent)

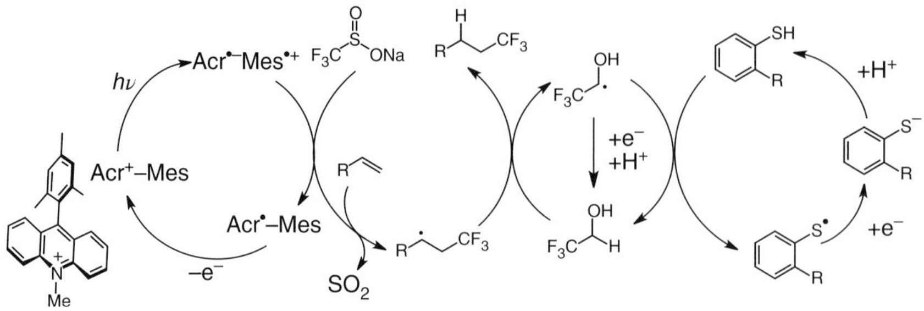

Scheme 7.16 Photocatalytic cycle of trifluoromethylation of alkenes. Source: Wilger et al. 2013 [252]. Reproduced with permission of Royal Society of Chemistry.

[253] results in the formation of the electrophilic triflurorомethyl radical ($CF_3^{\cdot}$) together with expulsion of SO_2. Addition of $CF_3^{\cdot}$ to the alkene occurs with anti-Markovnikov selectivity to produce the corresponding carbon-centered radical [252]. Alkyl-substituted alkenes provided hydrotrifluoromethylated products without the use of thiols as an H-atom donor [252]. In this case, trifluoroethanol used as a cosolvent acts as an H-atom donor. The trifluoromethylketyl radical produced oxidizes the Acr˙ moiety of Acr˙–Mes to regenerate Acr⁺–Mes. Methyl thiosalicylate is used as an H-atom donor for aliphatic alkenes, and thiophenol is used as an H-atom donor for styrenyl substrates [252]. The substrate scope for the photocatalytic trifluoromethylation is broad, including mono-, di-, and trisubstituted aliphatic and styrenyl alkenes, with high regioselectivity [252].

7.8 Photocatalytic Hydrogen Evolution

Hydrogen (H_2) is a clean energy source for the future to reduce dependence on fossil fuels and emissions of greenhouse gases in the long term [253]. A number of photocatalytic hydrogen-evolution systems have been developed in the past decades [254–261]. Such systems usually consist of an electron donor, a photosensitizer, and an electron mediator such as methyl viologen (MV^{2+}) and a hydrogen-evolution catalyst. A highly efficient photocatalytic hydrogen-evolution system *without an electron mediator* such as methyl viologen (MV^{2+}) has been constructed using 9-mesityl-10-methylacridinium ion (Acr^+–Mes), poly(*N*-vinyl-2-pyrrolidone)-protected platinum nanoclusters (Pt–PVP) and NADH (β-nicotinamide adenine dinucleotide, reduced form), used as a photocatalyst, a hydrogen-evolution catalyst, and an electron donor, respectively [262]. The hydrogen-evolution rate of the photocatalytic system in the absence of MV^{2+} is 300 times faster than that with MV^{2+} as shown in Figure 7.3 [263]. The photocatalytic efficiency of H_2 evolution has been much improved by using organic donor–acceptor linked molecules with a long-lived electron-transfer state [264], which can inject electrons directly to hydrogen-evolution catalysts without an electron mediator upon photoexcitation of the donor–acceptor linked dyads instead of using inorganic photocatalysts [263]. A very high quantum yield ($QY(H_2)$ = H_2 molecules produced/photons

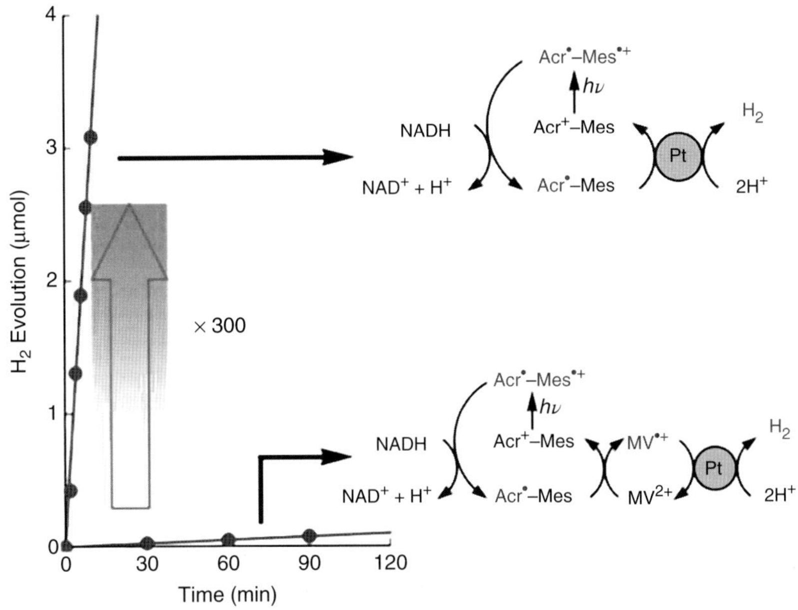

Figure 7.3 Time dependence of hydrogen evolution under steady-state irradiation ($\lambda > 390$ nm) of a deaerated phthalic acid buffer (pH 4.5, 50 mM) and MeCN [1 : 1 (v/v)] mixed solution (2.0 cm^3) containing Acr^+–Mes (0.10 mM), NADH (1.0 mM), and Pt–PVP (0.20 mg) in the absence and in the presence of MV^{2+} (5.0 mM) at 298 K. Source: Kotani et al. 2007 [263]. Reproduced with permission of Royal Society of Chemistry.

absorbed) × (2 electrons/H_2 molecule = 52%) and a high yield of H_2 (95%) based on an electron donor (NADH) have been achieved under photoirradiation of a deaerated phthalic acid buffer (pH 4.5, 50 mM) and acetonitrile (MeCN) [1 : 1 (v/v)] mixed solution containing 9-mesityl-10-methylacridinium ion (Acr^+–Mes) (0.10 mM), NADH (1.0 mM) and poly(*N*-vinyl-2-pyrrolidone)-protected platinum nanocluster (Pt–PVP) (0.20 mg) at 298 K [263]. Upon photoexcitation of Acr^+–Mes, the electron-transfer state ($Acr^{\cdot}$–$Mes^{\cdot +}$) is formed and electron transfer from NADH to the $Mes^{\cdot +}$ moiety of $Acr^{\cdot}$–$Mes^{\cdot +}$ occurs rapidly to produce $NADH^{\cdot +}$ and $Acr^{\cdot}$–Mes (Scheme 7.17) [263]. $NADH^{\cdot +}$ deprotonates to produce $NAD^{\cdot}$, followed by fast electron transfer from $NAD^{\cdot}$ to Acr^+–Mes to yield NAD^+ and $Acr^{\cdot}$–Mes as revealed by laser flash photolysis measurements [263]. Thus, the absorption of one photon resulted in formation of 2 equiv of $Acr^{\cdot}$–Mes (Scheme 7.17) [263]. Electron injection from $Acr^{\cdot}$–Mes to PtNPs with protons resulted in production of H_2. Cubic PtNPs with a diameter of 6.3 ± 0.6 nm exhibited the maximum catalytic activity for the photocatalytic H_2 production [265]. The observed photocatalytic H_2 production rate was virtually the same as the rate of electron injection from $Acr^{\cdot}$–Mes to PtNPs [265]. The rate constant of the electron injection (k_{et}) increased linearly with increasing proton concentration [265]. When H^+ was replaced by D^+, the inverse kinetic isotope effect was observed for the electron-transfer rate constant ($k_{et}(H)/k_{et}(D) = 0.47$) [265]. The linear dependence of k_{et} on proton concentration together with the

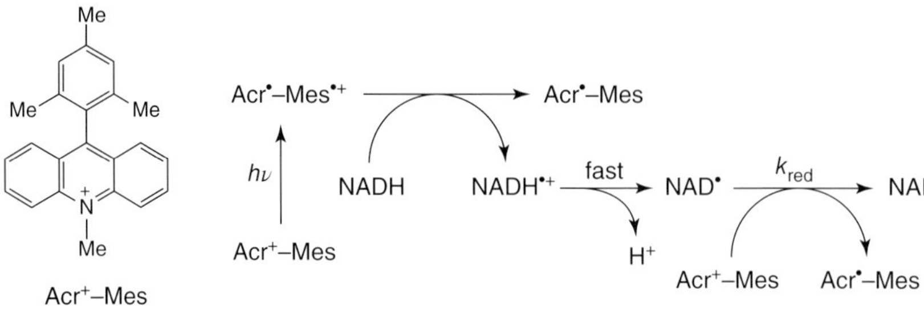

Scheme 7.17 Mechanism of photoinduced reduction of Acr^+–Mes with NADH. Source: Kotani et al. 2007 [263]. Reproduced with permission of Royal Society of Chemistry.

observed inverse kinetic isotope effect indicated that proton-coupled electron transfer (PCET) from $Acr^{\cdot}$–Mes to PtNPs to form the Pt—H bond is the rate-determining step for the photocatalytic H_2 production (Scheme 7.18) [265].

When Acr^+–Mes was replaced by 2-phenyl-4-(1-naphtyl)quinolinium ion ($QuPh^+$–NA), photocatalytic H_2 production with NADH and PtNPs was made possible under basic conditions (pH 10) [266]. PtNPs could be replaced by RuNPs with virtually the same catalytic activity for H_2 production [266]. Upon photoexcitation of $QuPh^+$–NA, the electron-transfer state of $QuPh^+$–NA ($QuPh^{\cdot}$–$NA^{\cdot+}$) is produced, followed by electron transfer from NADH to the $NA^{\cdot+}$ moiety of $QuPh^{\cdot}$–$NA^{\cdot+}$, resulting in the formation of 2 equiv of $QuPh^{\cdot}$–NA via deprotonation of $NADH^{\cdot+}$ and the subsequent electron transfer from $NAD^{\cdot}$ to $QuPh^+$–NA (Scheme 7.19) [266]. Electron transfer from the photogenerated $QuPh^{\cdot}$–NA to RuNPs resulted in H_2 evolution even under basic conditions. The rate of electron transfer from $QuPh^{\cdot}$–NA to RuNPs is much faster than the rate of hydrogen evolution. RuNPs with the size of 4.1 nm exhibited the highest H_2-evolution rate normalized by the weight of RuNPs [266]. The photocatalytic H_2 evolution with RuNPs (4.1 nm) under pH 10 conditions indicated that turnover frequency and turnover number normalized by the number of surface Ru atoms were as high as $1050\,h^{-1}$ and 300, respectively [266]. In the case of Acr^+–Mes, electron transfer from $Acr^{\cdot}$–Mes to PtNPs requires the assistance of a proton and the rate of the PCET increases with increasing proton concentration [265]. In contrast to this, electron transfer from $QuPh^{\cdot}$–NA to MNPs occurs without assistance of proton because of the much stronger reducing ability of $QuPh^{\cdot}$–NA as compared with $Acr^{\cdot}$–Mes judging from the significantly more negative oxidation potential of $QuPh^{\cdot}$–NA ($E_{ox} = -0.90$ V vs. SCE) than that of $Acr^{\cdot}$–Mes ($E_{ox} = -0.57$ V vs. SCE) [264]. Thus, the rate of

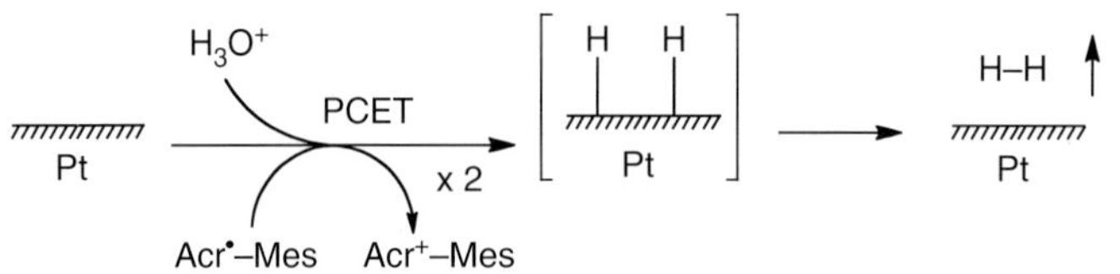

Scheme 7.18 PCET production of H_2 with $Acr^{\cdot}$–Mes on PtNPs. Source: Kotani et al. 2011 [265]. Reproduced with permission of John Wiley & Sons.

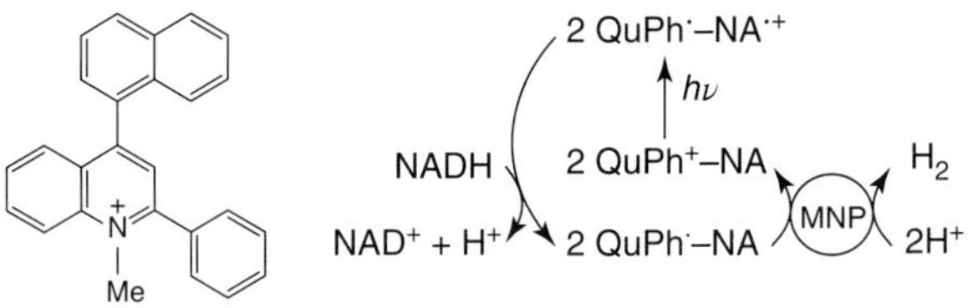

QuPh+–NA

Scheme 7.19 Structure of QuPh+–NA and the overall catalytic cycle for the photocatalytic hydrogen evolution with QuPh+–NA and metal nanoparticles (MNPs, *M* = Ru, Pt). Source: Yamada et al. 2011 [266]. Reproduced with permission of American Chemical Society.

electron transfer from QuPh˙–NA to MNPs remains about the same even under highly basic conditions (pH 10) [266]. RuNPs could be replaced by NiNPs, which are more earth abundant, although the catalytic reactivity was 40% of that with commercially available PtNPs using the same catalyst weight [267].

NADH used as an electron and proton donor in Scheme 7.19 can be replaced by oxalic acid for the photocatalytic H_2 production with QuPh+–NA in a deaerated mixed solution [1 : 1 (v/v)] of an aqueous buffer (pH 6.0) and acetonitrile [268]. Oxalate and its conjugate acid can act as a two-electron donor to produce 2 equiv of CO_2 and hydrogen (Eq. (7.1)). In this case as well, the catalytic reactivity of NiNPs was 32% of PtNPs [268].

$$(COOH)_2 \rightarrow 2CO_2 + H_2 \qquad (7.1)$$

A mixed solution of acetonitrile and water was used in the above reaction (Eq. (7.1)) because of the insolubility of QuPh+–NA in water. In order to perform the photocatalytic production of H_2 in water, QuPh+–NA was incorporated into nanosized mesoporous silica–alumina with a spherical shape (sAlMCM-41) by cation exchange and the composite (QuPh+–NA@sAlMCM-41) was employed as a photocatalyst in water [269]. Incorporation of cationic electron donor–acceptor dyads into nanosized mesoporous silica–alumina was reported to result in much longer lifetimes of the photogenerated electron-transfer states as compared with those in solution [73, 270]. The size and morphology of sAlMCM-41 were confirmed by transmission electron microscope (TEM) measurements and all particles were spherical and the particle sizes ranged from 200 to 700 nm [269]. In this case, K_2PtCl_6 was used as a precursor catalyst for PtNPS, because PtNPs could not be incorporated into sAlMCM-41 due to the larger size of PtNPs than the window size of sAlMCM-41 [269]. No H_2 evolution was observed under photoirradiation ($\lambda > 340$ nm) of a phthalate buffer (pH 4.5) dispersion containing QuPh+–NA@sAlMCM-41, sodium oxalate, and PtNPs as a photosensitizer, a sacrificial electron donor, and an H_2-evolution catalyst, respectively [269]. When K_2PtCl_6 was added to the reaction solution instead of PtNPs as a precursor of H_2-evolution catalyst, however, efficient H_2 production was observed [269]. $Cu(NO_3)_2$ also acted as a precursor catalyst for the photocatalytic H_2 production with oxalate in water (pH 4.5) [269]. Thus, hybrid H_2-evolution catalysts derived from earth-abundant metal salts in situ provide a

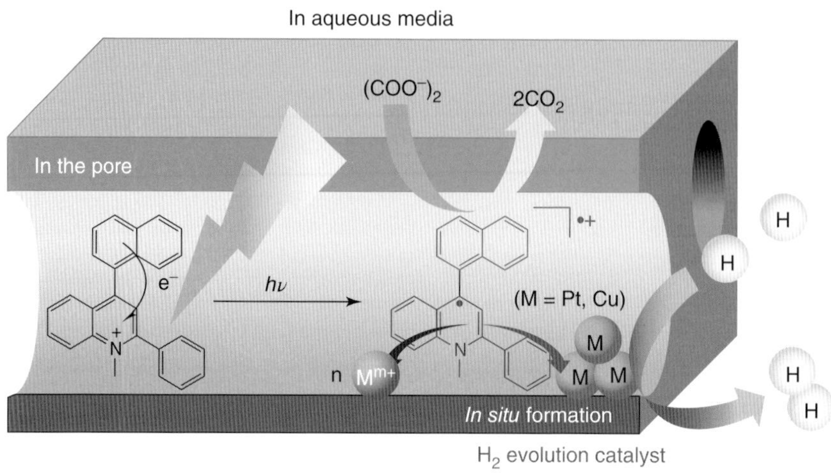

Figure 7.4 Photocatalytic production of H_2 from oxalate with QuPh⁺–N incorporated into nanosized mesoporous silica–alumina (sAlMCM-41) and Pt or Cu salts as precursor catalysts. Source: Yamada et al. 2013 [269]. Reproduced with permission of Royal Society of Chemistry.

convenient way to develop efficient water reduction catalysts working in water as shown in Figure 7.4 [269].

Under acidic conditions, oxalate anion is protonated to produce oxalic acid, which cannot act as an electron donor in the photocatalytic H_2 production system (vide supra). However, a composite photocatalytic system employing acidic oxalic acid as an electron donor has been successfully constructed by combining QuPh⁺–NA, Pt, and nanosheets prepared by exfoliation of $K_4Nb_6O_{17}$ (niobate-NS) as an organic photosensitizer, a hydrogen-evolution catalyst, and a semiconductor photocatalyst for oxidation of oxalic acid, respectively [271]. The composite photocatalyst, QuPh⁺–NA/niobate-NS (Pt), was prepared by a two-step route to locate a Pt catalyst near QuPh⁺–NA on the surfaces of niobate-NS [271]. QuPh⁺–NA was supported on niobate-NS by a cation exchange method and then Pt was supported on the QuPh⁺–NA/niobate-NS by a photodeposition method using $PtCl_4{}^{2-}$ as a precursor, which repulsively interacts with the negatively charged surfaces of niobate-NS [271]. The precursor of $PtCl_4{}^{2-}$ was reduced to metallic Pt by the photocatalysis of QuPh⁺–NA in the presence of oxalic acid. Photocatalytic H_2 evolution with the composite catalyst proceeds via photoexcitation of both niobate-NS and QuPh⁺–NA to produce an electron and a hole in the semiconductor and the electron-transfer state (QuPh$^{\cdot}$–NA$^{\cdot+}$), respectively [271].

Photocatalytic H_2 evolution also occurs via electron transfer from styrene to Acr$^{\cdot}$–Mes$^{\cdot+}$, produced upon photoexcitation of Acr⁺–Mes, followed by dimerization of styrene radical cation with neutral styrene to produce the dimer radical cation in the presence of $Co^{II}(dmgBF_2)_2$ (dmg = dimethylglyoximate) as shown in Scheme 7.20 [272]. The styrene dimer radical cation undergoes deprotonation to produce the neutral radical that reacts with $Co^{II}(dmgBF_2)_2$ to yield the dimerized product after deprotonation and the Co(III)-hydride

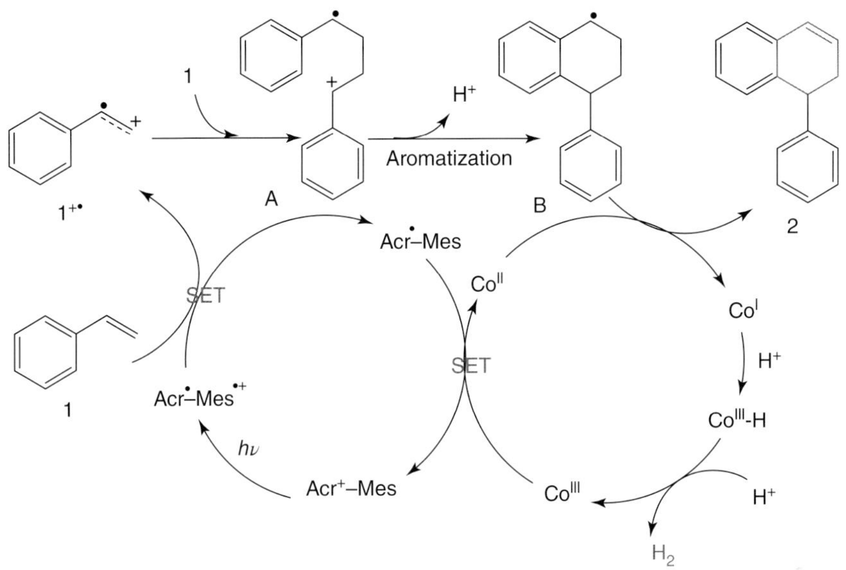

Scheme 7.20 The catalytic cycle for the photocatalytic hydrogen evolution with dimerization of styrenes in the presence of Acr^+–Mes and $Co^{II}(dmgH)_2py$. Source: Cao et al. 2018 [272]. Reproduced with permission of Elsevier.

complex [272]. The Co(III)-hydride complex reacts with proton to evolve H_2, accompanied by generation of $[Co^{III}(dmgBF_2)_2]^+$ that is reduced by $Acr^{\cdot}$–Mes to regenerate $Co^{II}(dmgBF_2)_2$ and Acr^+–Mes [272]. The maximum yield of 1-phenyl-1,2-dihydronaphthalene was 72% in the presence of NaH_2PO_4 that acts as a good proton acceptor. It should be noted that other photocatalysts such as rose bengal, rhodamine B, an Ir complex, eosin Y, and $[Ru(bpy)_3]Cl_2$ afforded no desired product [272]. In the control experiments, no desired product was observed without Acr^+–Mes, cobalt catalyst, or visible light, indicating that Acr^+–Mes and cobalt catalysts are essential for this transformation [272]. Photocatalytic generation of H_2 also occurs from alkenes, accompanied by dehydrogenative oxygenation of alkenes with Acr^+–Mes and $Co^{II}(dmgH)_2py$ to produce the carbonyl compounds [273].

8

Hydrogen Storage

In spite of the availability and usefulness of H_2, storage and transportation of H_2 are quite difficult, because H_2 gas is explosive and its volumetric energy density is quite low. Extensive efforts have so far been devoted to store hydrogen on-site, e.g. the improvement of the safety of high-pressure tank systems. In such systems, however, energy-consuming cryogenically coolable equipment is required to store pressurized hydrogen in a liquid form [274–276]. Other methods using metal hydrides (hydrogen storage alloy) [277–279], carbon materials [280–283] such as fullerene and carbon nanotubes, and metal-organic frameworks [284–294] can store and liberate only low amounts of hydrogen at high temperature and high pressure. Hence, low-cost, energy-efficient storage of hydrogen is definitely needed for stationary and portable applications in the hydrogen-delivery infrastructure. In this context, catalytic dehydrogenation and hydrogenation reactions of organic and inorganic molecules in a reversible way have recently attracted much attention from the viewpoint of a hydrogen donor as a hydrogen storage material, i.e. releasing hydrogen by dehydrogenation of a hydrogen donor and storing hydrogen by hydrogenation of the oxidized form of the hydrogen donor [295–305]. With reversible interconversion between hydrogen and hydrogen donors under ambient conditions, hydrogen donors can be regarded as sustainable and renewable energy resources. In particular, the utility of formic acid (HCOOH) as an organic hydrogen donor has merited significant attention [306–319], because HCOOH is liquid at room temperature with relatively high volumetric density (d = 1.22 g cm^{-3}) [320] and HCOOH can be formed by reduction of CO_2 with H_2 using various catalysts [321–340]. From the viewpoint of safety and cost-cutting, liquid form is suitable for transportation, handling, and storage as compared to the gaseous form. In addition, HCOOH is widely utilized as a preservative and antibacterial agent in livestock feed [341]. Thus, the combination of H_2 storage with the aid of CO_2 as a carrier, i.e. hydrogenation of CO_2 with H_2 to produce HCOOH and the reverse reaction, H_2 evolution in the decomposition of HCOOH to produce CO_2 as a sole by-product, is an ideal carbon-neutral process. The use of water as a solvent is preferred for the interconversion between H_2 and HCOOH, because the standard free energy change is slightly negative (−4 kJ mol^{-1} at 298 K) in water under the conditions that all the reactants and the products are soluble in water,

Electron Transfer: Mechanisms and Applications, First Edition. Shunichi Fukuzumi.

whereas the reaction between gaseous H_2 and CO_2 is accompanied by a change in free energy by +33 kJ mol^{-1} to yield liquid HCOOH [323, 342].

In the natural photosynthesis of plants, electrons and protons taken from water under irradiation of solar light are used to reduce nicotinamide adenine dinucleotide (phosphate): $NAD(P)^+$ with the aid of ferredoxin-NADP reductase, generating the 1,4-dihydro form, i.e. NAD(P)H, which is used to reductively fix CO_2 as carbohydrates through the Calvin cycle [343, 344]. On the other hand, in hydrogenase-containing photosynthetic organisms such as cyanobacteria and green algae, electrons and protons taken from water using solar energy are used for the reduction of protons to hydrogen (H_2) rather than the reduction of $NAD(P)^+$ to NAD(P)H [345–348]. Both the oxidation of H_2 with $NAD(P)^+$ and the reduction of protons with NAD(P)H are catalyzed by hydrogenases [345–348]. Thus, functional mimic of hydrogenase, i.e. interconversion between H_2 and NADH as an organic hydrogen donor, provides a sustainable biomimetic hydrogen storage system. Hydrogen evolution from 1,4-dihydronicotinamide adenine dinucleotide: NADH in water has so far been made possible under photoirradiation of an appropriate sensitizer via photoinduced electron transfer with the aid of metal nanoparticles at room temperature (vide supra) [307, 308, 349]. On the other hand, regeneration of NAD(P)H from $NAD(P)^+$ by the reduction with gaseous H_2 has been made possible with Ru and Rh catalysts under relatively high pressure of hydrogen (4.8 atm) [350].

From the viewpoint of solar energy utilization, biofuels such as the so-called bioethanol made from plants' carbohydrate are ideal hydrogen resources, since they can easily be produced from biomass by fermentation [351–359]. As biomass takes in carbon dioxide from the atmosphere in photosynthesis for its growth, consumption of biomass-derived ethanol to produce carbon dioxide is regarded as a carbon neutral process and does not adversely affect global warming. Bioethanol is usually produced in water with the aid of enzymes, thus, containing certain amount of water, at least 5 vol% [358, 360]. The mixture of water and ethanol is a well-known azeotrope and the use of the mixture as a biofuel to produce hydrogen is preferred as compared to that of pristine ethanol to avoid the costly isolation procedures. Dehydrogenation of bioethanol to produce hydrogen requires high temperatures (c. 573 K) generating CO as a waste by-product, which is a poison for the catalyst [360–364]. Thus, it is desired to produce hydrogen from bioethanol that contains water under ambient conditions without CO generation.

Given this state of development of hydrogen storage and evolution using the organic hydrogen donors described above, the following section is focused on efficient homogeneous catalysts, which are capable of interconverting between organic hydrogen donors and hydrogen (two electrons and two protons) in an aqueous solution under atmospheric pressure at ambient temperature. Two electrons and two protons thus obtained can be used for further redox reactions including electron-transfer reactions.

8.1 Interconversion Between Hydrogen and Formic Acid

In 2003, it was reported that a Ru–hydride complex [$Ru^{II}(\eta^6\text{-}C_6Me_6)$(bpy)H]$_2$($SO_4$) (where bpy = 2,2′-bipyridine), which was produced by the reduction of [$Ru^{II}(\eta^6\text{-}C_6Me_6)$(bpy)($OH_2$)]($SO_4$) (**1**) with $NaBH_4$, reduces CO_2 to HCOOH in water in a pH range of about 3–5 at ambient temperature [326, 365]. When [$Ru^{II}(\eta^6\text{-}C_6Me_6)$ (bpy)(OH_2)](SO_4) was replaced by [$Ru^{II}(\eta^6\text{-}C_6Me_6)$(4, 4′-MeO-bpy)($OH_2$)]($SO_4$) (**2**) (where 4,4′-OMe-bpy = 4,4′-dimethoxy-2,2′-bipyridine), the catalytic reduction of CO_2 (2.5 MPa) with H_2 (5.5 MPa) to HCOOH occurred under acidic conditions (pH 2.5–5.0) without any base at 313 K [326]. The use of an iridium complex was more effective on the catalytic reduction of CO_2 with H_2 under the same reaction conditions. The initial turnover frequency (TOF) for the catalytic reduction of CO_2 with H_2 to HCOOH using an Ir complex [$Ir^{III}(Cp^*)$(4,4′-OMe-bpy)(OH_2)]$^{2+}$ (**3**) (where $Cp^* = \eta^5\text{-}C_5Me_5$, the X-ray crystal structure is shown in Figure 8.1) was 27 h^{-1}, which is more than 10 times faster than that using a ruthenium aqua complex [$Ru^{II}(\eta^6\text{-}C_6Me_6)$(4,4′-MeO-bpy)($OH_2$)]$^{2+}$ (**2**) [326].

The catalytic cycle of the reduction of CO_2 with H_2 using [$Ir^{III}(Cp^*)$(4,4′-OMe-bpy)(OH_2)]$^{2+}$ is shown in Scheme 8.1 [326]. The reaction of the Ir–OH_2 (where Ir = $Ir^{III}(Cp^*)$(4,4′-OMe-bpy)) complex ([Ir–OH_2]$^{2+}$) with H_2 affords the Ir-hydride complex ([Ir–H]$^+$), which reacts with CO_2 to give the formate complex ([Ir–O(C=O)H]$^+$) [326].

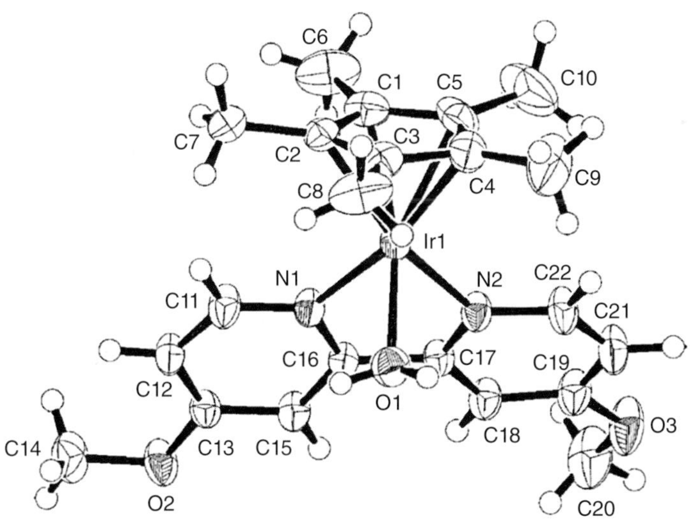

Figure 8.1 ORTEP drawing of [$Ir^{III}(Cp^*)$(4,4′-OMe-bpy)(OH_2)](OTf)$_2$ (**3**·(OTf)$_2$) with ellipsoids at 50% probability. Counter anions (OTf) {OTf = $CF_3SO_3^-$} are omitted for clarity. Source: Hayashi et al. 2004 [326]. Reproduced with permission of Royal Society of Chemistry.

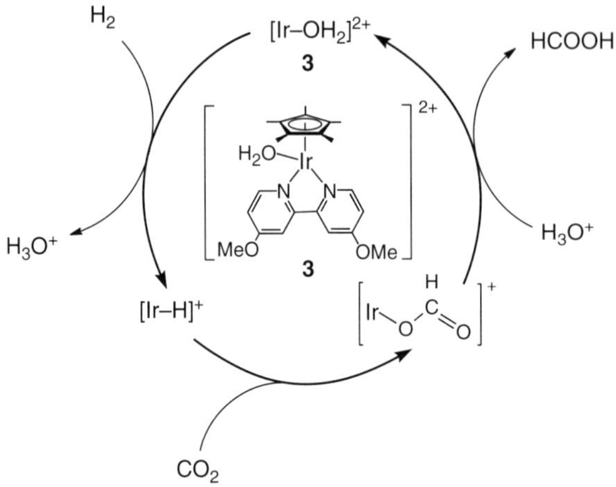

Scheme 8.1 Catalytic cycle of the reduction of CO_2 with H_2 using $[Ir^{III}(Cp^*)(4,4'\text{-OMe-bpy})(OH_2)](OTf)_2$ (**3**·$(OTf)_2$). Source: Hayashi et al. 2004 [326]. Reproduced with permission of Royal Society of Chemistry.

The catalytic efficiency for the reduction of CO_2 with H_2 to HCOOH is significantly improved by using a water-soluble [C,N] cyclometalated iridium aqua complex $[Ir^{III}(Cp^*)(4\text{–}(1H\text{–pyrazol–1–yl–}\kappa N^2)\text{benzoic acid–}\kappa C^3)(OH_2)]^+$ (**4**) as a catalyst, which enables the reaction under atmospheric pressure of CO_2 and H_2 [355]. The complex was obtained as a sulfate salt$[4]_2{\cdot}SO_4$ that was synthesized by the reaction of $[Ir^{III}(Cp^*)(OH_2)_3]SO_4$ with 4–(1*H*–pyrazol–1–yl)benzoic acid in H_2O under reflux conditions. The aqua complex **4** can release protons from the carboxyl group and the aqua ligand to form the corresponding benzoate complex **5** and hydroxo complex **6**, respectively (Scheme 8.2) [366]. The pK_a values of complexes **4** and **5** were determined from the spectral titration to be pK_{a1} = 4.0 and pK_{a2} = 9.5, respectively [366]. These values are consistent with those for benzoic acid [366] (pK_a = 4.19) and $[Ir^{III}(Cp^*)(4,4'\text{-OMe-bpy})(OH_2)](SO_4)$ [366] (pK_a = 9.2). The benzoate complex **5** is less soluble in water because of its neutral charge. The structure of **5** was determined by X-ray single crystal structure analysis [366, 368]. Figure 8.2 shows an ORTEP drawing of **5,** which is a neutral-charged mononuclear Ir complex with no counterions [366]. The complex has both a carboxylate group in its [C,N] cyclometalated ligand and an aqua ligand; thus, the intermolecular hydrogen bondings are formed in the crystal (Figure 8.2) [366].

A CO_2–saturated K_2CO_3 (0.1 M) aqueous solution of complex **5** was bubbled with H_2 and CO_2 at 1 : 1 volumetric ratio under atmospheric pressure (0.1 MPa) resulting in the catalytic formation of formate with high concentrations at pH 7.5 at 303 K (Scheme 8.3) [366]. The turnover number (TON) increases linearly with time to exceed over 100, which is significantly larger than the maximum TON for catalytic formation of formate with **2** (59) and **3** (64) at pH 3.0 at 313 K by the reduction of CO_2 (2.5 MPa) with H_2 (5.5 MPa) [366]. TOF was determined from the slope of the linear plot as 6.8 h^{-1} at 303 K and 22.1 h^{-1} at 333 K at pH 7.5 [366].

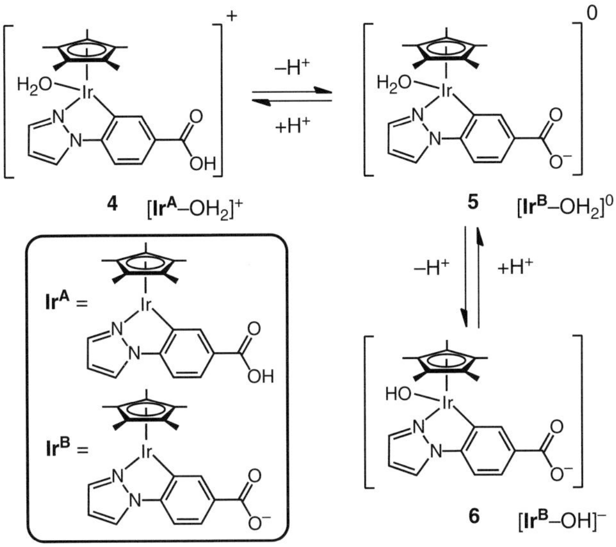

Scheme 8.2 Acid–base equilibria of iridium aqua complexes (**4**–**6**). Source: Maenaka et al. 2012 [366]. Reproduced with permission of Royal Society of Chemistry.

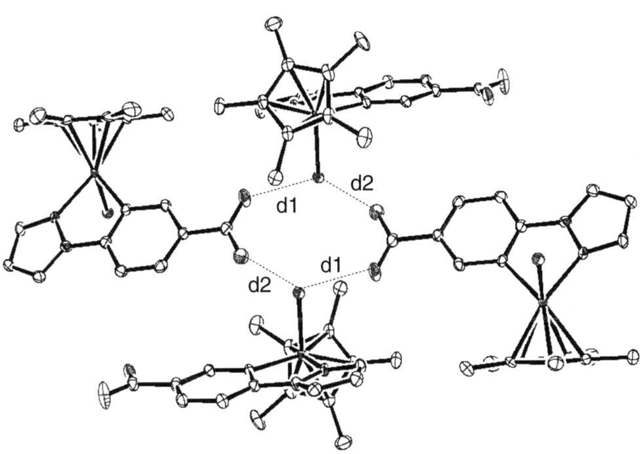

Figure 8.2 ORTEP drawing of 2. Hydrogen atoms are omitted for clarity. Selected bond length (Å): d1 = 2.590(4) and d2 = 2.560(4). Source: Maenaka et al. 2012 [366]. Reproduced with permission of Royal Society of Chemistry.

The TOF increased with increase in pH to afford the highest value ($36\,h^{-1}$) at pH 8.8 at 333 K (Figure 8.3) [366]. Further increase in pH resulted in a decrease in TOF to reach 0.0 at pH 10.4. On the other hand, under slightly acidic conditions, no formation of formate was confirmed at 333 K at pH 6.0. Judging from the similar pH dependence of TOF (black line in Figure 8.3) to that of the amount ratios of **5** and HCO_3^- (grey solid line and grey dashed line in Figure 8.3, respectively), hydrogenation of bicarbonate (HCO_3^-) rather than carbonate (CO_3^{2-}) or CO_2 is catalyzed mainly by **5** rather than **4** or **6** to produce

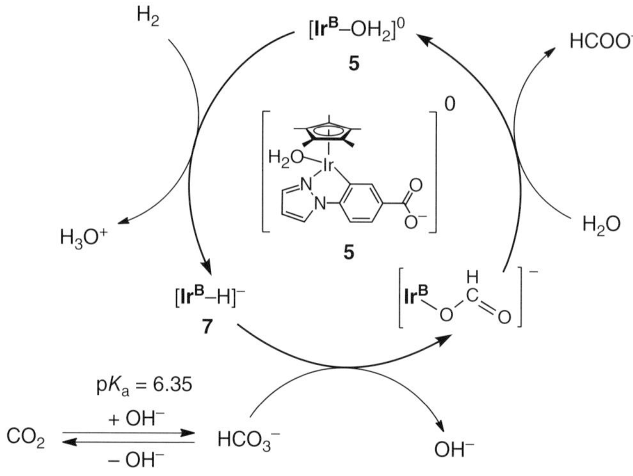

Scheme 8.3 Catalytic cycle of the reduction of CO_2 with H_2 using [Ir^{III}(Cp*)(4–(1H–pyrazol–1–yl–κN^2)benzoic acid–κC^3)(OH_2)]$_2SO_4$ ($\mathbf{4}_2 \cdot SO_4$). Source: Maenaka et al. 2012 [366]. Reproduced with permission of Royal Society of Chemistry.

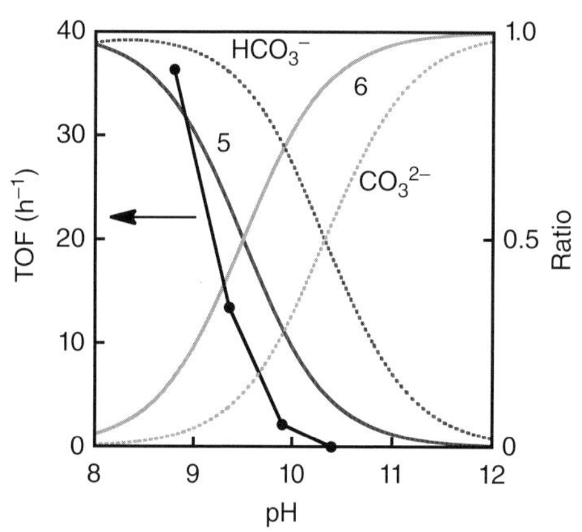

Figure 8.3 pH Dependence of the formation rate (TOF) of formate in the catalytic generation of formate from H_2, HCO_3^-, and CO_3^{2-} ([HCO_3^-] + [CO_3^{2-}] = 2.0 M) catalyzed by **5** and **6** ([**5**] + [**6**] = 0.18 mM) in deaerated H_2O at 333 K (black line). Source: Maenaka et al. 2012 [366]. Reproduced with permission of Royal Society of Chemistry.

formate at pH 8.8 [366]. Indeed, TOF linearly increased with concentrations of HCO_3^- and **5** at pH 8.8 [366]. These results indicate that both a catalyst **5** and a bicarbonate anion (HCO_3^-) are involved in the rate-determining step of the catalytic hydrogenation reaction in Scheme 8.3 [355]. The TOF for hydrogenation of HCO_3^- at pH 8.8 increased with increasing temperature with the activation energy of 11.3 kcal mol^{-1}, which is much smaller than that of the hydrogenation of CO_2 without catalysts (79 kcal mol^{-1}) [368].

In slightly basic water, the reaction of **5** with H_2 affords the corresponding hydride complex 7: [**Ir**$^{\mathbf{B}}$–H]$^-$, which was identified by ^{1}H nuclear magnetic resonance (NMR) spectrum, UV–vis absorption spectra, and electrospray ionization (ESI) mass spectrum [366]. The reaction of the hydride complex 7 with HCO_3^- gives the corresponding formate complex, which is the rate-determining step [366]. The formate complex is converted to regenerate the aqua complex **4** by releasing $HCOO^-$ in competition with the back reaction to form the hydride complex **7** and CO_2 [366].

In acidic water, the reverse reaction, i.e. the catalytic decomposition of HCOOH occurs with **4** to produce H_2 and CO_2 in 1 : 1 molar ratio, which was detected by GC [355]. No CO as a by-product was formed during the reaction [366]. The pH dependence of TOF is shown in Figure 8.4 in which the maximum TOF value 1880 h^{-1} is obtained at pH 2.8 at 298 K. The TOF value is more than four times larger than that (426 h^{-1}) obtained at pH 3.8 at 298 K in the catalytic decomposition of HCOOH using a heterodinuclear iridium–ruthenium complex $[Ir^{III}(Cp^*)(OH_2)(bpm)Ru^{II}(bpy)_2](SO_4)_2$ {bpm = 2,2′-bipyrimidine, bpy = 2,2′-bipyridine} under otherwise same experimental conditions [369]. The black line in Figure 8.4, which represents an increase in TOF with a decrease in pH in the region between 2.8 and 9.0, overlaps well with the curve of the ratio of complexes **4** : **5** (grey line in Figure 8.4), indicating that the complex **4** rather than **5** acts as a catalyst for the selective decomposition of HCOOH to H_2 and CO_2 [366].

The TOF value for hydrogen evolution from HCOOH at pH 2.8 increases with increasing concentration of [HCOOH] + [HCOOK] to approach a constant value (Figure 8.5) [366]. This indicates that the cationic aqua complex **4** reacts with $HCOO^-$ to afford the corresponding formate complex, which is then converted to the corresponding hydride complex **8**: $[\mathbf{Ir^A}\text{–H}]^0$ via β-hydrogen elimination. The hydride complex reacts with H_3O^+ to produce H_2, accompanied by regeneration of **4** (Scheme 8.4) [366]. Thus, the direction of the reaction under slightly basic conditions in Scheme 8.3 is reversed under acidic conditions in Scheme 8.4 [355]. Formation of the hydride complex was confirmed by the ^{1}H NMR spectrum independently obtained in DMSO-d_6 [366]. At lower pH, formation of a hydride complex **8** via β-hydrogen elimination from the formate complex is the rate-determining step in the catalytic cycle in Scheme 8.4 because of relatively high concentration of protons. This was confirmed by observing the kinetic deuterium isotope effect (KIE = 4.0) for the catalytic hydrogen evolution from formic acid-*d* (DCOOH) [366]. Thus, interconversion between H_2 and HCOOH in water has been made possible using the Ir complexes **4** and **5** at ambient

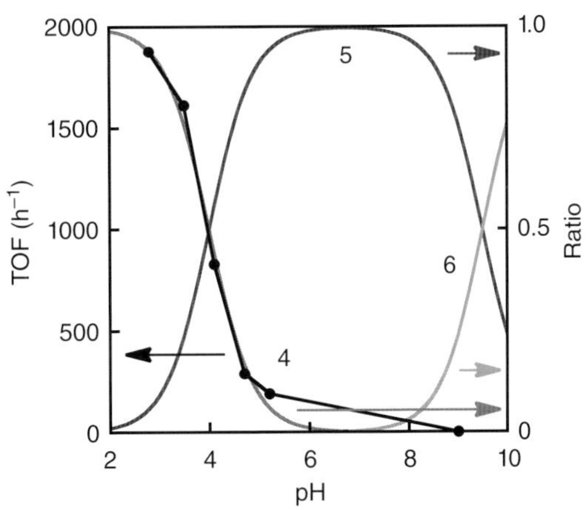

Figure 8.4 pH Dependence of the H_2 evolution rate (TOF) in the catalytic hydrogen generation from formic acid and formate ([HCOOH] + [HCOOK] = 3.3 M) catalyzed by **4** (0.20 mM) in deaerated H_2O at 298 K (black line). Source: Maenaka et al. 2012 [366]. Reproduced with permission of Royal Society of Chemistry.

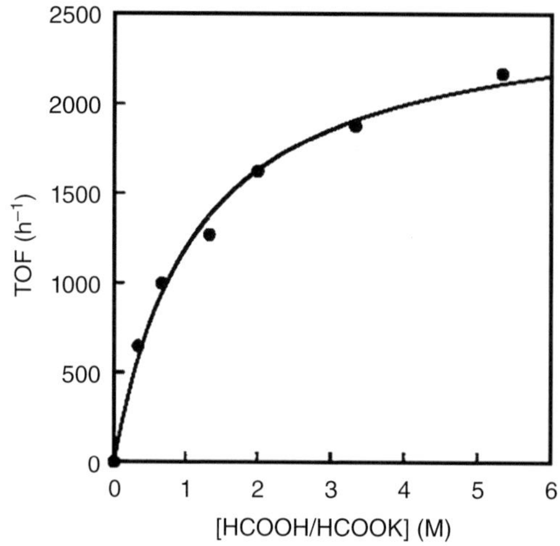

Figure 8.5 Plot of TOF vs. concentration of HCOOH and HCOOK mixture (HCOOH/HCOOK), i.e. [HCOOH] + [HCOOK] in the decomposition reaction of HCOOH/HCOOK catalyzed by **4** (0.20 mM) in deaerated H_2O at pH 2.8 at 298 K. Source: Kumar et al. 2009 [355]. Reproduced with permission of John Wiley & Sons.

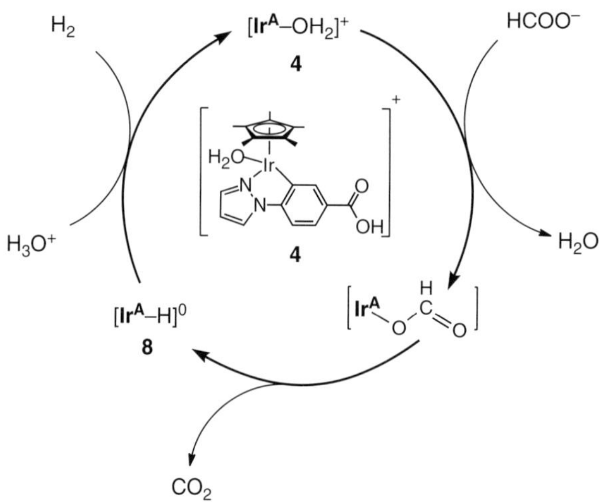

Scheme 8.4 Catalytic cycle of the decomposition of HCOOH using [Ir^{III}(Cp^*)(4-(1*H*-pyrazol-1-yl-κN^2)benzoic acid-κC^3)(OH_2)]$_2SO_4$ ($\mathbf{4}_2\cdot SO_4$). Source: Maenaka et al. 2012 [366]. Reproduced with permission of Royal Society of Chemistry.

temperature and pressure by changing pH: the Ir benzoate complex **5** catalyzes hydrogenation of bicarbonate in slightly basic water (Scheme 8.3), whereas the reverse reaction, that is, the decomposition of formic acid to H_2 and CO_2, is also catalyzed by **4** in acidic water (Scheme 8.4) [366]. In the same manner, a dinuclear Ir(Cp^*) complex with 4,4′,6,6′-tetrahydroxy-2,2′-bipyrimidine (thbpym) as a bridging ligand, [{Ir(Cp^*)(Cl)}$_2$(thbpym)]$^{2+}$, acts as a pH-modulated catalyst for interconversion between H_2 and HCOOH by the deprotonation equilibrium of the phenolic ligand in Scheme 8.5 [370]. The X-ray crystal structure of [{Ir(Cp^*)(Cl)}$_2$(thbpym)]$^{2+}$ is shown in Figure 8.6 [370]. At pH 8.4, the deprotonated form of [{Ir(Cp^*)(Cl)}$_2$(thbpym)]$^{2+}$ acts as an efficient catalyst for hydrogenation of bicarbonate at 0.1 MPa at 298 K to produce formate with

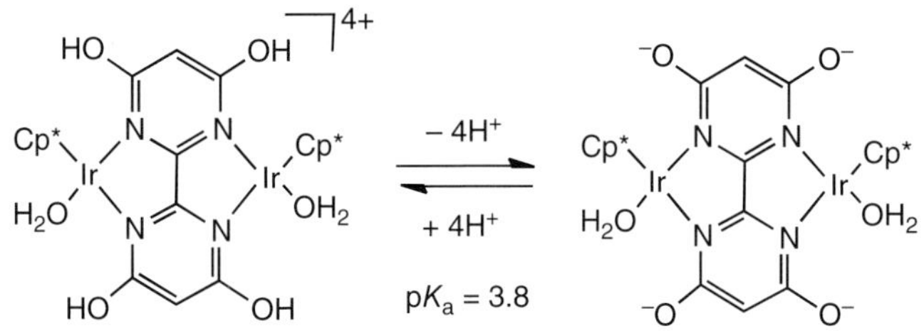

Scheme 8.5 A dinuclear Ir(Cp*) complex with 4,4′,6,6′-tetrahydroxy-2,2′-bipyrimidine (thbpym) as a bridging ligand, $[\{Ir(Cp^*)(Cl)\}_2(thbpym)]^{2+}$, and the protonation and deprotonation equilibrium. Source: Hull et al. 2012 [370]. Reproduced with permission of Springer Nature.

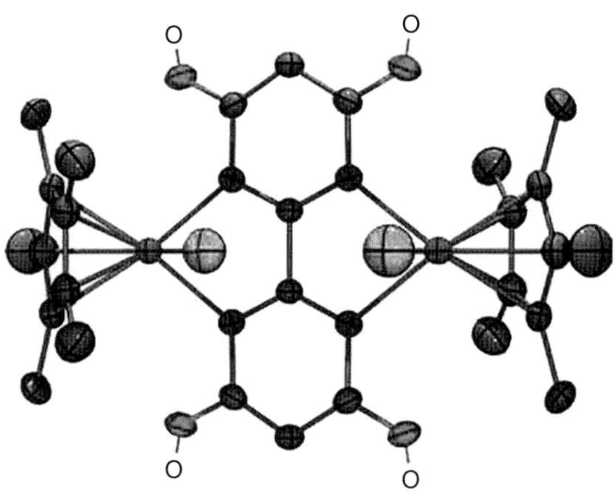

Figure 8.6 Crystal structure of $[\{Ir(Cp^*)(Cl)\}_2(thbpym)]^{2+}$ [370]. C—H bonds are omitted for clarity. Source: Hull et al. 2012 [370]. Reproduced with permission of Springer Nature.

TOF of 64 h^{-1} [370]. At pH 3.5, the reverse reaction, i.e. the decomposition of HCOOH to H_2 and CO_2 is efficiently catalyzed by $[\{Ir(Cp^*)(Cl)\}_2(thbpym)]^{2+}$, which is partially deprotonated, with a TOF value of 228 000 h^{-1} at 388 K. This is the highest TOF value ever reported for the production of H_2 from HCOOH. The mechanism of the pH-dependent catalytic interconversion between HCOOH and H_2 with $[\{Ir(Cp^*)(Cl)\}_2(thbpym)]^{2+}$ may be similar to the case of the Ir complexes **5** and **4** in Schemes 8.3 and 8.4, respectively. However, the Ir–hydride and formate complexes have yet to be detected [370].

8.2 Interconversion Between Hydrogen and NADH

When HCOOH in Scheme 8.3 was replaced by 1,4-dihydro-β-nicotinamide adenine dinucleotide (NADH), H_2 evolved from NADH in the presence of a catalytic amount of **4** in an acidic aqueous solution (Scheme 8.6) [371]. The oxidized product of NADH was confirmed to be an oxidized form of β-nicotinamide adenine dinucleotide (NAD^+) by ^{1}H NMR [371]. The yield and TON reached up to 96 % and 6.9 (20 minutes), respectively, both of which were determined by integrating the characteristic ^{1}H NMR signals of NAD^+ and NADH [371]. The reaction

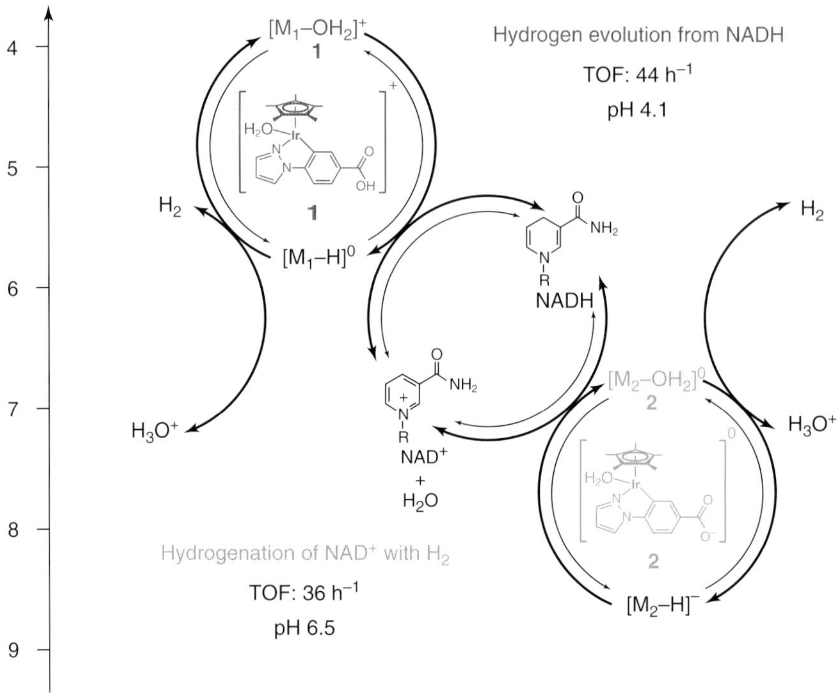

Scheme 8.6 Catalytic mechanism for interconversion between NADH and H_2 with water soluble [C,N] cyclometalated Ir complexes (**1** and **2**) depending on pH. Source: Maenaka et al. 2011 [371]. Reproduced with permission of American Chemical Society.

of **4** and the deprotonated complex **5** with NADH afforded the corresponding anionic Ir–hydride complex **7** as indicated by the negative-ion ESI mass spectrum ($m/z = 515.2$) [371]. The reaction of the hydride complex **7** with proton in water yields H_2, which is the rate-determining step in the catalytic cycle in Scheme 8.6 (top left-hand side) as indicated by the saturation behavior of TOF with increasing concentration of NADH (Figure 8.7) [371].

At pH 8, the reverse reaction of H_2 production from NADH, i.e. hydrogenation of NAD+ by H_2 with the Ir complex **5,** occurs to produce NADH [371]. Only the 1,4-isomer of NADH was selectively produced by the reaction of NAD+ with H_2 in the presence of a catalytic amount of **5** [371]. The yield based on the amount of NAD+ and TON reached up to 97% and 9.3 (90 minutes), respectively [371].

The pH dependence of TOF for both H_2 evolution from NADH and formation of NADH from NAD+ with **4** and **5** is shown in Figure 8.8 [371]. An increase in TOF of H_2 evolution from NADH with a decrease in pH in the region between 4.1 and 7.0 (black line in Figure 8.8) overlaps well with the curve of the ratio of **4** (grey line in Figure 8.8), whereas the pH dependence of TOF for hydrogenation of NAD+ (dashed line in Figure 8.8) overlaps well with the curve of the ratio of **5**

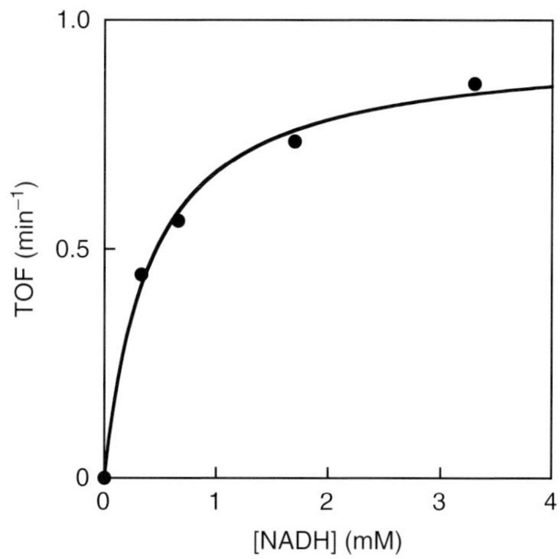

Figure 8.7 Plot of TOF for catalytic hydrogen evolution vs. the concentration of NADH in the reaction of NADH with proton in the presence of **4** and **5** ([**4**] + [**5**] = 65 μM) in deaerated phthalic buffer (2.0 ml) at 298 K at pH 4.1. Source: Maenaka et al. 2011 [371]. Reproduced with permission of American Chemical Society.

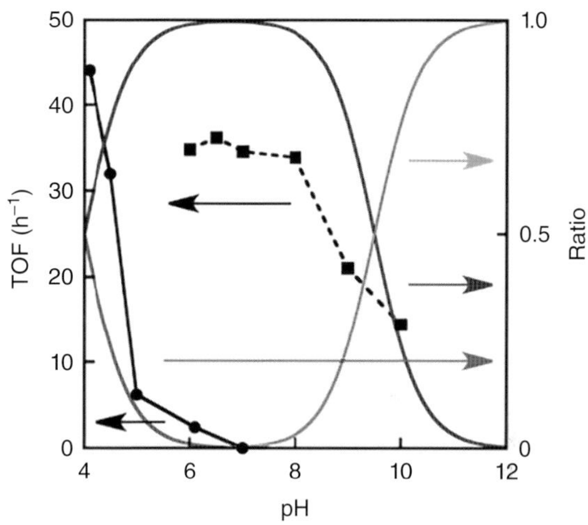

Figure 8.8 pH Dependence of the rate (TOF) of H_2 evolution in the oxidation of NADH (3.3 mM) catalyzed by **4** and **5** ([**4**] + [**5**] = 0.18 mM) in deaerated H_2O at 298 K (•) and pH dependence of the rate (TOF) of formation of NADH in the reduction of NAD^+ (0.78 mM) by H_2 catalyzed by **4** and **5** ([**4**] + [**5**] = 15 μM) in deaerated H_2O at 298 K (■). Source: Maenaka et al. 2011 [371]. Reproduced with permission of American Chemical Society.

(grey line in Figure 8.8) [371]. This indicates that the complex **4** reacts with NADH to produce H_2 and the complex **5** reacts with H_2 to reduce NAD^+ to NADH as in the case of interconversion between HCOOH and H_2. At pH 6.5, the TOF for the formation of NADH is maximized (36 h^{-1}), whereas the TOF for the hydrogen evolution reaches 44 h^{-1} at pH 4.1. Further decrease in pH resulted in the decomposition of NADH due to acid-catalyzed hydration [372].

The rate-determining step in the catalytic hydrogenation of NAD^+ with H_2 is the formation of the Ir–H complex, which reacts with NAD^+ rapidly to produce NADH (the right-hand catalytic cycle in Scheme 8.6) [371]. Formation of the Ir–H complex under an atmospheric pressure of H_2 was confirmed by ESI mass spectrum, ^{1}H NMR, and UV–vis absorption spectra [371]. The TOF value

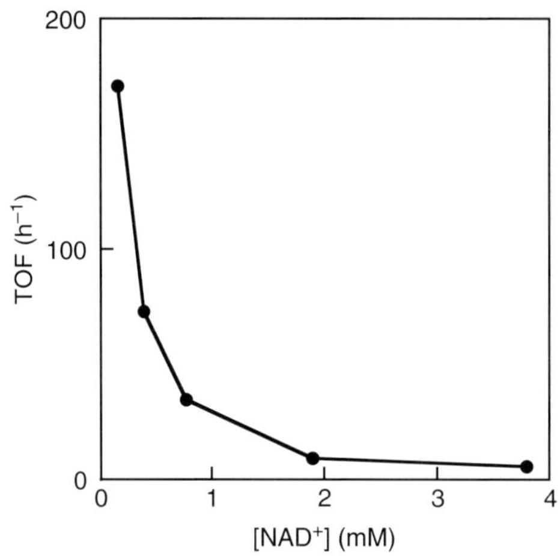

Figure 8.9 Plot of TOF vs. the concentration of NAD^+ in the catalytic hydrogenation of NAD^+ with H_2 to produce NADH under atmospheric pressure in the presence of **5** (15 μM) in deaerated phosphate buffer (pH 7.0) at 298 K. Source: Maenaka et al. 2011 [371]. Reproduced with permission of American Chemical Society.

of the catalytic hydrogenation of NAD^+ decreased with increasing concentration of NAD^+ (Figure 8.9) due to the coordination of NAD^+ to **5**, which prohibits the formation of the hydride complex via H_2 coordination [371]. The binding constant of NAD^+ with **5** was determined to be $1.1 \times 10^4\ M^{-1}$ from the absorption change due to the binding of NAD^+ with **5** [371]. This is the first example of a hydrogenase functional mimic using a water-soluble iridium aqua complex that can catalyze the oxidation of H_2 with NAD^+ to produce protons and NADH and also the reduction of protons with NADH to produce H_2 and NAD^+ in water under atmospheric pressure at room temperature.

8.3 Hydrogen Evolution from Alcohols

Alcohols can also be used as hydrogen donors for the catalytic hydrogen evolution with the Ir complex **4** [373]. At pH 13.6, the aqua ligand of the Ir complex **4** is deprotonated to produce $[\mathbf{Ir^B}\text{–OH}]^-$ complex **6** (Scheme 8.7) [373]. When ethanol was added to a heavy water solution of **6**-*d*, i.e. $[\mathbf{Ir^B}\text{–OD}]^-$ at pD 13.6 in D_2O, **6**-*d* was reduced to produce the corresponding Ir–hydride complex **7** as confirmed by the ^1H NMR spectral changes at high yield (98%) [373]. The second-order rate constants (k) of formation of the hydride complex were determined in the reaction of **6** with various aliphatic alcohols in an aqueous

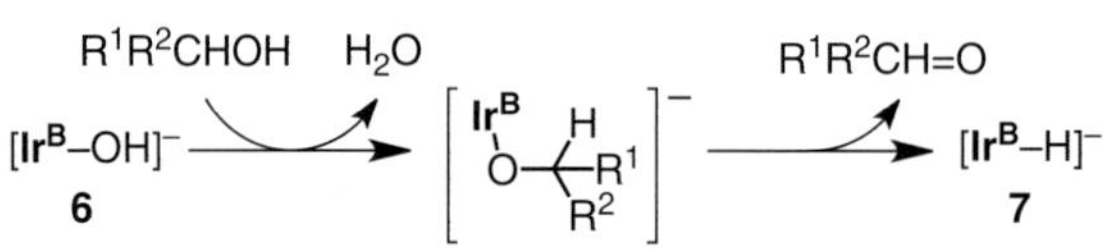

Methanol; $R^1 = R^2 = H$ 1–Propanol; $R^1 = H$, $R^2 = C_2H_5$
Ethanol; $R^1 = H$, $R^2 = CH_3$ 2–Propanol; $R^1 = CH_3$, $R^2 = CH_3$

Scheme 8.7 Formation of a hydride complex **7** in the hydrogenation of 6 with aliphatic alcohols. Source: Maenaka et al. 2012 [373]. Reproduced with permission of American Chemical Society.

solution (pH 13.7) at 298 K [373]. When CH_3OH and C_2H_5OH were replaced by CD_3OH and CD_3CD_2OH, respectively, in the reaction of **6** at pH 13.7, kinetic deuterium isotope effects (KIE's) were observed to be 3.2 and 2.1, respectively [373]. Thus, the β-hydrogen elimination of alkoxy complexes, which are produced by the replacement of a hydroxy (OH) ligand of **6** by an alkoxy ligand, is the rate-determining step for the formation of **7**: $[\mathbf{Ir}^B\text{–H}]^-$ (Scheme 8.7) [373]. The k value increases in accordance with the electron donating ability of the alkyl groups bonded to the carbon at β-position of aliphatic alcohols, $CH_3(CH_2)_2$ for entry 9 > two CH_3's for entry 8 > CH_3CH_2 for entry 7 > CH_3 for entry 4 > no substituent alkyl group for entry 1 [373]. An alcohol having no β-hydrogen (*t*-butyl alcohol, entry 10) shows no hydride donating ability to **6** [373].

The hydride complex **7** formed in the reaction of **6** with ethanol in water is stable at pH 14; however, when the pH was decreased to 0.8 by adding H_2SO_4, **7** was immediately converted to an aqua complex **4** accompanied by evolution of hydrogen (H_2) [373]. The yield of H_2 was determined by GC to be 82% [373]. The conversion between the hydride complex and the aqua complex **4** accompanied by H_2 evolution was repeated by alternate change in pH between c. 12 and c. 2 in the presence of excess amount of ethanol as shown in Figure 8.10 [373]. The conversion from **4** to **7** in the reaction of ethanol with **6** and that from **7** to **4** accompanied by evolution by changing the pH is shown in Scheme 8.8 [373]. Thus, hydrogen derived from aliphatic alcohols can be stored as a form of an

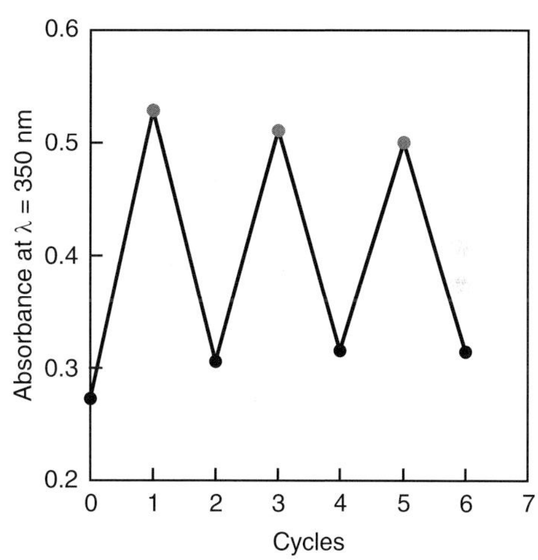

Figure 8.10 Change in absorbance at $\lambda = 350$ nm due to formation of the hydride complex **7** (grey closed circle) in the reaction of **6** (0.12 mM) with ethanol (0.82 M) in water (pH 11.8–12.2) and that due to the hydrogen evolution (black closed circle) in the reaction of the hydride complex **7** with proton in water at 298 K (pH 2.0–3.3) by adding an aqueous solution of H_2SO_4 (5.0 M) or NaOH (5.0 M). Source: Maenaka et al. 2012 [373]. Reproduced with permission of American Chemical Society.

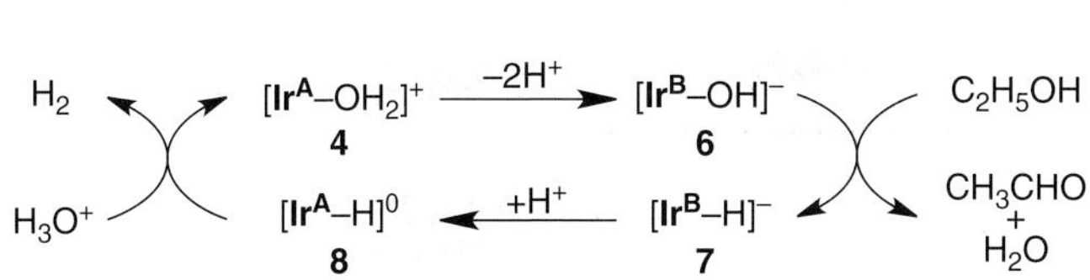

Scheme 8.8 Catalytic cycle of hydrogen evolution from ethanol using $[\text{Ir}^{III}(\text{Cp}^*)(4\text{–}(1H\text{–}$pyrazol-1-yl-$\kappa N^2$)benzoic acid-$\kappa C^3)(OH_2)]_2SO_4$ ($\mathbf{4}_2\cdot SO_4$). Source: Maenaka et al. 2012 [373]. Reproduced with permission of American Chemical Society.

Ir–hydride complex at higher pH and can be provided whenever it is needed at lower pH simply by changing the pH [373].

When an aqueous solution of the Ir–hydride complex 7 produced by the reaction of **6** with ethanol in D_2O at pD 13.6 was photoirradiated for seven hours with a Xe lamp through a colored glass filter transmitting $\lambda > 340$ nm, the signal of a proton at $\delta = 8.25$ ppm disappeared completely and the signal of a hydride proton exhibited upfield shift from $\delta = -14.34$ to $\delta = -17.48$ [373]. The ESI mass spectrum of the CO bubbled aqueous solution of the photoproduct agrees with the calculated isotopic distribution for the corresponding CO complex, i.e. $[Ir(Cp^*)(4\text{–}(1H\text{–pyrazol–1–yl–}\kappa C^5)\text{benzoate–}\kappa C^3)(CO)]^-$ [373]. This indicates that the mass number of the photoproduct remained the same as 7 and the [C,N] cyclometalated Ir–hydride complex **7**: $[\mathbf{Ir^B}\text{–H}]^-$ was converted to the corresponding [C,C] cyclometalated complex **9**: $[\mathbf{Ir^C}\text{–H}]^{2-}$ (Scheme 8.9) [362]. The quantum yield of this photoreaction was determined to be 1.7% [373]. The intersystem crossing of the photoexcited 7 and the subsequent formation of **9** were monitored by femtosecond and nanosecond laser flash photolysis, respectively [373].

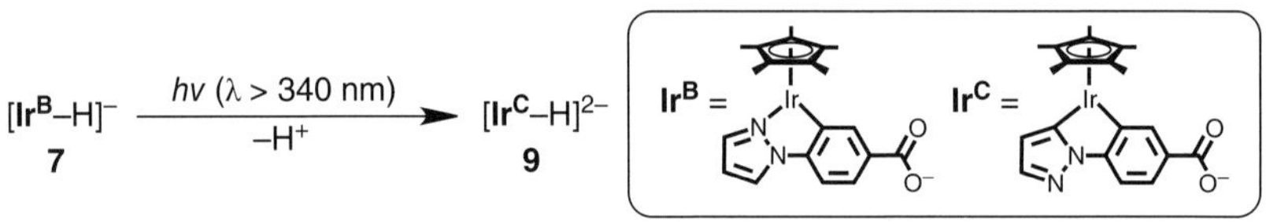

Scheme 8.9 Photoconversion from a hydride complex **7** to **9**. Source: Maenaka et al. 2012 [373]. Reproduced with permission of American Chemical Society.

In contrast to the [C,N] cyclometalated Ir–hydride complex (7), the [C,C] cyclometalated Ir–hydride complex (**9**) can react with proton in water to produce H_2 even under basic conditions as shown in Figure 8.9 [373]. The TON of H_2 evolution with **9** increases linearly with time to reach 3.3 (2.5 hours), whereas 7 has no catalytic reactivity even at elevated temperature at 323 K (Figure 8.11)

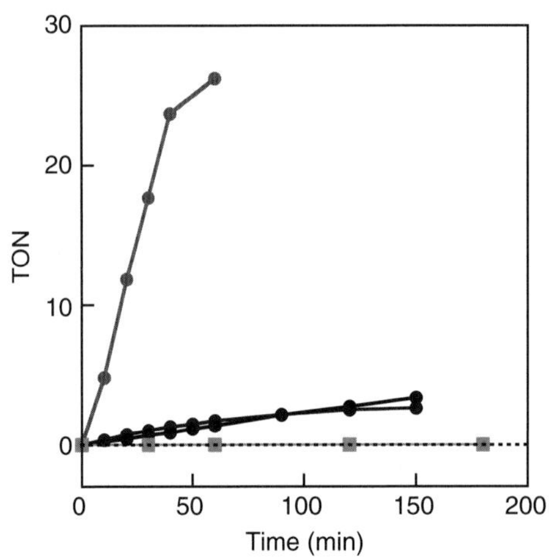

Figure 8.11 Time course of hydrogen evolution from 2-propanol (4.3 M) catalyzed by **9**: $[\mathbf{Ir^C}\text{–H}]^{2-}$ (55 μM) in water (pH 11.9) at 353 K (top line) and 323 K and that from ethanol (5.7 M) catalyzed by **9** (55 μM) in water (pH 11.9) at 323 K. The complex **9** was formed from **7**: $[\mathbf{Ir^B}\text{–H}]^-$ under photoirradiation ($\lambda > 340$ nm) for 30 minutes. Time course of hydrogen evolution from 2-propanol (4.3 M) catalyzed by **7** (55 μM) in water (pH 11.9) at 323 K is shown by the bottom line. Source: Maenaka et al. 2012 [373]. Reproduced with permission of American Chemical Society.

[373]. The TON for hydrogen evolution with **9** increases with increasing temperature to be 26 (1.0 hour) at 353 K [373]. This catalytic reactivity of **9** may be attributed to the electron-donating effect of the phenylpyrazole ligand on the metal center in **9** with the [C,C] cyclometalated iridium as indicated by the upfield shift of the signal of a hydride ligand bonded to Ir^{III} center [373].

8.4 Hydrogen Evolution from Paraformaldehyde

The same Ir catalyst (**1**) can also act as an efficient catalyst for decomposition of paraformaldehyde to produce H_2 and CO_2 in a 2 : 1 mol ratio under basic conditions at room temperature [374]. At pH 11, **1** is converted to the hydroxo complex **3**, which reacts with paraformaldehyde $HO(CH_2O)_nH$ via the monomerized and hydrated equivalent, methanediol [375], to produce the formaldehyde adduct ($[Ir–OCH_2OH]^-$) and $HO(CH_2O)_{n-1}H$ [374].

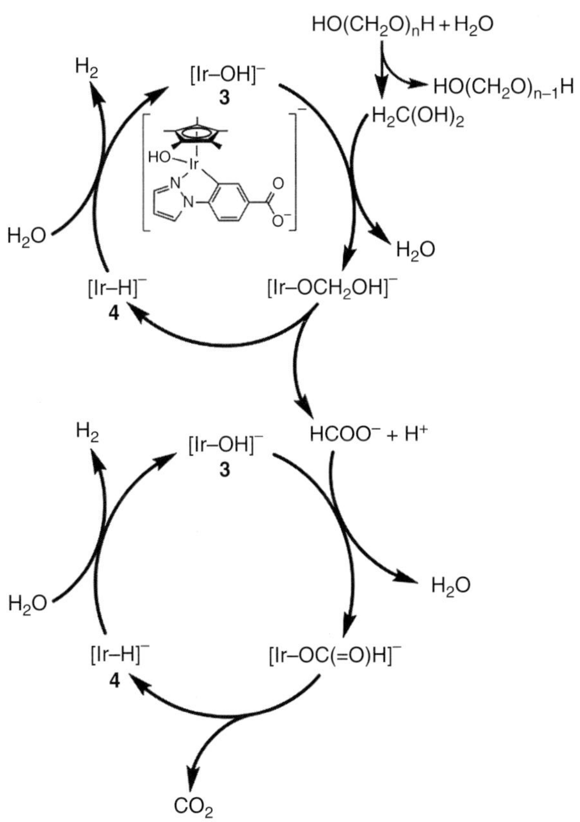

Scheme 8.10 Catalytic cycles for decomposition of paraformaldehyde to H_2 and formate that is further decomposed to H_2 and CO_2 with **3**. Source: Suenobu et al. 2015 [374]. https://pubs.rsc.org/en/content/articlelanding/2015/CC/C4CC06581F#!divAbstract; http://creativecommons.org/licenses/by/3.0/. Licensed under CCBY 3.0

The β-hydrogen elimination from $[Ir{-}OCH_2OH]^-$ occurs to produce the hydride complex (**4**) and formic acid. The hydride complex (**4**) reacts with H_2O to produce H_2, accompanied by regeneration of **3** (upper-side catalytic cycle in Scheme 8.10) [374]. The hydroxo complex **3** also reacts with formate to produce the hydride complex (**4**) and CO_2 by β-hydrogen elimination. The hydride complex (**4**) also reacts with H_2O to produce H_2, accompanied by regeneration of **3** (lower-side catalytic cycle in Scheme 8.10) [374]. Thus, paraformaldehyde can be regarded as a convenient solid H_2 carrier, which has a higher energy density (6.7%) than HCOOH (4.4%) [374].

9

Metal Ion-Coupled Electron Transfer (MCET)

9.1 MCET of O_2

Binding of metal ions to radical anions of electron acceptors results in significant positive shifts of the one-electron reduction potentials of electron acceptors (vide supra) [24–27, 194–200]. In the case of back electron transfer (ET) in the charge-separated (CS) states, the strong binding of metal ions to the CS states decelerates the electron transfer because of a large increase in the reorganization energy of electron transfer [203–205]. Such examples were presented in the previous sections. In contrast, uphill ET reactions, which are thermodynamically infeasible to occur, are made possible by the presence of metal ions provided that the strong binding of metal ions to radical anions of electron acceptors changes the energetics of electron transfer from uphill to downhill. In such a case, the binding of metal ions is coupled with electron transfer to promote metal ion-coupled electron transfer (MCET) reactions, which would otherwise be impossible to occur [194]. This is particularly important for the intermolecular MCET of dioxygen (O_2), because electron transfer is only the spin-allowed process of the reactions of O_2, which is triplet in the ground state, with singlet molecules. The promoting effects of metal ions on the electron transfer are certainly related to the Lewis acidity of metal ions. Charges and ion radii are important factors to determine the Lewis acidity of metal ions. The binding energies of a variety of metal ions with superoxide ion ($O_2^{\bullet-}$) can be readily derived from the g_{zz}-values of the EPR spectra of the superoxide-metal ion complexes ($O_2^{\bullet-}/M^{n+}$), providing the quantitative measure of Lewis acidity of the metal ions (vide infra) [201].

The $O_2^{\bullet-}/M^{n+}$ complex is produced by the photoinduced ET reduction of O_2 by dimeric 1-benzyl-1,4-dihydronicotinamide [$(BNA)_2$] in MeCN [376]. When an oxygen-saturated MeCN solution containing $(BNA)_2$ (1.0×10^{-4} M) was irradiated with a high-pressure mercury lamp, $O_2^{\bullet-}$ is detected by the ESR spectrum in frozen MeCN at 143 K [376]. The ESR spectrum shows a typical anisotropic signal with $g_{\parallel} = 2.090$ and $g_{\perp} = 2.005$ [377]. The ESR spectra of $O_2^{\bullet-}$ produced in the presence of a variety of closed shell metal ions were also measured at 143 K [201]. The anisotropic ESR signals are changed significantly in the presence of each metal ion as compared to that in its absence [201]. In particular, the g_{zz}-values of $O_2^{\bullet-}/M^{n+}$ complexes become significantly smaller

Electron Transfer: Mechanisms and Applications, First Edition. Shunichi Fukuzumi.

than the g_{zz}-value of $O_2^{\bullet-}$ due to the binding of M^{n+} to $O_2^{\bullet-}$ ($O_2^{\bullet-}/M^{n+}$, $n = 1-3$) [201]. Figure 9.1 shows typical examples of ESR spectra of $O_2^{\bullet-}/M^{2+}$ complexes (M^{2+} = Mg^{2+}, Ca^{2+}, Sr^{2+} and Ba^{2+}) measured in frozen MeCN at 143 K [201]. The g_{zz}-value of $O_2^{\bullet-}/M^{n+}$ gives valuable information concerning the binding strength of $O_2^{\bullet-}/M^{n+}$. The deviation of the g_{zz}-value from the free spin value

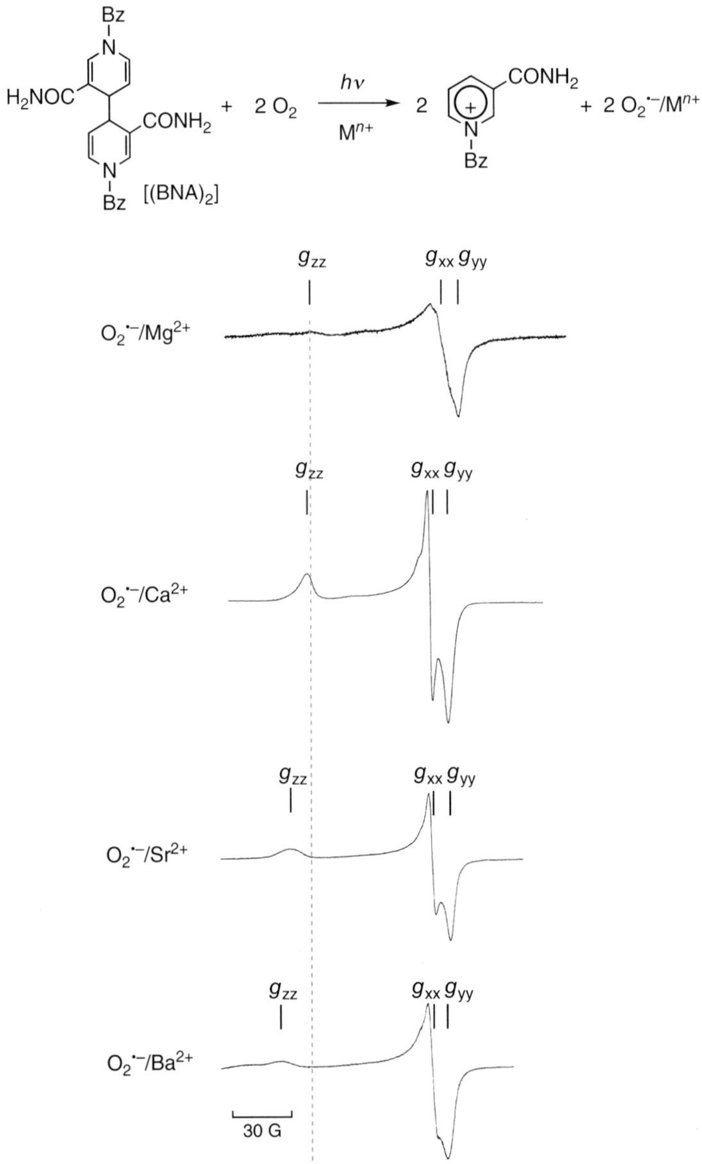

Figure 9.1 ESR spectra of $O_2^{\bullet-}/M^{2+}$ complexes (M^{2+} = Mg^{2+}, Ca^{2+}, Sr^{2+} and Ba^{2+}) in frozen MeCN at 143 K. Source: Fukuzumi et al. 2000 [201]. Reproduced with permission of John Wiley & Sons.

(g_e = 2.0023) is caused by the spin–orbit interaction as given by Eq. (9.1) [378, 379], where λ is the spin–orbit coupling constant

$$\Delta E = (g_{zz} - g_e)/2\lambda \tag{9.1}$$

(0.014 eV) [380] and ΔE is the energy splitting of π_g levels due to the biding of M^{n+} to $O_2^{\bullet-}$. The ΔE value is readily obtained from the deviation of the g_{zz}-value from the free spin value. The ΔE value increases generally in the order: monovalent cations (M^+) < divalent cations (M^{2+}) < trivalent cations (M^{3+}) [201]. The ΔE value also increases with decreasing ion radius when the oxidation state of the metal ion is the same. The same trend has been reported for $O_2^{\bullet-}$ adsorbed on the surface of various metal oxides [381, 382]. Scandium ion, which has the smallest ion radius among the trivalent metal cations, gives the largest ΔE value, and this indicates that the binding energy between Sc^{3+} and $O_2^{\bullet-}$ is the strongest [201].

When 3 equiv of hexamethylphosphoric triamide (HMPA) ligand is added to the $(BNA)_2$–O_2–Sc^{3+} system, the $O_2^{\bullet-}/Sc^{3+}$ complex becomes significantly stable and the ESR spectrum of the $O_2^{\bullet-}/Sc^{3+}$(HMPA) complex can be detected even at 60 °C under irradiation of the light [376]. The lanthanide complexes of radical anions of aromatic ketones are stabilized by the presence of HMPA ligand [383, 384]. Oxygen enriched in ^{17}O can provide valuable information about the inequivalency of oxygen nuclei in the $O_2^{\bullet-}/Sc^{3+}$–HMPA complex. Figure 9.2a shows the EPR spectrum of $O_2^{\bullet-}/Sc^{3+}$(HMPA) with ^{17}O enriched oxygen [376]. There are two sets of six lines due to the hyperfine splitting of two different ^{17}O atoms ($I = 5/2$), although the center lines are masked by the strong eight-line signal of $^{16}O_2^{\bullet-}/Sc^{3+}$(HMPA) with $g = 2.0165$ and $a_{Sc} = 3.82$ G [376]. The two inequivalent $a(^{17}O)$ values are determined as 21 and 14 G by comparison between the observed signal (Figure 9.2a) and the computer simulation lines (Figure 9.2b) [376]. Such inequivalent $a(^{17}O)$ values are fully consistent with an "end-on" coordination form of $^{\bullet}O$–O^-/Sc^{3+}(HMPA) in which the electron spin is more localized at the terminal oxygen. This is confirmed by the DFT (density function theory) calculation using the spin-restricted B3LYP functional and the 6-311++G(3d,3p) basis set for the open shell $O_2^{\bullet-}/Sc^{3+}$, which gives more localized spin density at the terminal oxygen (65%) [376]. The calculated O–O distance decreases in the order: $O_2^{\bullet-}$ (1.343 Å) > $O_2^{\bullet-}/Li^+$ (1.309 Å) > $O_2^{\bullet-}/Mg^{2+}$ (1.297 Å) > $O_2^{\bullet-}/Sc^{3+}$ (1.211 Å) as the ΔE value increases [376].

Metal ions such as Mg^{2+} have been reported to promote electron transfer from (TPP)Co (TPP^{2-} = dianion of tetraphenylporphyrin) to *p*-benzoquinone (Q), although no reaction between (TPP)Co and Q occurs in the absence of metal ions in MeCN [385]. Such promoting effects of metal ions in ET reduction of substrates have been ascribed to the binding of metal ions to the radical anions produced in the ET reactions [24–27, 194, 195]. The promoting effects of metal ions on electron transfer from (TPP)Co to O_2 are examined in the presence of a series of metal ions (M^{n+}, $n = 1$–3) by the UV–vis spectral change for the decay of (TPP)Co (λ_{max} = 411 nm) and the formation of $[(TPP)Co]^+$ (λ_{max} = 434 nm) in MeCN at 298 K [201]. No electron transfer from (TPP)Co (E_{ox} = 0.35 V vs. saturated calomel electrode [SCE] in MeCN) [386] to O_2 (E_{red} = −0.86 V vs. SCE in MeCN) [387] has occurred in MeCN at 298 K. In the presence of

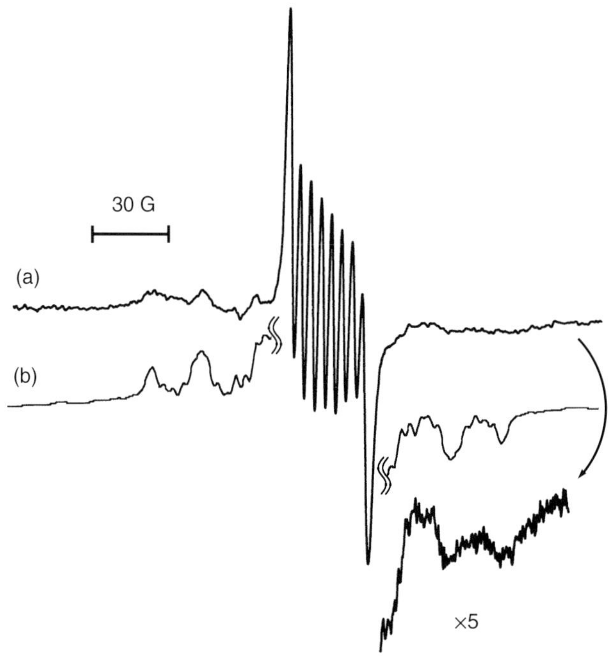

Figure 9.2 (a) ESR spectrum observed after irradiation of an ^{17}O (40%) oxygen-saturated propionitrile solution containing $(BNA)_2$ (6.9×10^{-3} M), $Sc(OTf)_3$ (8.1×10^{-2} M), and HMPA (2.5×10^{-1} M) with a high-pressure mercury lamp at room temperature. (b) Computer simulation spectrum with $g = 2.0165$, $a_{Sc} = 3.82$ G, $a(^{17}O1) = 21$ G, $a(^{17}O2) = 14$ G, $\Delta H_{msl} = 3.5$ G. The center lines are omitted for clarity. Source: Fukuzumi et al. 1999 [376]. Reproduced with permission of American Chemical Society.

M^{n+}, however, efficient electron transfer from (TPP)Co to O_2 occurs to yield $[(TPP)Co]^+$ (Scheme 9.1) [201].

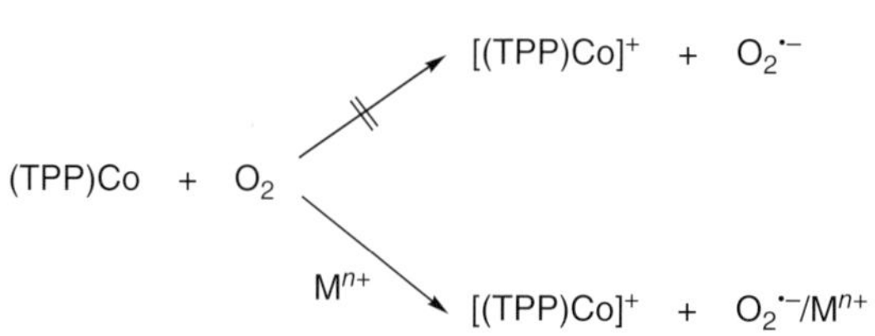

Scheme 9.1 M^{n+}-promoted electron transfer from (TPP)Co to O_2. Source: Fukuzumi et al. 2000 [201]. Reproduced with permission of John Wiley & Sons.

The ET rates obeyed second-order kinetics, showing a first-order dependence on the concentration of each reactant [O_2 and (TPP)Co] [201]. The observed second-order rate constant (k_{obs}) for the M^{n+}-promoted electron transfer increases linearly with increasing metal ion concentration [201]. This indicates that the binding of M^{n+} to $O_2^{\bullet-}$ is coupled with electron transfer when M^{n+}-promoted ET occurs in a *concerted manner* (MCET) rather than a stepwise manner. If the rate-determining step were an uphill electron transfer from

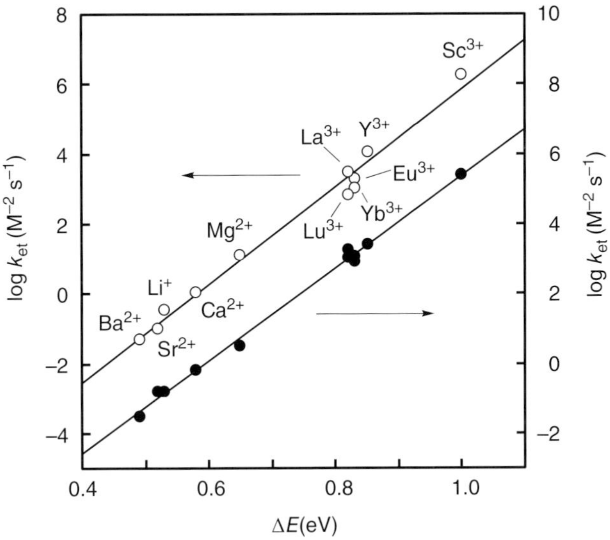

Figure 9.3 Plots of log k_{et} vs. ΔE in M^{n+}-promoted electron transfer from (TPP)Co to O_2 (o) and *p*-benzoquinone (•). Source: Fukuzumi et al. 2000 [201]. Reproduced with permission of John Wiley & Sons.

(TPP)Co to O_2, followed by rapid binding of M^{n+} to $O_2^{\bullet-}$, the ET rate would be independent of metal ion concentration.

The rate constants of M^{n+}-promoted electron transfer (k_{et}) were determined from the slopes of the linear plots of k_{obs} vs. $[M^{n+}]$. There is a *striking linear correlation* between log k_{et} and the ΔE values of $O_2^{\bullet-}/M^{n+}$ derived from the g_{zz}-values as shown in Figure 9.3 [201]. The remarkable correlation spans a range of almost 10^7 in the rate constant. The slope of the linear correlation between log k_{et} and ΔE is determined to be 14.0, which is close to the value of $1/(2.3RT)$ (=16.9, $T = 298$ K) [201]. This means that the variation of ΔE is directly reflected in the difference in the activation free energy for the M^{n+}-promoted electron transfer from (TPP)Co to O_2. The stronger the binding of M^{n+} with $O_2^{\bullet-}$, the faster will be the M^{n+}-promoted electron transfer. Thus, the ΔE values can be regarded as a good measure of the binding energies in the $O_2^{\bullet-}/M^{n+}$ complexes [201].

Electron transfer from (TPP)Co to Q is also promoted by M^{n+}, whereas no electron transfer occurs from (TPP)Co to Q without M^{n+} (Scheme 9.2) [201]. There is also an excellent linear correlation between log k_{et} and ΔE as shown in Figure 9.3 (closed circles) [201]. More importantly, the slope (13.3) for Q (closed circles) is nearly the same as the slope (14.0) for O_2 (open circles). Such an agreement indicates that the ΔE values of $O_2^{\bullet-}/M^{n+}$ are in parallel with those of $Q^{\bullet-}$–M^{n+} [201].

The effects of ligands on the Lewis acidity of metal ions were also examined using scandium ion containing different ligands such as TTP^{2-} and HMPA in relation with the Lewis acid-promoted electron transfer from (TPP)Co to O_2 [388]. Organotin compounds constitute another useful class of Lewis acids, which are frequently used for organic synthesis. The effects of counterions were also examined using organotin halides, triflates, and organotin dinuclear

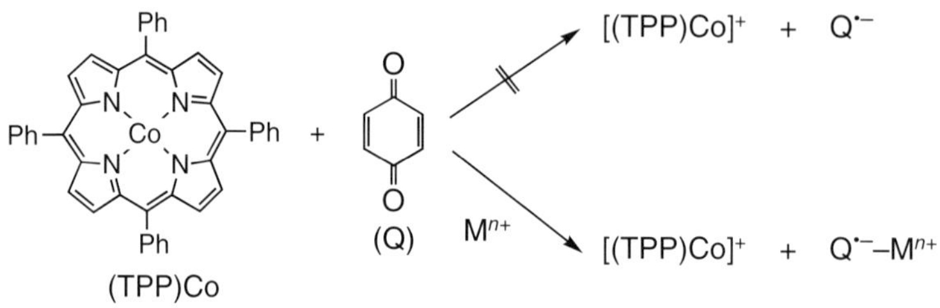

Scheme 9.2 M^{n+}-promoted electron transfer from (TPP)Co to Q. Source: Fukuzumi et al. 2000 [201]. Reproduced with permission of John Wiley & Sons.

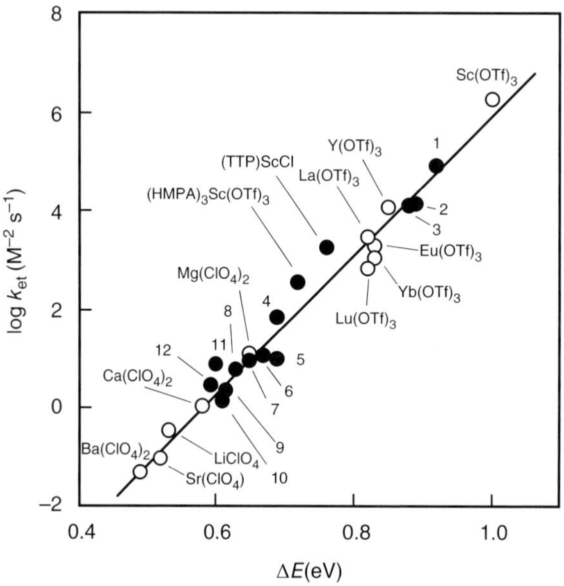

Figure 9.4 Plots of log k_{et} vs. ΔE in electron transfer from (TPP)Co to O_2, promoted by metal ions (triflate or perchlorate salts) (○) and organotin compounds and scandium complexes (•) in MeCN at 298 K. Numbers refer to a variety of organotin compounds. Source: Ohkubo et al. 2003 [388]. Reproduced with permission of American Chemical Society.

clusters, which have been used as unique and useful Lewis acids to catalyze a variety of nucleophilic reactions of silyl and stannyl substrates under mild conditions [389, 390]. There is again a striking single linear correlation between log k_{et} and ΔE of the $O_2^{\bullet -}$/Lewis acid complexes as shown in Figure 9.4, where the k_{et} and ΔE values of various metal ions with different ligands and organotin halides, triflates, and organotin dinuclear clusters are included [388].

9.2 Binding Modes of Metal Ions

As described above, metal ions acting as Lewis acids accelerate ET reactions, when metal ions bind with the product radical anions. In such a case, the binding mode of metal ions plays an important role in the activation of electron transfer by metal ions, because the binding mode of metal ions directly influences the redox potentials of substrates and the reorganization energy of electron transfer, both of which are important parameters to determine the rate of electron transfer. The binding mode of metal ion complexes is a key factor to characterize the redox function of metalloenzymes [391–394]. In fact, there are several possible coordination sites for metal ions in the case of *o*-quinones such as PQQ (pyrroloquinoline quinone) involved in biological redox systems [395, 396]. Semiquinone radical anions derived from *o*-quinones are known to form not only simple 1 : 1 complexes with metal ions but also more intricate complexes with metal ions, i.e. 1 : 2 complexes [397]. A variety of binding modes are possible depending on the types of metal ion complexes of quinones and semiquinone radical anions derived from both *o*-quinones and *p*-quinones, affecting the metal ion-promoted ET reduction of quinones (vide infra) [398].

o-Quinones such as 9,10-phenanthrenequinone (PQ) form 1 : 1 complexes with metal ions ($M^{n+} = Sc^{3+}$, Y^{3+}, and Ca^{2+}) as indicated by the UV–vis spectral changes of PQ in the presence of various concentrations of metal ions in Figure 9.5 [398]. The formation constants (K_{ox}) are determined from linear plots of $(A - A_0)^{-1}$ ($=\Delta Abs^{-1}$) vs. $[M^{n+}]^{-1}$ (insets of Figure 9.5). The K_{ox} values of other metal ions were also determined together with the absorption maxima (λ_{max}) of PQ–M^{n+} complexes [398]. The greater the K_{ox} values [1.3 M^{-1} (Mg^{2+}), 4.9 M^{-1} (Ca^{2+}), 7.2 M^{-1} (Ba^{2+}), 20 M^{-1} (Y^{3+}), and 29 M^{-1} (Sc^{3+})], the larger the red shifts of λ_{max} values [319 nm in the absence of M^{n+}, 321 nm (Mg^{2+}), 328 nm (Ca^{2+}), 331 nm (Ba^{2+}), 350 nm (Y^{3+}), and 360 nm (Sc^{3+})] [398]. This indicates that the stronger the binding in the PQ–M^{n+} complex, the more stabilized is the excited state of the PQ–M^{n+} complex as compared with the ground state.

In contrast to the 1 : 1 complex formation between PQ and M^{n+}, PTQ forms a 2 : 1 complex with M^{n+} (Figure 9.5) as indicated by the UV–vis spectral changes of PTQ in the presence of various concentrations of M^{n+} in Figure 9.6 [398]. The formation of a 2 : 1 complex is confirmed by the titration of the absorption spectra (insets of Figure 9.6). Such a 2 : 1 complex formation has also been reported between PTQ and transition metal ions [399].

The ^{13}C NMR signal of carbonyl carbons in the PQ–Sc^{3+} complex shows a downfield shift, whereas the ^{13}C NMR signal of carbonyl carbons of PTQ exhibits an upfield shift in the presence of Sc^{3+} [^{13}C NMR (CD_3CN): δ 180.2 in the absence of Sc^{3+}; δ 176.0 in the presence of 3.0×10^{-1} M of Sc^{3+}] [398]. The B3LYP theoretical calculation of the PTQ indicates that a larger negative charge is located on nitrogens as compared with carbonyl oxygens [398]. Thus, Sc^{3+} binds with the nitrogens of PTQ rather than with carbonyl oxygens.

The radical anions of PQ and PTQ also form complexes with metal ions. The ESR spectra provide valuable information on the binding modes of metal ions. Figure 9.7 shows the ESR spectra of $PQ^{\bullet-}$ and metal ion complexes, which were formed by photoinduced electron transfer from $(BNA)_2$ to PQ or electron transfer from decamethylferrocene to PQ in the presence of metal ions [398]. The *g* values of $PQ^{\bullet-}$–M^{n+} complexes are smaller than that of $PQ^{\bullet-}$ (2.0048) [398]. The smaller *g* values of $PQ^{\bullet-}$–M^{n+} complexes than that of $PQ^{\bullet-}$ (2.0048) indicate that the spin density on oxygen nuclei in $PQ^{\bullet-}$ is reduced by the complexation with M^{n+}. The hyperfine patterns of $PQ^{\bullet-}$in the presence of metal ions (Figure 9.7c,e,g,i) are obviously different from that in the absence of metal ions (Figure 9.7a) [398]. In the case of La^{3+} (Figure 9.7e), the superhyperfine splitting due to two La^{3+} ions [$a(2La^{3+})$ = 1.80 G] is clearly observed in addition to

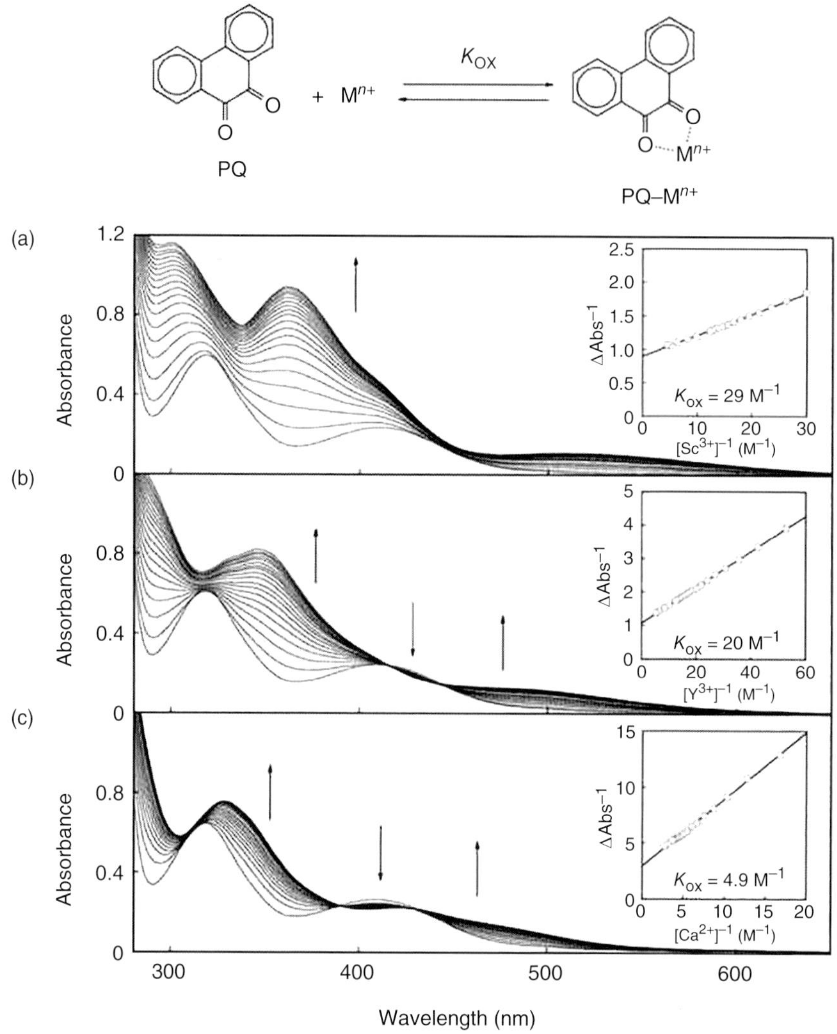

Figure 9.5 UV–visible absorption spectra of PQ (1.2×10^{-4} M) in the presence of various concentrations of (a) Sc^{3+} (0 to 8.2×10^{-2} M), (b) Y^{3+} (0 to 8.0×10^{-2} M), and (c) Ca^{2+} (0 to 2.7×10^{-1} M) in MeCN at 298 K [394]. Insets: Plot of ΔAbs^{-1} at (a) 360 nm vs. $[Sc^{3+}]^{-1}$, (b) 350 nm vs. $[Y^{3+}]^{-1}$, and (c) 328 nm vs. $[Ca^{2+}]^{-1}$. Source: Yuasa et al. 2006 [398]. Reproduced with permission of John Wiley & Sons.

the hyperfine splitting due to protons [a(4H) = 0.46 G and a(4H) = 1.80 G] (94) [398]. The complete agreement of the observed ESR spectrum (Figure 9.7e) with the computer simulation spectrum (Figure 9.7f) indicates that $PQ^{\bullet-}$ forms a 1 : 2 complex with La^{3+} ($PQ^{\bullet-}$–$2La^{3+}$). Similarly, Y^{3+} forms a 1 : 2 complex with $PQ^{\bullet-}$ ($PQ^{\bullet-}$–$2Y^{3+}$) as indicated by the superhyperfine structure due to two Y^{3+} ions in Figure 9.7g together with the computer simulation spectrum in Figure 9.7h [398]. In the case of Sc^{3+}, which has much stronger Lewis acidity and the smaller ion radius than La^{3+} and Y^{3+}, the hyperfine and superhyperfine patterns of $PQ^{\bullet-}$ in

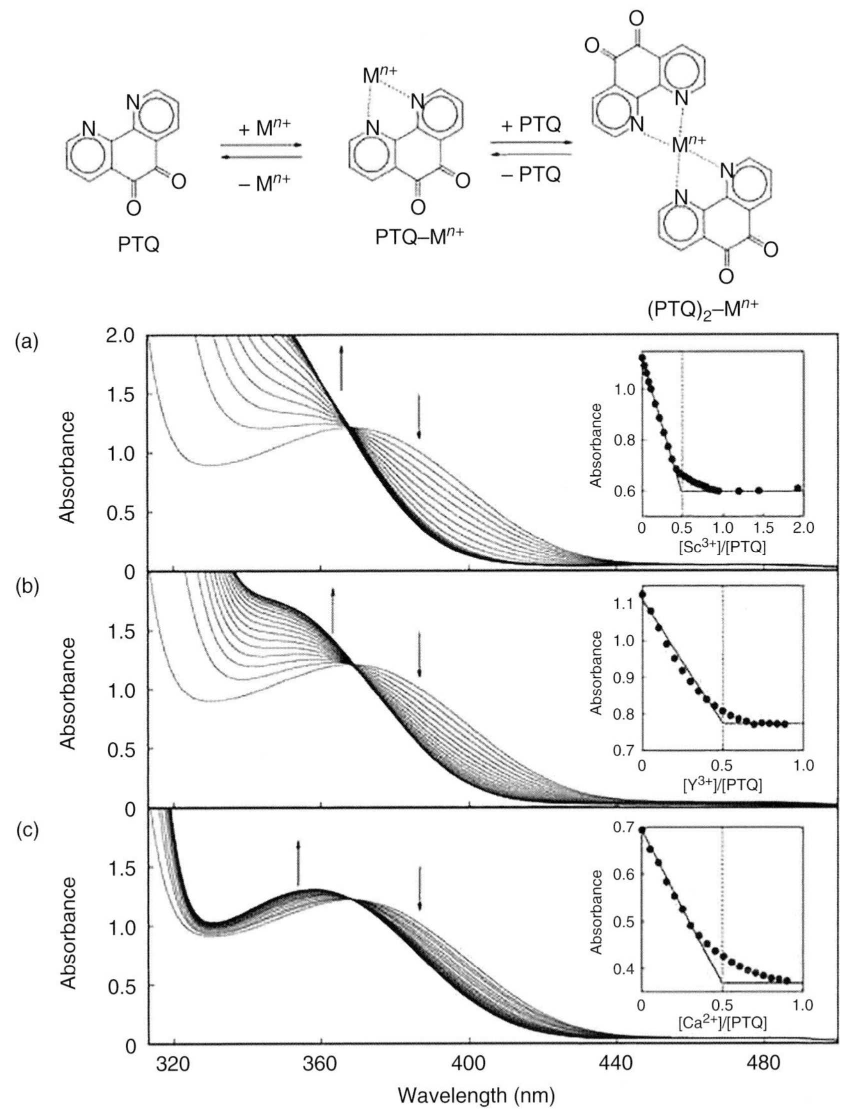

Figure 9.6 (a) UV–visible absorption spectra of PTQ (2.0×10^{-3} M) in the presence of various concentrations of (a) Sc^{3+} (0 to 1.9×10^{-3} M), (b) Y^{3+} (0 to 1.9×10^{-3} M), and (c) Ca^{2+} (0 to 1.8×10^{-3} M) in MeCN at 298 K [394]. Insets: Plot of absorbance at (a) 380 nm vs. [Sc^{3+}]/[PTQ], (b) 380 nm vs. [Y^{3+}]/[PTQ], and (c) 400 nm vs. [Ca^{2+}]/[PTQ]. Source: Yuasa et al. 2006 [398]. Reproduced with permission of John Wiley & Sons.

Figure 9.7i together with the computer simulation spectrum (Figure 9.7j) clearly indicate that $PQ^{\cdot-}$ forms a 1 : 1 complex with Sc^{3+} [398]. The small ion radius and high Lewis acidity of Sc^{3+} result in the formation of a 1 : 1 complex with $PQ^{\cdot-}$ rather than a 1 : 2 complex. Although the absence of superhyperfine patterns in the case of Mg^{2+} and Ca^{2+} has precluded determining the binding mode of metal ion complexes with $PQ^{\cdot-}$, the 1 : 1 binding mode of $PQ^{\cdot-}$ with Mg^{2+} is confirmed

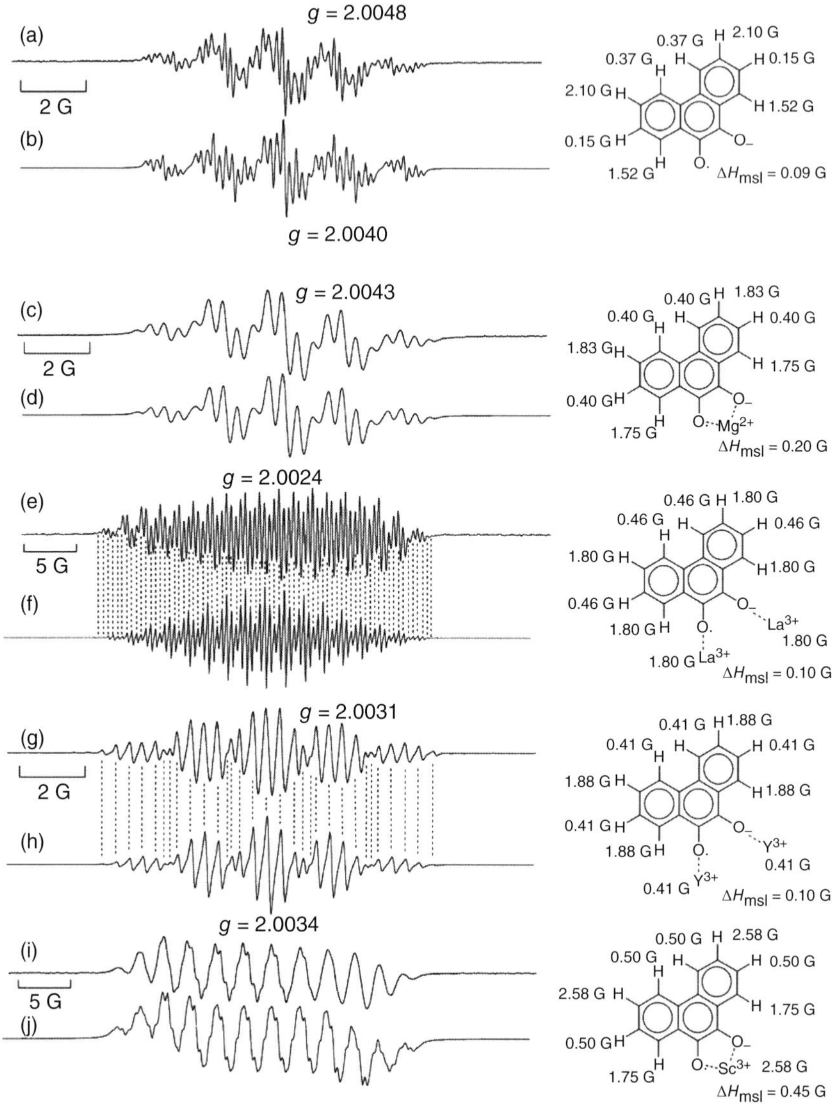

Figure 9.7 (a) EPR spectrum of $PQ^{\bullet -}$ produced by photoinduced electron transfer from $(BNA)_2$ (1.0×10^{-3} M) to PQ (1.1×10^{-3} M) in deaerated MeCN at 243 K and (b) the computer simulation spectrum. (c) EPR spectrum of $PQ^{\bullet -}–Mg^{2+}$ produced by electron transfer from decamethylferrocene (Fc^*: 1.0×10^{-3} M) to PQ (2.0×10^{-2} M) in the presence of Mg^{2+} (2.0×10^{-2} M) in deaerated MeCN at 298 K and (d) the computer simulation spectrum. (e) EPR spectrum of $PQ^{\bullet -}–(La^{3+})_2$ produced by electron transfer from Fc^* (2.1×10^{-4} M) to PQ (2.1×10^{-4} M) in the presence of La^{3+} (2.5×10^{-3} M) in deaerated MeCN at 298 K and (f) the computer simulation spectrum. (g) EPR spectrum of $PQ^{\bullet -}–(Y^{3+})_2$ produced by electron transfer from Fc^* (2.1×10^{-4} M) to PQ (2.1×10^{-4} M) in the presence of Y^{3+} (2.7×10^{-3} M) in deaerated MeCN at 298 K and (h) the computer simulation spectrum. (i) EPR spectrum of $PQ^{\bullet -}–Sc^{3+}$ produced by photoinduced electron transfer from $(BNA)_2$ (1.0×10^{-3} M) to PQ (1.2×10^{-2} M) in the presence of Sc^{3+} (8.1×10^{-1} M) in deaerated MeCN at 243 K and (j) the computer simulation spectrum. Source: Yuasa et al. 2006 [398]. Reproduced with permission of John Wiley & Sons.

by the electrochemical measurements (vide infra) [398]. Thus, the binding modes of metal ions with $PQ^{\bullet-}$ are changed depending on the Lewis acidity and the ion radius of metal ions.

The EPR spectra of $PTQ^{\bullet-}$ and metal ion complexes of $PTQ^{\bullet-}$ are shown together with the computer simulation spectra in Figure 9.8 [398]. In contrast to the 2 : 1 complex formation of PTQ with M^{n+} (Figure 9.8), the EPR spectra of metal ion complexes of $PTQ^{\bullet-}$, which exhibit superhyperfine splitting due to one metal ion (Y^{3+} or Sc^{3+}), indicate that $PTQ^{\bullet-}$ forms 1 : 1 complexes with the metal ions [398]. The hyperfine splittings due to six protons [a(4H) = 1.32 G and a(2H) = 0.24 G] and 2 equiv nitrogens [a(2N) = 0.57 G] of $PTQ^{\bullet-}$ are little changed by the complex formation with M^{n+} [398]. This indicates that the interaction between $PTQ^{\bullet-}$ and M^{n+} is largely electrostatic. In contrast to the case of PTQ (vide supra), the B3LYP theoretical calculation of the $PTQ^{\bullet-}$ indicates that a larger negative charge is located on carbonyl oxygens rather than on nitrogens [398]. In addition, the superhyperfine splitting constant due to Sc^{3+}

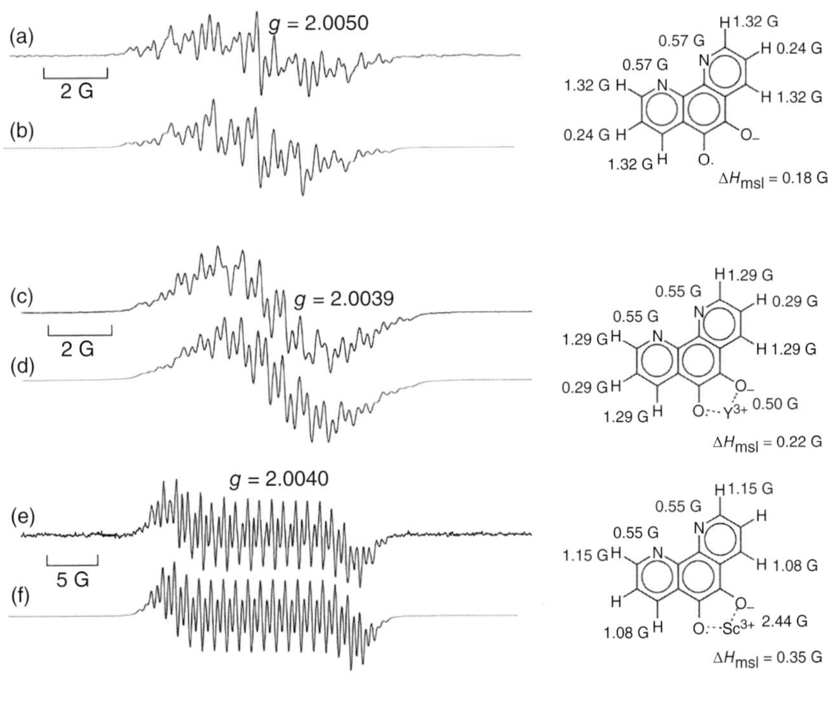

Figure 9.8 (a) EPR spectrum of $PTQ^{\bullet-}$ produced by photoinduced electron transfer from $(BNA)_2$ (1.0×10^{-3} M) to PTQ (1.0×10^{-3} M) in deaerated MeCN at 298 K and (b) the computer simulation spectrum. (c) EPR spectrum of $PTQ^{\bullet-}$–Y^{3+} generated by electron transfer from Fc^{*} (1.0×10^{-3} M) to PTQ (1.0×10^{-3} M) in the presence of Y^{3+} (1.7×10^{-1} M) in deaerated MeCN at 298 K and (d) the computer simulation spectrum. (e) EPR spectrum of $PTQ^{\bullet-}$–Sc^{3+} generated by electron transfer from dimethylferrocene (3.6×10^{-4} M) to PTQ (3.9×10^{-4} M) in the presence of Sc^{3+} (4.0×10^{-1} M) and H_2O (11 M) in MeCN at 298 K and (f) the computer simulation spectrum. Source: Yuasa et al. 2006 [398]. Reproduced with permission of John Wiley & Sons.

of the $PTQ^{\bullet-}$–Sc^{3+} complex [$a(Sc^{3+}) = 2.44$ G] (Figure 9.8e) is virtually the same as that of the $PQ^{\bullet-}$–Sc^{3+} complex [$a(Sc^{3+}) = 2.58$ G] (Figure 9.7i) [398]. This indicates that metal ions bind with carbonyl oxygens of $PTQ^{\bullet-}$ rather than with nitrogens as in the case of PTQ.

The complex formation of PQ and PTQ with M^{n+} results in an increase in the electron acceptor ability of PQ and PTQ. The positive shifts of one-electron reduction potentials (E_{red}) of *o*-quinones caused by the complex formation with various metal ions were verified by electrochemical measurements [398]. The CVs of PQ in the absence and presence of various metal ions exhibit a reversible redox wave as shown in Figure 9.9 [398]. The E_{red} value of PQ without metal ions is determined from the half-wave potential as −0.65 V (vs. SCE) as shown in Figure 9.9a [398]. The addition of 2.0×10^{-2} M Ba^{2+}, Ca^{2+}, Mg^{2+}, Y^{3+}, and Sc^{3+} to a deaerated MeCN solution of PQ results in large positive shifts of the E_{red} value of PQ as shown in Figure 9.9b–f [398]. The largest positive shift of the E_{red} value of PQ (+1.12 V) was observed in the presence of Sc^{3+}, which indicates a strong interaction of Sc^{3+} with $PQ^{\bullet-}$ [398]. Large positive shifts of E_{red} were also observed in the PTQ-M^{n+} [398].

According to the Nernst equation, the positive shift of E_{red} (ΔE_{red}) of PQ due to the complex formation with M^{n+} is determined by the difference in the binding constants between metal ion complexes with PQ (K_{ox}) and $PQ^{\bullet-}$ (K_{red}) as given by Eq. (9.2), which is rewritten by Eq. (9.3) [398]. The 1 : 1 complex formation of $PQ^{\bullet-}$ with Mg^{2+} is confirmed by the dependence of ΔE_{red} of PQ in the presence of Mg^{2+} on the concentration of Mg^{2+} (Figure 9.10) [398].

$$\Delta E_{red} = (RT/F)\ \ln[(K_{red}[M^{n+}])/(K_{ox}[M^{n+}] + 1)] \tag{9.2}$$

$$\exp(-F\Delta E_{red}/RT) = K_{ox}K_{red}^{-1} + K_{red}^{-1}[M^{n+}]^{-1} \tag{9.3}$$

According to Eq. (9.3), a linear correlation is obtained between $\exp(-F\Delta E_{red}/RT)$ and $[Mg^{2+}]^{-1}$ [398]. From the slope and intercept of the linear plot the formation constant is determined as $K_{ox} = 1.1\ M^{-1}$ in MeCN at 298 K, which agrees with the K_{ox} value determined independently from UV–vis spectral change of PQ in

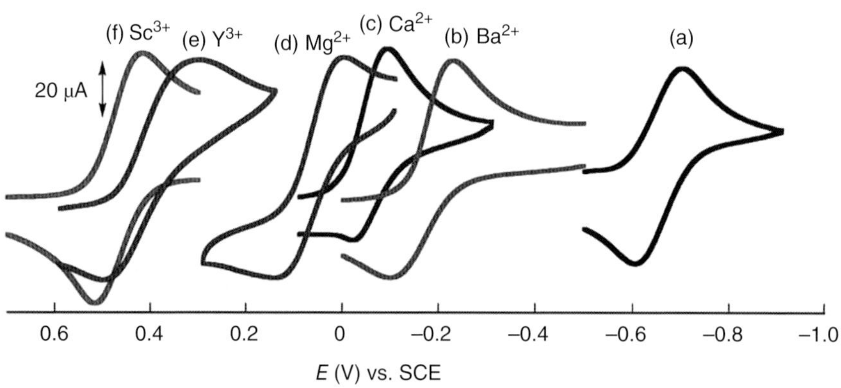

Figure 9.9 CVs of PQ (1.5×10^{-3} M) (a) in the absence of metal ion, and in the presence of 2.0×10^{-2} M of (b) Ba^{2+}, (c) Ca^{2+}, (d) Mg^{2+}, (e) Y^{3+}, and (f) Sc^{3+} in deaerated MeCN containing TBAP (0.10 M) with a Pt working electrode at 298 K. Source: Yuasa et al. 2006 [398]. Reproduced with permission of John Wiley & Sons.

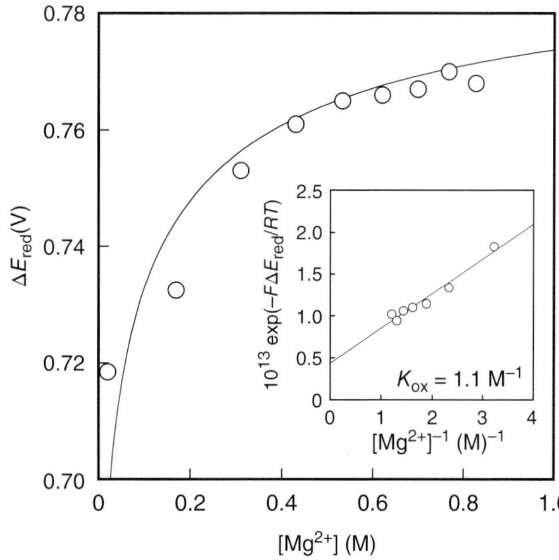

Figure 9.10 Dependence of the positive shift in E_{red} (ΔE_{red}) of PQ in the presence of Mg^{2+} on [Mg^{2+}]. Inset: Plot of $\exp(-F\Delta E_{red}/RT)$ vs. $[Mg^{2+}]^{-1}$. Source: Yuasa et al. 2006 [398]. Reproduced with permission of John Wiley & Sons.

the presence of various concentrations of Mg^{2+} ($K_{ox} = 1.3\,M^{-1}$) [398]. The K_{red} values are determined from the K_{ox} values using Eq. (9.3) as 9.3×10^9, 5.6×10^{11}, 7.5×10^{13}, and $6.3 \times 10^{20}\,M^{-1}$ for $PQ^{\cdot-}$–Ba^{2+}, $PQ^{\cdot-}$–Ca^{2+}, $PQ^{\cdot-}$–Mg^{2+}, and $PQ^{\cdot-}$–Sc^{3+}, respectively [398]. These values are much larger than that of neutral PQ.

The K_{red} values are correlated well with the ΔE values of $O_2^{\cdot-}/M^{n+}$ derived from the g_{zz}-values [201], as indicated by a linear correlation between $\log K_{red}$ and ΔE (closed circles in Figure 9.11) [398]. A good linear correlation observed

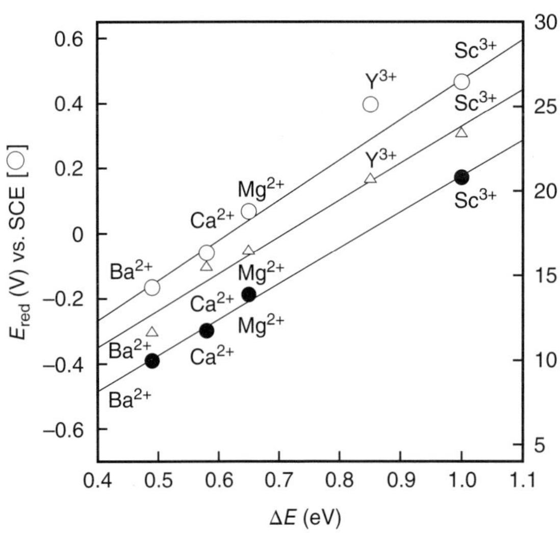

Figure 9.11 Plots of E_{red} of PQ (o) and PTQ (Δ) in the presence of M^{n+} (2.0×10^{-2} M) and plot of $\log K_{red}$ of PQ (•) vs. ΔE. Source: Yuasa et al. 2006 [398]. Reproduced with permission of John Wiley & Sons.

in the plot of the E_{red} values of PQ in the presence of Ba^{2+}, Ca^{2+}, Mg^{2+}, and Sc^{3+} vs. ΔE indicates that the binding strength of metal ion with $PQ^{\bullet -}$ increases with increasing Lewis acidity of the metal ion, which leads to large positive shifts of E_{red} of PQ in the presence of M^{n+} [398]. A good linear correlation is also observed in the relationship between E_{red} values of PTQ in the presence of M^{n+} and ΔE in Figure 9.11 (open triangles) [398]. It should be noted that the slopes of the linear correlation of the E_{red} values of PQ (1.2) and PTQ (1.1) vs. ΔE are close to unity. Such linear correlations with the slope being close to unity strongly indicate that the binding strength of metal ion complexes of $PQ^{\bullet -}$ and $PTQ^{\bullet -}$ is determined mainly by the Lewis acidity of M^{n+} rather than by the ion radius.

The large positive shifts of the E_{red} values of PQ in the presence of M^{n+} (Figure 9.11) result in enhancement of the rate of ET reduction of PQ. No electron transfer from (TPP)Co ($E^0{}_{ox}$ = 0.35 V vs. SCE) to PQ has occurred in the presence of Ba^{2+}, Ca^{2+}, and Mg^{2+} in deaerated MeCN at 298 K, as expected from the highly positive free energy change of electron transfer ($\Delta G_{et} > 0.28$ eV) [398]. In the presence of Sc^{3+}, however, the free energy change of electron transfer becomes negative ($\Delta G_{et} = -0.12$ eV), when efficient electron transfer from (TPP)Co to PQ occurs to yield $(TPP)Co^+$ and the $PQ^{\bullet -}$–Sc^{3+} complex as shown in Scheme 9.3 [398]. The second-order rate constant of electron transfer (k_{obs}) increases with increasing concentration of Sc^{3+} to approach a limited value as shown in Figure 9.12 [398]. Such saturated dependence of k_{obs} with respect to the concentration of Sc^{3+} results from the 1 : 1 complex formation between PQ and Sc^{3+}, which enhances the electron acceptor ability of PQ [398]. If the ET reaction proceeds via the PQ–Sc^{3+} complex (Scheme 9.3), the dependence of k_{obs} on $[Sc^{3+}]$ is expressed by Eq. (9.4), which is rewritten by a linear relation between $k_{obs}{}^{-1}$ and $[Sc^{3+}]^{-1}$ (Eq. (9.5)), where K_{ox} is the formation constant of the PQ–Sc^{3+} complex, and k_{et} is the second-order rate constant of

$$k_{obs} = k_{et}K_{ox}[Sc^{3+}]/(1 + K_{ox}[Sc^{3+}]) \tag{9.4}$$

$$k_{obs}{}^{-1} = k_{et}{}^{-1} + k_{et}{}^{-1}K_{ox}{}^{-1}[Sc^{3+}]^{-1} \tag{9.5}$$

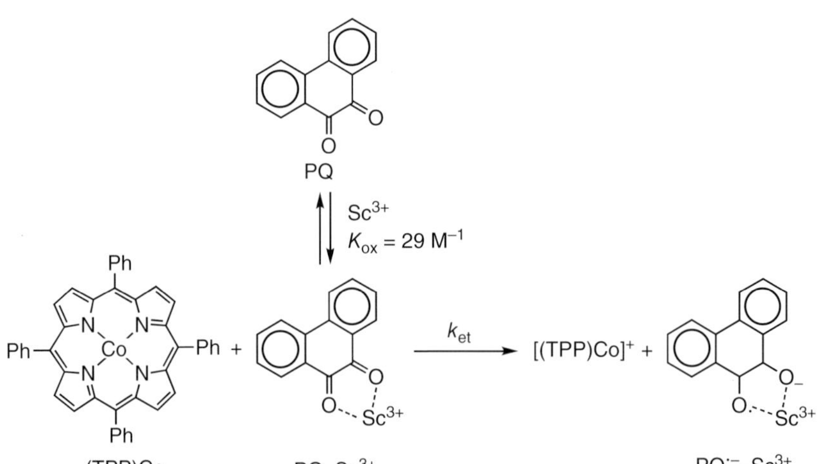

Scheme 9.3 Sc^{3+}-promoted electron transfer from (TPP)Co to PQ. Source: Yuasa et al. 2006 [398]. Reproduced with permission of John Wiley & Sons.

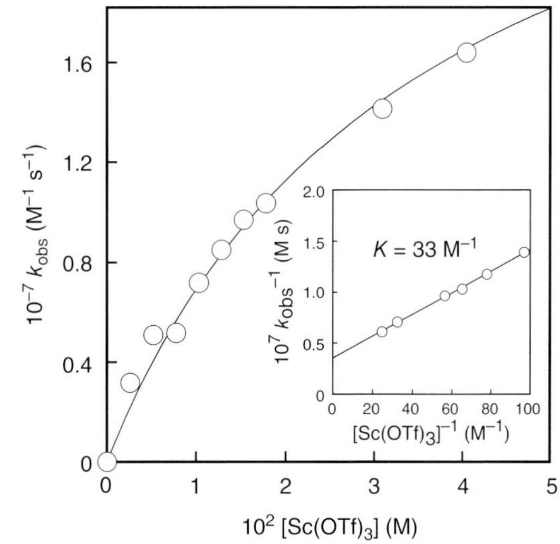

Figure 9.12 Dependence of k_{obs} on [Sc(OTf)$_3$] for electron transfer from (TPP)Co (2.0×10^{-7} M) to PQ in the presence of Sc^{3+} in deaerated MeCN at 298 K. Source: Yuasa et al. 2006 [398]. Reproduced with permission of John Wiley & Sons.

electron transfer from (TPP)Co to the PQ–Sc^{3+} complex [398]. From the slope and intercept of the linear plot of $k_{obs}{}^{-1}$ vs. $[Sc^{3+}]^{-1}$ (inset of Figure 9.12) the K_{ox} value ($33\,M^{-1}$) is determined, which agrees well with the value determined from the UV–vis spectral change of PQ in the presence of various concentrations of Sc^{3+} ($K_{ox} = 29\,M^{-1}$) [398]. Such agreement indicates that the PQ–Sc^{3+} complex is indeed a reactive intermediate in the ET reaction.

In the case of PTQ, the binding mode is changed from $(PTQ)_2$–M^{n+} to $PTQ^{\cdot-}$–M^{n+} upon the ET reduction (vide supra). Electron transfer from 10,10′-dimethyl-9,9′-biacridine [$(AcrH)_2$] to $(PTQ)_2$–M^{n+} (M^{n+} = Mg^{2+}, Y^{3+}, and Sc^{3+}) occurs to yield 10-methylacridinium ion ($AcrH^+$) and $PTQ^{\cdot-}$–M^{n+} in MeCN at 298 K, whereas no electron transfer occurs when the weaker Lewis acid (Ca^{2+} or Ba^{2+}) is employed [398]. In this case, the initial electron transfer from $(AcrH)_2$ to $(PTQ)_2$–M^{n+} is followed by the facile C—C bond cleavage to give $AcrH^+$ and $AcrH^{\cdot}$ (Scheme 9.4). Since $AcrH^{\cdot}$ (E_{ox} vs. SCE = −0.46 V) [400]

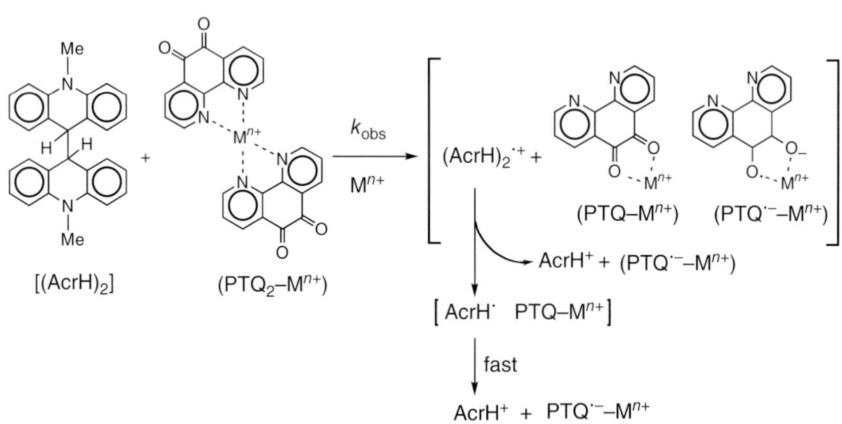

Scheme 9.4 M^{n+}-promoted electron transfer from $(AcrH)_2$ to PTQ in MeCN. Source: Yuasa et al. 2006 [398]. Reproduced with permission of John Wiley & Sons.

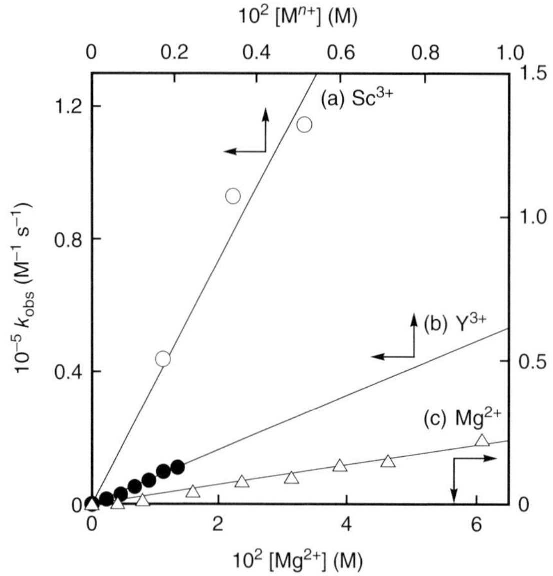

Figure 9.13 Dependence of k_{obs} on [M^{n+}] for the electron transfer from $(AcrH)_2$ (3.0×10^{-6} M) to $(PTQ)_2$–M^{n+} complexes in the presence of (a) $Sc(OTf)_3$ (o), (b) $Y(OTf)_3$ (•), (c) $Mg(ClO_4)_2$ (Δ) in deaerated MeCN at 298 K. Source: Yuasa et al. 2006 [398]. Reproduced with permission of John Wiley & Sons.

is a much stronger reductant than $(AcrH)_2$ (E_{ox} vs. SCE = 0.62 V) [401], the rapid electron transfer from $AcrH^{\cdot}$ to the PTQ–M^{n+} complex occurs to yield the final product, $AcrH^+$ and $PTQ^{\cdot-}$–M^{n+} (Scheme 9.4) [398]. The observed second-order rate constant (k_{obs}) increases proportionally with increasing M^{n+} concentration (Figure 9.13) [398]. The first-order dependence of k_{obs} on [M^{n+}] results from the formation of 1 : 1 complexes of $PTQ^{\cdot-}$–M^{n+} in the ET reactions, which requires M^{n+} to be consumed in the electron transfer (Scheme 9.4) [398]. This shows a sharp contrast to the case of PQ in which the 1 : 1 binding mode of the PQ–Sc^{3+} complex remains the same in the $PQ^{\cdot-}$–Sc^{3+} complex and the rate exhibits saturated dependence on the concentration of Sc^{3+} (Figure 9.12) [398].

9.3 Self-Organized MCET

Rates of M^{n+}-promoted electron transfer from electron donors to acceptors normally increase linearly or approach limited values with increasing concentration of M^{n+} (vide infra). In contrast, self-organized MCET systems involving multiple molecular environments can lead to decreases of entropy equivalent to an increase of molecular electronic order for the activated complex, resulting in a substantial increase in the rate of electron transfer [402, 403]. In such a case, the rate of electron transfer is no longer linearly related with concentrations of reactants and promoting molecules for electron transfer. New frontiers of electron transfer may be exploited in such nonlinear dynamic and self-organized MCET systems that nature has developed to a high degree of perfection [403]. In the case of Sc^{3+}, which is the strongest Lewis acid among metal ions (vide supra), high kinetic order is indeed observed in Sc^{3+}-promoted ET reduction of Q in propionitrile (EtCN): third order with respect to concentration of Sc^{3+} and second order with respect to concentration of Q (vide infra) [402].

When tris(2-phenylpyridine)iridium [$Ir(ppy)_3$] [404, 405] is employed as an electron donor, no electron transfer from $Ir(ppy)_3$ (E_{ox} = 0.77 V vs. SCE) [405] to Q (E_{red} = −0.50 V vs. SCE) [400] occurs, in agreement with the highly positive free energy change of electron transfer (ΔG_{et} = 1.27 eV) [402]. In the presence of $Sc(OTf)_3$, however, efficient electron transfer from $Ir(ppy)_3$ to Q occurs as in the case of ET from (TPP)Co to Q [201]. Under the pseudo-first-order kinetic conditions in the presence of a large excess of Q and Sc^{3+} as compared with $Ir(ppy)_3$, the rate obeys pseudo-first-order kinetics [402]. The dependence of observed pseudo-first-order rate constant (k_{obs}) on [Sc^{3+}] is shown in Figure 9.14a, where the k_{obs} value increases exhibiting a second-order dependence on [Sc^{3+}] at lower concentrations, changing to a third-order dependence at higher concentrations [402]. The dependence of k_{obs} on [Q] is also unusual: the k_{obs} value increases with [Q] to show a second-order dependence on [Q] as shown in Figure 9.14b [402]. Such a second-order dependence of k_{obs} on [Q] is ascribed to self-organized behavior in the Sc^{3+}-promoted electron transfer from $Ir(ppy)_3$ to Q to produce the π-dimer semiquinone radical anion complex that is triply bridged by three Sc^{3+} ions ($Q^{\cdot-}$–$(Sc^{3+})_3$–Q) [402].

Formation of the π-dimer semiquinone radical anion complexes with Sc^{3+} in the Sc^{3+}-promoted electron transfer from $Ir(ppy)_3$ to Q was confirmed by their detection by EPR. The EPR spectrum obtained at 298 K exhibits 23 hyperfine lines (Figure 9.15a) [402]. The computer simulation spectrum with hyperfine splitting due to 8 equiv protons (a(8H) = 1.12 G) and superhyperfine splitting

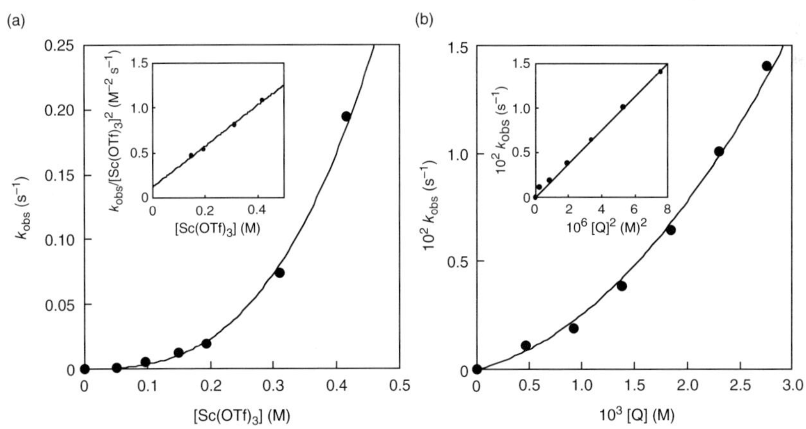

Figure 9.14 (a) Dependence of k_{obs} on [$Sc(OTf)_3$] for electron transfer from $Ir(ppy)_3$ (1.5×10^{-5} M) to Q (4.8×10^{-4} M) in the presence of $Sc(OTf)_3$ in EtCN at 233 K [402]. Inset: Plot of k_{obs}/[$Sc(OTf)_3$]$_2$ vs. [$Sc(OTf)_3$]. (b) Dependence of k_{obs} on [Q] for electron transfer from $Ir(ppy)_3$ (1.5×10^{-5} M) to Q in the presence of $Sc(OTf)_3$ (5.1×10^{-2} M) in EtCN at 233 K [402]. Inset: Plot of k_{obs} vs. $[Q]^2$. Source: Yuasa et al. 2003 [402]. Reproduced with permission of American Chemical Society.

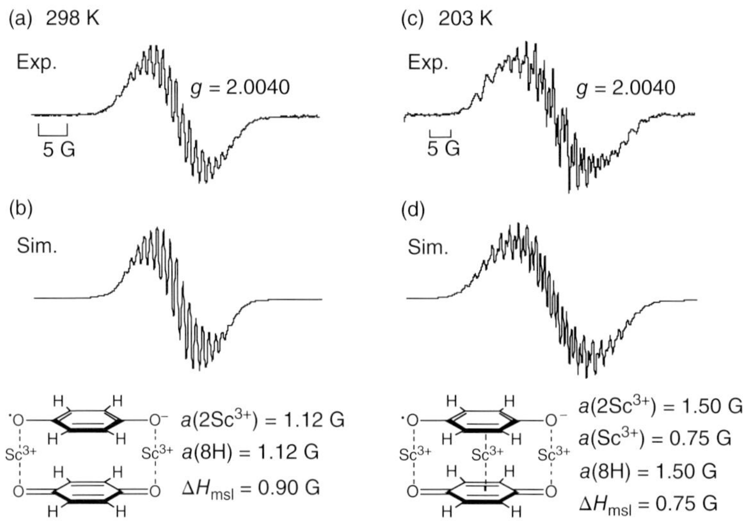

Figure 9.15 EPR spectra of an EtCN solution of $Ir(ppy)_3$ (2.3×10^{-4} M) and Q (3.4×10^{-1} M) in the presence of $Sc(OTf)_3$ (4.8×10^{-1} M) at (a) 298 K and (c) 203 K. Their computer simulation spectra are shown in (b) and (d), respectively. ΔH_{msl} stands for the maximum slope linewidth. Source: Yuasa et al. 2003 [402]. Reproduced with permission of American Chemical Society.

due to 2 equiv Sc^{3+} ions ($a(2Sc^{3+}) = 1.12$ G) is shown in Figure 9.15b [402]. The complete agreement of the observed EPR spectrum (Figure 9.15a) with the computer simulation spectrum (Figure 9.15b) indicates that $Q^{\bullet-}$ forms a π-dimer complex with Q, which is bridged by 2 equiv. Sc^{3+} ions ($Q^{\bullet}$–$(Sc^{3+})_2$–Q) [402]. When temperature is lowered to 203 K, the EPR spectrum is changed to exhibit further superhyperfine splitting due to additional Sc^{3+} ion (Figure 9.15c) [402]. This is well reproduced by computer simulation spectrum with hyperfine splitting due to 8 equiv protons ($a(8H) = 1.50$ G), superhyperfine splitting due to 2 equiv Sc^{3+} ions ($a(2Sc^{3+}) = 1.50$ G) and an additional Sc^{3+} ion ($a(Sc^{3+}) = 0.75$ G) as shown in Figure 9.15d [402]. The larger $a(2Sc^{3+})$ value and additional superhyperfine splitting with the smaller $a(Sc^{3+})$ value indicate that the dimer radical anions $(Q)_2^{\bullet-}$ are more strongly bridged by 2 equiv Sc^{3+} ions with an additional Sc^{3+} ion, which may be placed between the π planes of $(Q)_2^{\bullet-}$ to produce the $Q^{\bullet-}$–$(Sc^{3+})_3$–Q complex as shown in the structure in Figure 9.15d [402].

Such complex formation of $Q^{\bullet-}$ with Q and Sc^{3+} results in a positive shift of the one-electron reduction potential of Q in accordance with the Nernst equation (Eq. (9.6)),

$$E_{red} = E^0{}_{red} + (2.3RT/F)\,\log\{K_1[Q][Sc^{3+}]^2(1 + K_2[Sc^{3+}])\} \tag{9.6}$$

where $E^0{}_{red}$ is the one-electron reduction potential of Q in the absence of Sc^{3+}, K_1 is the formation constant for $Q^{\bullet-}$–$(Sc^{3+})_2$–Q and K_2 is the formation constant for $Q^{\bullet-}$–$(Sc^{3+})_3$–Q [402]. Since Sc^{3+} has no effect on the one-electron oxidation potential of $Ir(ppy)_3$, the free energy change of electron transfer from $Ir(ppy)_3$ to Q in the presence of Sc^{3+} decreases due to the positive shift in the E_{red} value of Q. If such a change in the energetics is directly reflected in the transition state of self-organized electron transfer, the dependence of k_{obs} on $[Sc^{3+}]$ is derived as given by Eq. (9.7), where k_0 is the rate constant in the absence of Sc^{3+} [391]. This equation agrees with the

$$k_{obs} = k_0 K_1 [Q]^2 [Sc^{3+}]^2 (1 + K_2 [Sc^{3+}]) \tag{9.7}$$

second- and third-order dependence of k_{obs} on $[Sc^{3+}]$ (Figure 9.14a) and also the second-order dependence of k_{obs} on [Q] (Figure 9.14b) [402]. The linear plot of $k_{obs}/[Sc^{3+}]^2$ vs. $[Sc^{3+}]$ (inset in Figure 5.23a) affords the K_2 value as 18 M^{-1} [402]. This K_2 value is consistent with the value (20 M^{-1}) obtained independently from the absorbance change at 604 nm due to the $Q^{\cdot-}–(Sc^{3+})_3–Q$ complex in the electron transfer from $Ir(ppy)_3$ to Q in the presence of a large excess of Sc^{3+} in EtCN at 233 K [402]. Thus, the rates of self-organized electron transfer from $Ir(ppy)_3$ to Q promoted by Sc^{3+} exhibit unusually high-order kinetics due to the formation of the triply bridged π-dimer semiquinone radical anion complex with three Sc^{3+} ions ($Q^{\cdot-}–(Sc^{3+})_3–Q$).

The ($Q^{\cdot-}–(Sc^{3+})_3–Q$) complex can also be formed by proportionation reaction between Q and QH_2 in the presence of Sc^{3+} (Scheme 9.5) [406]. Addition of a high concentration of $Sc(OTf)_3$ (5.4×10^{-1} M) to a deaerated EtCN solution of QH_2 (1.2×10^{-3} M) and Q (3.9×10^{-3} M) affords a less dark solution at 298 K (inset of Figure 9.16a) [406]. When the temperature is lowered to 203 K, there is a change in color (darker color in inset of Figure 9.16b) [406]. The color change depending on temperature is completely reversible and it can be repeated many times [406].

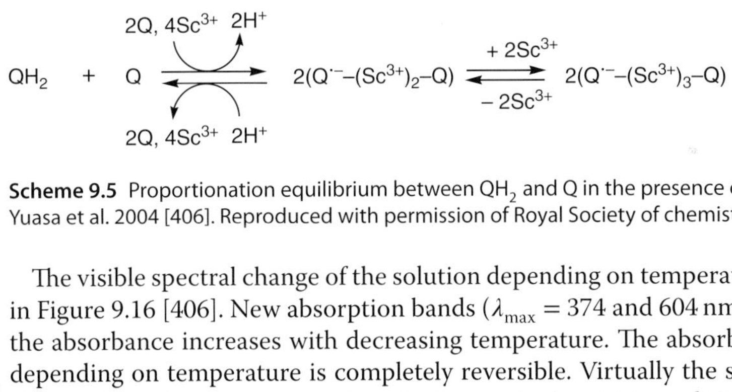

Scheme 9.5 Proportionation equilibrium between QH_2 and Q in the presence of Sc^{3+}. Source: Yuasa et al. 2004 [406]. Reproduced with permission of Royal Society of chemistry.

The visible spectral change of the solution depending on temperature is shown in Figure 9.16 [406]. New absorption bands (λ_{max} = 374 and 604 nm) appear and the absorbance increases with decreasing temperature. The absorbance change depending on temperature is completely reversible. Virtually the same spectral change depending on temperature is observed when QH_2 and Q are replaced by hydronaphthoquinone (NQH_2) and 1,4-naphthoquinone (NQ), respectively [406]. Such temperature-induced color change is known as thermochromism for a long time [407, 408]. Thermochromism results from many types of thermally driven changes in the energy of electronic excitations such as the proton tautomerization of organic compounds involving different proton-accepting moieties [409], valence tautomerism of metal complexes involving a redox-active metal center [410–412], conformational change of the polymer backbone [413], and formation of supramolecular complexes or aggregates [414]. The thermochromism observed in Figure 9.14 is unique, because this is caused by the change in the paramagnetic species associated with the change in the EPR spectrum (vide infra).

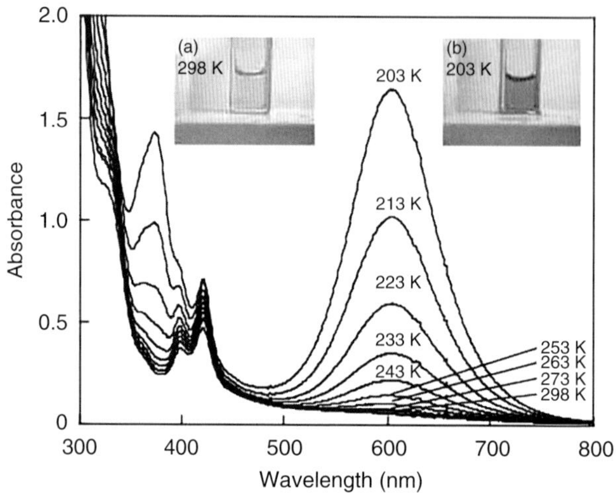

Figure 9.16 Absorption spectral changes of a deaerated EtCN solution of QH_2 (3.2×10^{-3} M) and Q (1.8×10^{-2} M) in the presence of $Sc(OTf)_3$ (3.2×10^{-1} M) at various temperatures (1 mm path length). Inset: Visible color change of a deaerated propionitrile solution of QH_2 (1.2×10^{-3} M) and Q (3.9×10^{-3} M) in the presence of $Sc(OTf)_3$ (5.4×10^{-1} M) at (a) 298 K and (b) 203 K. Source: Yuasa et al. 2004 [406]. Reproduced with permission of Royal Society of Chemistry.

The species responsible for the color change depending on temperature are observed by EPR as shown in Figure 9.17 [416]. The EPR spectrum observed at 298 K with the hyperfine splitting due to 8 equiv protons [a(8H) = 1.12 G] and superhyperfine splitting due to 2 equiv Sc^{3+} ions [$a(2Sc^{3+})$ = 1.12 G] agrees with that of the π-dimer formed between $Q^{\bullet-}$ and Q, which is bridged by 2 equiv Sc^{3+} ions ($Q^{\bullet-}–(Sc^{3+})_2–Q$) in Sc^{3+}-promoted electron transfer from iridium complex to Q (Figure 9.15a) [402]. When temperature is lowered to 203 K, the EPR spectrum is changed to exhibit further superhyperfine splitting due to additional Sc^{3+} ion [$a(Sc^{3+})$ = 0.75 G] [406]. The EPR spectrum agrees with that of $Q^{\bullet-}–(Sc^{3+})_3–Q$ in which additional Sc^{3+} is placed between the π-planes of $(Q)_2^{\bullet-}$ to produce $Q^{\bullet-}–(Sc^{3+})_3–Q$ (Figure 9.13b). The change in the EPR spectrum between $Q^{\bullet-}–(Sc^{3+})_2–Q$ and $Q^{\bullet-}–(Sc^{3+})_3–Q$ was reversible as in the case of the color change in Figure 9.16 [406].

The decrease in the number of binding Sc^{3+} ions from three to two is clearly identified as the disappearance of the EPR signal at the lowest or highest magnetic field region due to the loss of additional superhyperfine splitting due to one Sc^{3+} ion with increasing temperature (Figure 9.17), where the change in the EPR spectra (solid line) at the higher magnetic field region is shown together with the computer simulation spectra (broken line) under the same experimental conditions as employed for the color change in Figure 9.14 [406]. The decrease in the EPR signal intensity with temperature agrees with the accompanied decrease in absorbance at 604 nm (inset of Figure 9.17) [406]. Thus, the unique thermochromism in Figure 9.17 involves the change in the number of binding scandium ions with dimer radical anions in the disproportionation equilibrium depending on temperature (Scheme 9.5), which causes the drastic color change, associated with the change in the EPR spectrum [406].

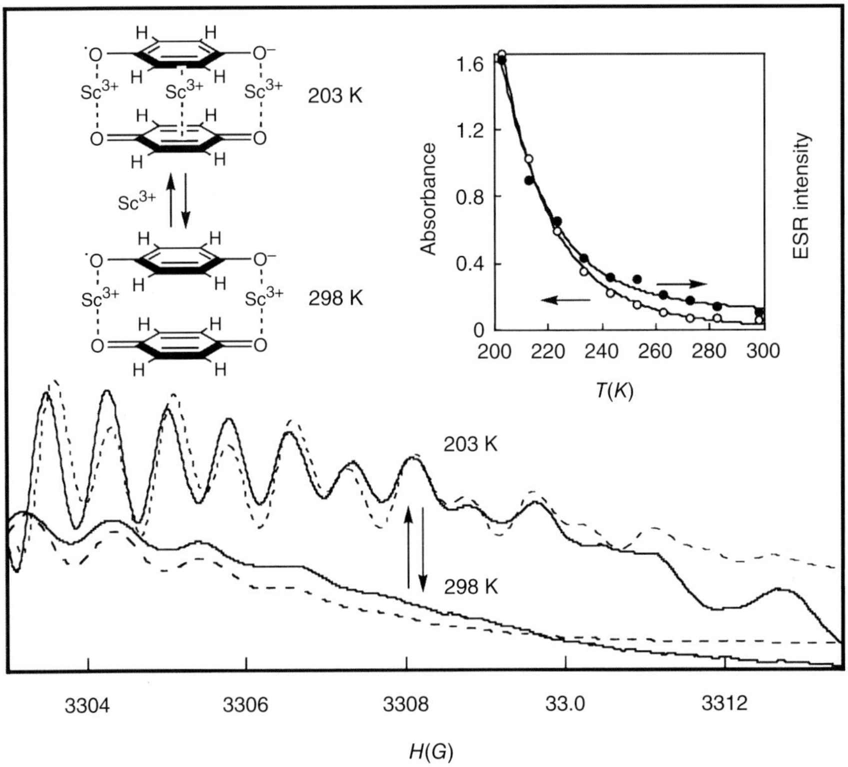

Figure 9.17 EPR spectra of a deaerated propionitrile solution of QH_2 (3.2×10^{-3} M) and Q (1.8×10^{-2} M) in the presence of $Sc(OTf)_3$ (3.2×10^{-1} M) at the magnetic range 3303.0–3313.5 G at 203 K and 298 K. Broken lines are simulation EPR spectra of $Q^{\bullet-}$–$(Sc^{3+})_3$–Q [a(8H) = $a(2Sc^{3+})$ = 1.50 G, $a(Sc^{3+})$ = 0.75 G, and ΔH_{msl} = 0.64 G] and $Q^{\bullet-}$–$(Sc^{3+})_3$–Q [a(8H) = $a(2Sc^{3+})$ = 1.12 G and ΔH_{msl} = 0.93 G]. Inset: Dependence of absorbance at 604 nm and EPR intensity at 3312.8 G on temperature [406]. Source: Yuasa et al. 2004 [406]. Reproduced with permission of Royal Society of Chemistry.

When a chiral scandium complex of 2,6-bis-(oxazolinyl)pyridine, [Sc(R)-pybox]$(OTf)_3$ [OTf = OSO_2CF_3] is employed instead of $Sc(OTf)_3$, a 2 : 2 chiral π-dimer complex of 1,4-napthosemquinone radical anion ($NQ^{\bullet-}$) with Sc^{3+}(R)-pybox, [$(NQ^{\bullet-})_2$–$(Sc^{3+}(R)\text{-pybox})_2$], the chiral π-dimer complex is formed by the ET reduction of 1,4-naphtoquinone (NQ) by electron donors such as 1,1′-dimethylferrocene (Me_2Fc) in the presence of Sc^{3+}(R)-pybox (Scheme 9.6) [415]. The chiral π-dimer complex is also formed by the proportionation reaction between NQH_2 and NQ in the presence of Sc^{3+}(R)-pybox (Scheme 9.6) [415]. The formation of $(NQ^{\bullet-})_2$–$(Sc^{3+}(R)\text{-pybox})_2$ was confirmed by positive-ion electrospray ionization (ESI) mass spectrum (Figure 9.18a), which exhibits a signal at m/z 1591.4 [415]. This corresponds to $[(NQ^{\bullet-})_2\text{–}(Sc^{3+}(R)\text{-pybox})_2]^+$. The signal has a characteristic distribution of isotopomers (Figure 9.18a) that matches well with the calculated isotopic distribution for $[(NQ^{\bullet-})_2\text{–}(Sc^{3+}(R)\text{-pybox})_2]^+$ (Figure 9.18b) [415]. The stoichiometry in Scheme 9.6 was confirmed by UV–vis and ^{1}H NMR titration [415]. The chiral π-dimer complex $[(Q^{\bullet-})_2\text{–}(Sc^{3+}(R)\text{-pybox})_2]^+$ is also formed by the proportionation reaction between QH_2 and Q in the presence of Sc^{3+}(R)-pybox [415].

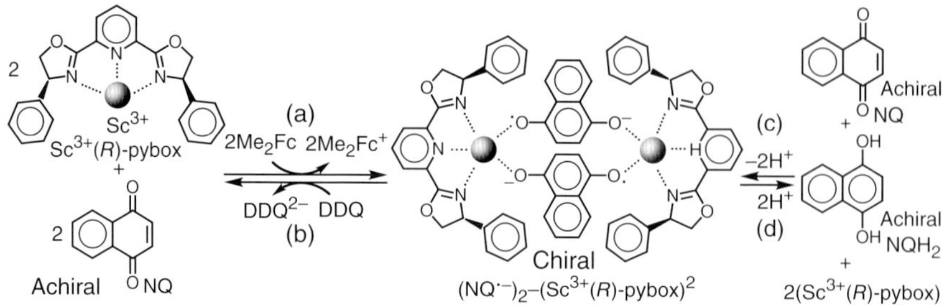

Scheme 9.6 Formation of a chiral assembly $[(NQ^{\cdot-})_2\text{–}(Sc^{3+}(R)\text{-pybox})_2]^+$ in electron-transfer reduction of NQ or the proportionation equilibrium between QH_2 and Q in the presence of $(Sc^{3+}(R)\text{-pybox})_2$ in EtCN. Source: Yuasa et al. 2007 [415]. Reproduced with permission of American Chemical Society.

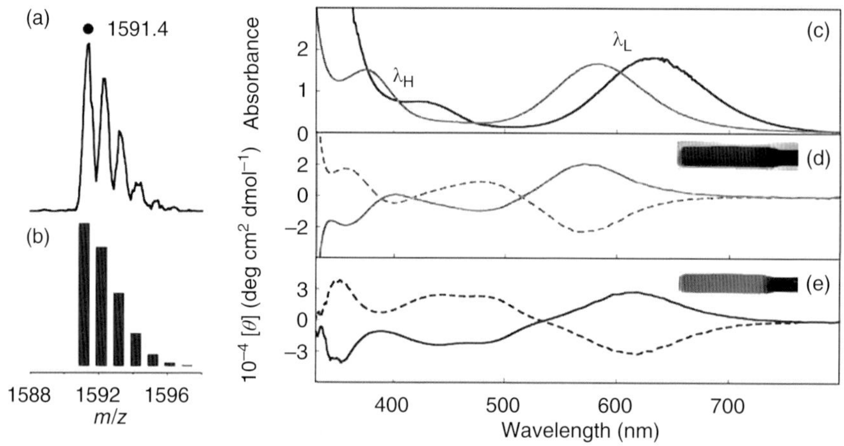

Figure 9.18 (a) Positive-ion ESI mass spectrum of an MeCN solution of NQ (3.0×10^{-4} M) and NQH_2 (3.0×10^{-4} M) in the presence of $[Sc(R)\text{-pybox}](OTf)_3$ (6.0×10^{-4} M). The signal at m/z 1591.4 corresponds to $[(NQ^{\cdot-})_2\text{–}(Sc^{3+}(R)\text{-pybox})_2]^+$. (b) Calculated isotopic distributions for $[(NQ^{\cdot-})_2\text{–}(Sc^{3+}(R)\text{-pybox})_2]^+$. (c) UV–vis spectra of a deaerated MeCN solution of Q (3.4×10^{-2} M) and QH_2 (3.4×10^{-2} M) in the presence of $Sc^{3+}(R)$-pybox (6.7×10^{-2} M) (light grey solid line), NQ (7.0×10^{-3} M), and NQH_2 (7.0×10^{-3} M) in the presence of $Sc^{3+}(R)$-pybox (1.3×10^{-2} M) (dark grey solid line) (1 mm path-length). Corresponding CD spectra of (d) Q–QH_2 and (e) NQ–NQH_2 systems in the presence of $Sc^{3+}(R)$-pybox (solid lines) and $Sc^{3+}(S)$-pybox (dashed lines). Insets: Photographs of a deaerated MeCN solution of (d) Q (1.0×10^{-1} M) and QH_2 (1.0×10^{-1} M) in the presence of $Sc^{3+}(R)$-pybox (1.0×10^{-1} M), and (e) NQ (1.0×10^{-1} M) and NQH_2 (1.0×10^{-1} M) in the presence of $Sc^{3+}(R)$-pybox (1.0×10^{-1} M). Source: Yuasa et al. 2007 [415]. Reproduced with permission of American Chemical Society.

The absorption spectrum of a 2 : 2 chiral π-dimer complex of $Q^{\cdot-}$ with $Sc^{3+}(R)$-pybox, $[(Q^{\cdot-})_2\text{–}(Sc^{3+}(R)\text{-pybox})_2]^+$, produced by the proportionation reaction between QH_2 and Q (light grey solid line in Figure 9.18c) shows strong new absorption bands (denoted as $\lambda_H = 376$ nm and $\lambda_L = 583$ nm) [415]. When Q and QH_2 are replaced by NQ and NQH_2, the absorption maxima ($\lambda_H = 376$ nm and $\lambda_L = 583$ nm) are shifted to $\lambda_H = 422$ nm and $\lambda_L = 633$ nm, respectively (dark grey solid line in Figure 9.18c) [415]. Such UV–vis spectral changes are associated with remarkable color changes (shown in photographs in Figure 9.18d,e) [415]. The circular dichroism (CD) spectra of the quinone–hydroquinone systems in the presence of Sc^{3+}(*R*)-pybox and Sc^{3+}(*S*)-pybox are shown in Figure 9.18d,e, where the Cotton effects of the CD bands with complete mirror images for their enantiomer pairs are observed [415]. This indicates the chiral organization of the quinone–hydroquinone systems with Sc^{3+}-pybox.

The absorption bands due to $[(NQ^{\cdot -})_2–(Sc^{3+}(R)\text{-pybox})_2]^+$ completely disappear upon addition of 2,3-dichloro-5,6-dicyano-*p*-benzoquinone [DDQ (8.5×10^{-4} M)] to the MeCN solution of $[(NQ^{\cdot -})_2–(Sc^{3+}(R)\text{-pybox})_2]^+$ [415]. Since neutral NQ has virtually no interaction with Sc^{3+}(*R*)-pybox, the formation and dispersion of $[(NQ^{\cdot -})_2–(Sc^{3+}(R)\text{-pybox})_2]^+$ are controlled by the ET reduction and oxidation (Scheme 9.6) [415]. The ET reduction of 2 equiv of NQ by 2 equiv of Me_2Fc in the presence of 2 equiv of Sc^{3+}(*R*)-pybox yields 1 equiv of $[(NQ^{\cdot -})_2–(Sc^{3+}(R)\text{-pybox})_2]^+$, while the two-electron oxidation of $[(NQ^{\cdot -})_2–(Sc^{3+}(R)\text{-pybox})_2]^+$ by 1 equiv of DDQ results in reproducing NQ and Sc^{3+}(*R*)-pybox [415]. Such formation and dispersion cycles of the chiral π-dimer complex in response to ET reduction and oxidation are highly reversible and it can be repeated many times as shown in Figure 9.19 [415]. Thus, the effective redox control on building affinity of the chiral supramolecules affords reversible formation and dispersion of chiral assemblies in response to a simple external signal such as an electron, giving achiral–chiral switchability [416, 417].

The optimized structure of $[(Q^{\cdot -})_2–(Sc^{3+}(R)\text{-pybox})_2]^+$ is calculated by using density functional theory (DFT) at the B3LYP/6-31G* basis (Figure 9.20a [top view] and b [front view]) [415]. There are different sets of protons termed H_A and H_B in $(Q^{\cdot -})_2$, where H_B is shielded as compared with H_A by phenyl rings of (*R*)-pybox [415]. This is consistent with two doublet peaks in the ^{1}H NMR spectrum (Figure 9.20d) [415]. Nuclear overhauser effects (NOEs) are detected between H_B (or H_A) protons and phenyl protons of Sc^{3+}(*R*)-pybox (termed *a*) when irradiated at H_B as shown in Figure 9.20f [415]. This indicates that Sc^{3+}(*R*)-pybox is located near $(Q^{\cdot -})_2$, supporting the suggested structure of $[(Q^{\cdot -})_2–(Sc^{3+}(R)\text{-pybox})_2]^+$. The enantiomer pairs of $(Q^{\cdot -})_2$ in $(Q^{\cdot -})_2–(Sc^{3+}\text{-pybox})_2$ without Sc^{3+}-pybox are shown Figure 9.20c [415]. Those

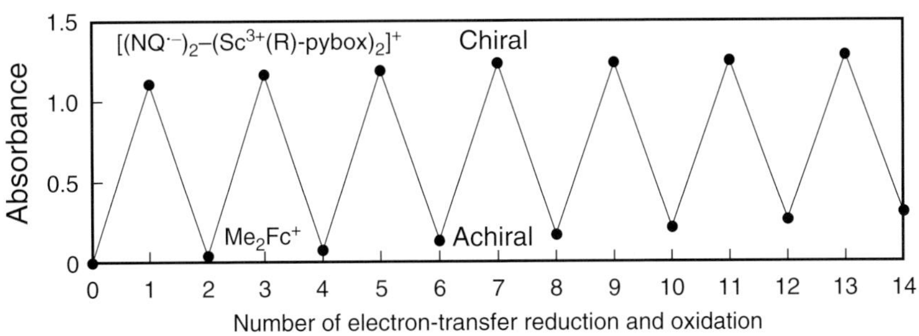

Figure 9.19 Absorbance at 633 nm for cycles of ET reduction of a deaerated MeCN solution of NQ (6.9×10^{-3} M) and Sc^{3+}(R)-pybox (1.0×10^{-1} M) by Me_2Fc (1.7×10^{-3} M) (chiral) and oxidation by DDQ (8.5×10^{-4} M) (achiral) at 298 K (1 mm path-length). Source: Yuasa et al. 2007 [415]. Reproduced with permission of American Chemical Society.

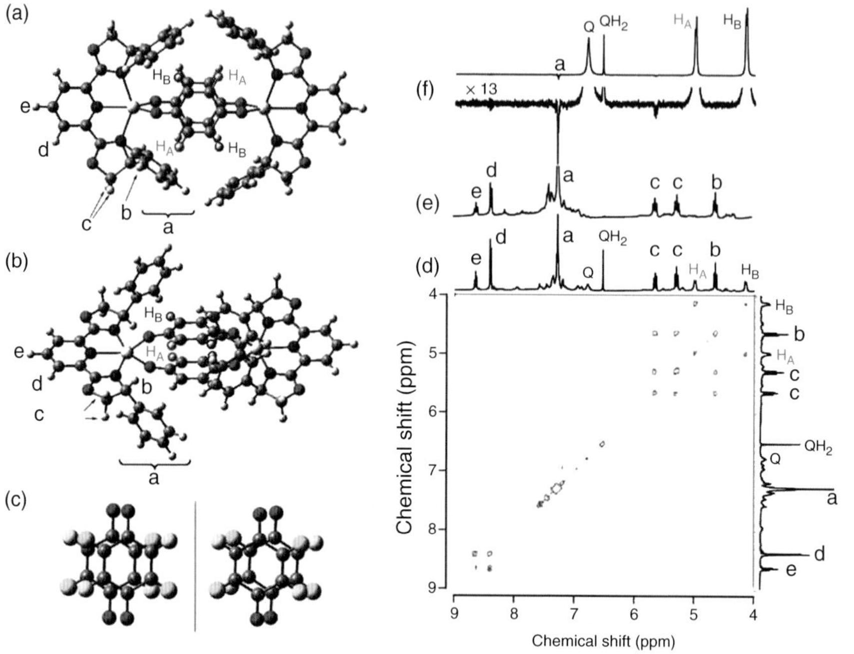

Figure 9.20 The optimized structure of $[(Q^{\cdot-})_2-(Sc^{3+}(R)\text{-pybox})_2]^+$ calculated by using a DFT at the B3LYP/6-31G*: top view (a) and front view (b). (c) An enantiomer pair of the $(Q^{\cdot-})2$ unit in $[(Q^{\cdot-})_2-(Sc^{3+}(R)\text{-pybox})_2]^+$. (d) ^1H, ^1H COSY NMR and (f) NOE NMR spectra of a deaerated CD_3CN solution of Q (5.0×10^{-2} M) and Me_2Fc (5.0×10^{-2} M) in the presence of Sc^{3+}(R)-pybox (5.0×10^{-2} M) at 298 K. (e) ^1H NMR spectrum of a deaerated CD_3CN solution of Q-d_4 (5.0×10^{-2} M) and Me_2Fc (5.0×10^{-2} M) in the presence of Sc^{3+}(R)-pybox (5.0×10^{-2} M) at 298 K. Source: Yuasa et al. 2007 [415]. Reproduced with permission of American Chemical Society.

enantiomer pairs are non-superimposable mirror images of each other. Such mirror symmetry breaking in the $(Q^{\cdot-})_2$ unit through the chiral π-dimer complex formation should cause the induced circular dichroism (ICD) in the long wavelength region (Figure 9.20d,e) [415]. Thus, the dimer formation of $Q^{\cdot-}$ plays a crucial role in such a supramolecular chirogenesis.

9.4 Accelerating and Decelerating Effects of Metal Ions

When electron transfer from an electron donor (D) to an electron acceptor (A) is coupled with the binding of metal ions (M^{n+}) to the product radical anion ($A^{\cdot-}$) (vide supra), the $\Delta S^{\neq}$ value becomes largely negative because of restricted geometry in the binding of metal ions in the transition state, whereas the $\Delta H^{\neq}$ value becomes smaller because of the metal ion binding that thermodynamically stabilizes the product radical anion. In contrast, the $\Delta S^{\neq}$ value of electron transfer without binding of metal ions is normally close to zero, when the reactivity is determined solely by the $\Delta H^{\neq}$ value [418, 419]. If there is a change in the binding mode of the metal ion complex from a 1 : 1 complex ($A^{\bullet-}$–M^{n+}) to a 1 : 2 complex [$A^{\bullet-}$–$(M^{n+})_2$] with increasing metal ion concentration, the $\Delta H^{\neq}$ and $\Delta S^{\neq}$ values would be different depending on the binding modes. In such a case, the rate of the ET pathway to afford a 1 : 2 complex [$A^{\bullet-}$–$(M^{n+})_2$] with a smaller $\Delta H^{\neq}$ value and a more negative $\Delta S^{\neq}$ value would be faster than that to afford a 1 : 1 complex ($A^{\bullet-}$–M^{n+}) at lower temperature, whereas this would be reversed at higher temperature. This indicates that the ET rate increases with increasing concentrations of metal ions at lower temperature, but the rate decreases with increasing concentrations of metal ions at higher temperature. Such an example is shown as the accelerating and decelerating effects of metal ions on the ET reduction of 1-(*p*-tolylsulfinyl)-2,5-benzoquinone (TolSQ) and 9,10-phenanthrenequinone (PQ) by electron donors with different electron-donor abilities depending on temperature (vide infra) [420].

No electron transfer from 10,10′-dimethyl-9,9′-biacridine [$(AcrH)_2$] (E_{ox} = 0.62 V vs. SCE) [401] to 1-(*p*-tolylsulfinyl)-2,5-benzoquinone (TolSQ) [E_{red} = −0.26 V vs. SCE] occurs in MeCN at 298 K, in agreement with the highly positive free energy change of ET (ΔG_{et} = 0.88 eV) [420]. In the presence of scandium triflate [$Sc(OTf)_3$] (OTf = OSO_2CF_3) [1.0 M], however, the reduction potential of TolSQ is shifted to 0.70 V (vs. SCE) [421]. In such a case, efficient electron from $(AcrH)_2$ to TolSQ occurs to yield 2 equiv of $AcrH^+$ (Scheme 9.7) [420]. The initial electron transfer from $(AcrH)_2$ to TolSQ is followed by the facile C—C bond cleavage to give $AcrH^+$ and $AcrH^{\bullet}$ (E_{red} = −0.46 V vs. SCE) [400], which is a much stronger electron donor than $(AcrH)_2$. Thus, the subsequent electron transfer from $AcrH^{\bullet}$ to TolSQ occurs rapidly to yield 2 equiv of $AcrH^+$ and the 1 : 1 complex ($TolSQ^{\bullet-}$–Sc^{3+}) and the 1 : 2 complex [$TolSQ^{\bullet-}$–$(Sc^{3+})_2$] depending on Sc^{3+} concentration (Scheme 9.7) [420].

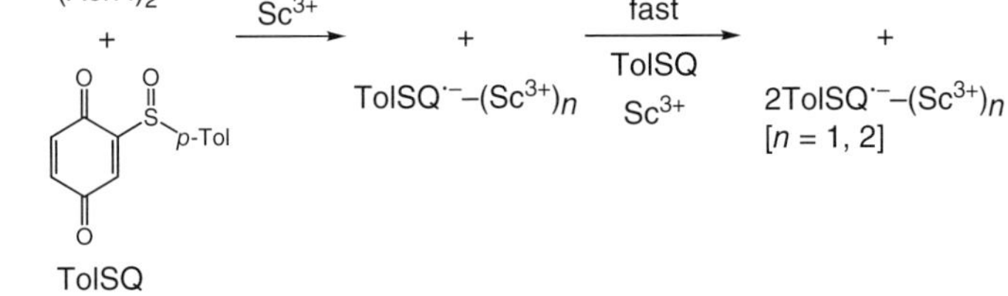

Scheme 9.7 Sc^{3+}-promoted electron transfer from $(AcrH)_2$ to TolSQ in MeCN. Source: Yuasa et al. 2008 [420]. Reproduced with permission of John Wiley & Sons.

The formation of $TolSQ^{\bullet-}$–Sc^{3+} and $TolSQ^{\bullet-}$–$(Sc^{3+})_2$ depending on Sc^{3+} concentration was detected by EPR in electron transfer from $(AcrH)_2$ to TolSQ in the presence of low and high concentrations of Sc^{3+} as shown in Figure 9.21a,c,

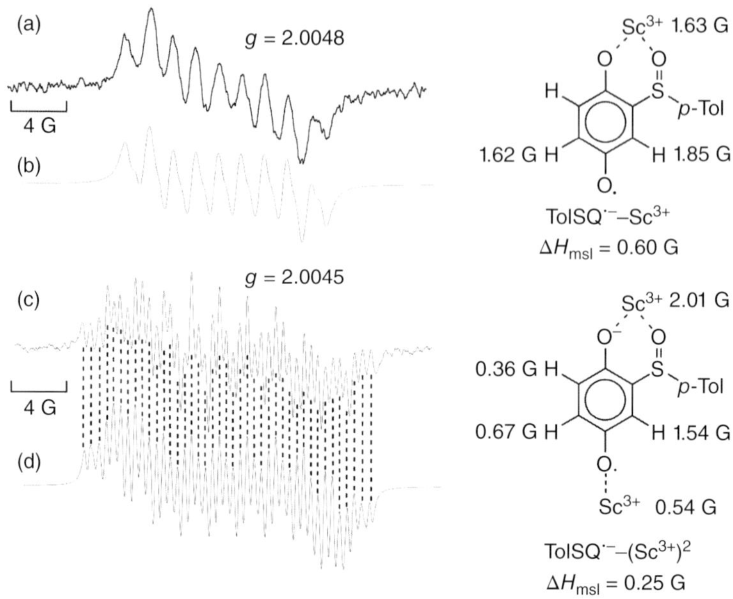

Figure 9.21 (a) EPR spectrum of TolSQ$^{•-}$–Sc^{3+} produced by electron transfer from (AcrH)$_2$ (1.6×10^{-2} M) to TolSQ (4.2×10^{-2} M) in the presence of Sc^{3+} (4.2×10^{-3} M) and H_2O (4.6 M) in deaerated MeCN at 298 K and (b) the computer simulation spectrum. (c) EPR spectrum of TolSQ$^{•-}$–(Sc^{3+})$_2$ produced by electron transfer from (AcrH)$_2$ (1.6×10^{-2} M) to TolSQ (4.3×10^{-2} M) in the presence of Sc^{3+} (2.1×10^{-1} M) and H_2O (2.4 M) in deaerated MeCN at 298 K and (d) the computer simulation spectrum. The *hfc* values determined by the computer simulation with the maximum slope linewidth (ΔH_{msl}) are shown together with the structures of TolSQ$^{•-}$–Sc^{3+} and TolSQ$^{•-}$–(Sc^{3+})$_2$. Source: Yuasa et al. 2008 [420]. Reproduced with permission of John Wiley & Sons.

respectively [420]. The EPR spectrum of TolSQ$^{•-}$–Sc^{3+} (Figure 9.21a) is well reproduced by the computer simulation spectrum with the *hfc* values of a(2H) = 1.85, 0.62 G, and superhyperfine splitting due to one Sc^{3+} ion [a(Sc^{3+}) = 1.63 G] (Figure 9.21b) [420]. In the presence of a high concentration of Sc(OTf)$_3$ (2.1×10^{-1} M), the hyperfine pattern is changed to exhibit splitting due to the additional Sc^{3+} ion (Figure 9.21c) [420]. This is also well reproduced by the computer simulation spectrum (Figure 9.21d) [420]. Thus, the TolSQ$^{•-}$–Sc^{3+} complex is converted to the 1 : 2 complex with Sc^{3+} [TolSQ$^{•-}$–(Sc^{3+})$_2$] in the presence of high concentrations of Sc^{3+}.

The dependence of the observed second-order rate constant (k_{et}) on [Sc^{3+}] for electron transfer from (AcrH)$_2$ to TolSQ at 233 and 298 K is affected by the formation of TolSQ$^{•-}$–Sc^{3+} and TolSQ$^{•-}$–(Sc^{3+})$_2$ depending on the Sc^{3+} concentration. The k_{et} value increases with increasing Sc^{3+} concentration exhibiting a saturated behavior at low concentrations of Sc^{3+} ([Sc^{3+}] $< 5.0 \times 10^{-3}$ M) at both 233 and 298 K as shown in Figure 9.22a,b, respectively [420]. Such saturated dependence of k_{et} on [Sc^{3+}] is ascribed to a 1 : 1 complex formation between TolSQ and Sc^{3+} (TolSQ–Sc^{3+}), which enhances the electron-acceptor ability of TolSQ [420]. The formation constants (K_1) of the TolSQ–Sc^{3+} complex at 298 and 233 K were

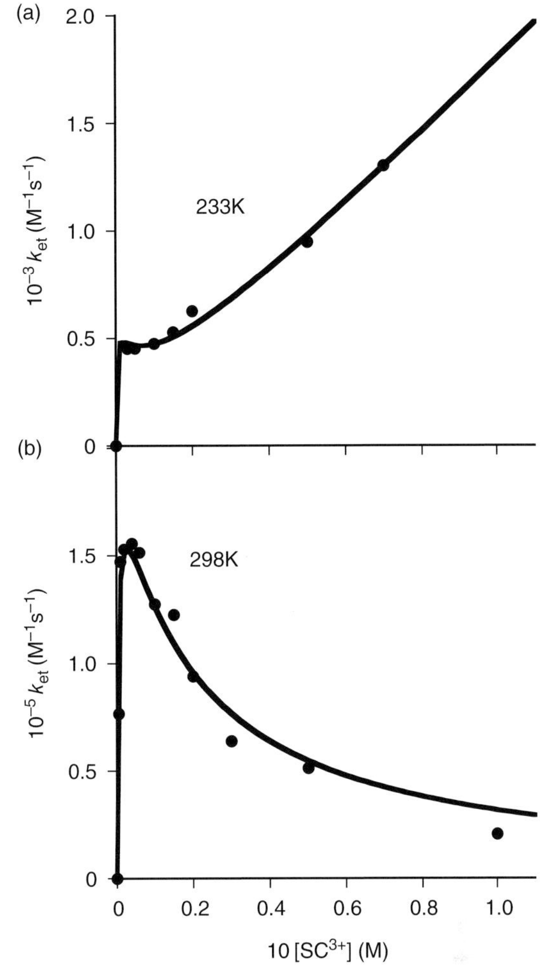

Figure 9.22 Dependence of k_{et} on [Sc^{3+}] for electron transfer from $(AcrH)_2$ (1.0×10^{-5} M) to TolSQ in the presence of Sc^{3+} in deaerated MeCN at (a) 233 K and (b) 298 K. Source: Yuasa et al. 2008 [420]. Reproduced with permission of John Wiley & Sons.

determined from UV–vis spectral changes of TolSQ in the presence of various concentrations of Sc^{3+} in MeCN as $(2.5 \pm 0.1) \times 10^3\ M^{-1}$ and $(9.7 \pm 0.1) \times 10^3\ M^{-1}$ at 298 and 233 K, respectively [420].

The k_{et} value increases further with an increase in [Sc^{3+}] at high concentrations of Sc^{3+} ([Sc^{3+}] $> 5.0 \times 10^{-3}$ M) at 233 K (Figure 9.20a) [420]. In sharp contrast to this, the k_{et} value decreases with increasing [Sc^{3+}] at high concentrations of Sc^{3+} ([Sc^{3+}] $> 5.0 \times 10^{-3}$ M) at 298 K (Figure 9.20b) [420]. The decelerating effect of metal ions (M^{n+}) on the rate of ET normally often results from the complex formation between electron donor and metal ion (D–M^{n+}), which reduces the electron-donor ability to decelerate the ET reaction [422]. However, it was confirmed that Sc^{3+} has no effect on the oxidation potential of $(AcrH)_2$ [420].

The accelerating effect of Sc^{3+} on k_{et} in Figure 9.20 results from two ET pathways to produce the 1 : 1 complex ($TolSQ^{\cdot-}$–Sc^{3+}) and the 1 : 2 complex [$TolSQ^{\cdot-}$–$(Sc^{3+})_2$] as shown in Scheme 9.8 (pathway A and pathway B, respectively) [420]. No ET from $(AcrH)_2$ to TolSQ occurs without Sc^{3+} (vide supra).

In the presence of Sc^{3+}, ET becomes possible by the 1 : 1 complex formation between TolSQ and Sc^{3+} (TolSQ–Sc^{3+}) to afford the TolSQ$^{\bullet -}$–Sc^{3+} complex [420]. With increasing concentration of Sc^{3+}, the 1 : 1 complex (TolSQ$^{\bullet -}$–Sc^{3+}) is converted to the 1 : 2 complex [TolSQ$^{\bullet -}$–$(Sc^{3+})_2$] (pathway B) [420]. In such a case, the ET rate increases with increasing concentration of Sc^{3+}, because additional Sc^{3+} is involved in the transition state of ET to afford the 1 : 2 complex [TolSQ$^{\bullet -}$–$(Sc^{3+})_2$].

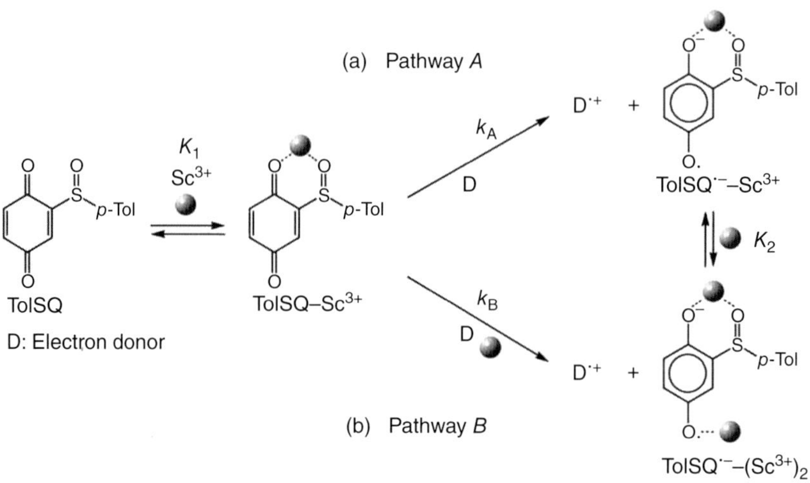

Scheme 9.8 Two pathways in electron transfer from electron donors to TolSQ–Sc^{3+} to produce (a) TolSQ$^{\bullet -}$–Sc^{3+} and (b) TolSQ$^{\bullet -}$–$(Sc^{3+})_2$. Source: Yuasa et al. 2008 [420]. Reproduced with permission of John Wiley & Sons.

According to Scheme 9.8, the activation parameters are different between the pathway A and the pathway B, because the pathway B to afford the 1 : 2 complex [TolSQ$^{\bullet -}$–$(Sc^{3+})_2$] is expected to have a smaller activation enthalpy ($\Delta H^{\neq}$) and a more negative activation entropy ($\Delta S^{\neq}$) due to the second binding of Sc^{3+} with more restricted geometry in the ET transition state as compared to the pathway A to afford the 1 : 1 complex (TolSQ$^{\bullet -}$–Sc^{3+}). This expectation is verified by the Eyring plots in Figure 9.23 (open and closed circles for the pathways A and B, respectively) [420]. The activation enthalpies ($\Delta H^{\neq}$) and entropies ($\Delta S^{\neq}$) are determined from the slopes and the intercepts of the Eyring plots as $\Delta H^{\neq} = 11.6 \pm 0.4$ kcal mol^{-1} and $\Delta S^{\neq} = 3.2 \pm 1.5$ cal mol^{-1} K^{-1} at a low Sc^{3+} concentration (1.0×10^{-2} M) and $\Delta H^{\neq} = 8.3 \pm 0.3$ kcal mol^{-1} and $\Delta S^{\neq} = -9.4 \pm 1.1$ cal mol^{-1} K^{-1} at a high Sc^{3+} concentration (5.0×10^{-2} M) [420]. The larger $\Delta H^{\neq}$ value with a positive $\Delta S^{\neq}$ value at a low Sc^{3+} concentration (1.0×10^{-2} M) corresponds to pathway A [420]. At a high concentration of Sc^{3+} (5.0×10^{-2} M), the contribution of pathway B becomes predominant with the smaller $\Delta H^{\neq}$ value and the more negative $\Delta S^{\neq}$ value because of the second binding of Sc^{3+} to TolSQ$^{\bullet -}$–Sc^{3+} to give TolSQ$^{\bullet -}$–$(Sc^{3+})_2$ (pathway B) [420].

There is a crossing point in the two plots at 263 K (Figure 9.23). As a result, the k_{et} value increases with increasing Sc^{3+} concentration below 263 K, and

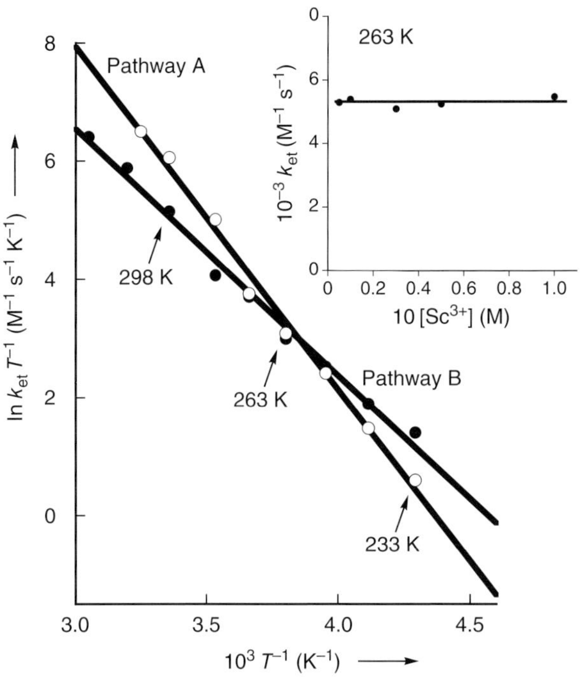

Figure 9.23 Plots of ln($k_{et}T^{-1}$) vs. T^{-1} for electron transfer from $(AcrH)_2$ (1.0×10^{-5} M) to TolSQ in the presence of $Sc(OTf)_3$ (1.0×10^{-2} M (○), 5.0×10^{-2} M (•)) in deaerated MeCN. Inset: Dependence of k_{et} (•) on [Sc^{3+}] for ET from $(AcrH)_2$ (1.0×10^{-5} M) to TolSQ in the presence of Sc^{3+} in deaerated MeCN at 263 K. Source: Yuasa et al. 2008 [420]. Reproduced with permission of John Wiley & Sons.

decreases with increasing Sc^{3+} concentration above 263 K. At the crossing point (263 K), the k_{et} values remain constant with increasing Sc^{3+} concentration (inset of Figure 9.23). Thus, metal ions exhibit both acceleration and deceleration effects on the ET reduction of TolSQ depending on the difference in temperature in relation with binding modes of metal ions to $TolSQ^{\bullet-}$ [420].

9.5 Driving Force Dependence of MCET Rate Constants

In the case of metal ion-promoted ET reactions between D and A, the driving force of electron transfer is altered by addition of metal ions that can bind with $A^{\bullet-}$ (vide supra). In addition, the reorganization energy of electron transfer is expected to be altered in the presence of metal ions, since the binding of metal ions associated with electron transfer certainly requires much larger reorganization energy (λ) than the λ value without metal ion. The ET dynamics between D and A at a fixed distance without metal ion is now well understood in the light of the Marcus theory of electron transfer [6]. Once the driving force ($-\Delta G_{ET}$) and reorganization energy (λ_{DA}) of electron transfer between D and A are determined, the activation free energy of electron transfer ($\Delta G^{\neq}{}_{ET}$) is well predicted using the Marcus equation (Eq. (9.8)). The reorganization energy of electron transfer between D and A (λ_{DA}) is obtained as the

$$\Delta G^{\neq}{}_{ET} = (\lambda_{DA}/4)(1 + \Delta G_{ET}/\lambda_{DA})^2 \tag{9.8}$$

average of the reorganization energy for the electron self-exchange between D and $D^{\cdot+}$ (λ_D) and that between A and $A^{\cdot-}$ (λ_A): Eq. (9.9) [6]. Thus, knowledge of the fundamental redox

$$\lambda_{DA} = (\lambda_D + \lambda_A)/2 \tag{9.9}$$

properties of D (E_{ox} and λ_D) and A (E_{red} and λ_A) is sufficient to predict the $\Delta G^{\neq}{}_{ET}$ values in Eq. (9.8). In other words, the ET reactivity is automatically determined once the combination of D and A is fixed [423]. When the third component such as metal ions are involved in electron transfer from D to A, how can the Marcus analysis be applied to MCET?

The applicability of the Marcus equation (Eq. (9.8)) to MCET reactions has been examined using a ferrocene–naphthoquinone dyad (Fc–NQ) as discussed below (Scheme 9.9) [424, 425]. No electron transfer from ferrocene (Fc) to naphthoquinone (NQ) moiety occurs in Fc–NQ with flexible methylene spacer including an amide linkage thermally in MeCN at 298 K. However, the addition of scandium triflate ($Sc(OTf)_3$: 1.0×10^{-3} M) to an MeCN solution of Fc–NQ (2.0×10^{-3} M) results in the formation of Fc^+ as indicated by the appearance of the absorption band due to Fc^+ at 860 nm with the absorption band at $\lambda_{max} = 420$ nm (Scheme 9.9) [424].

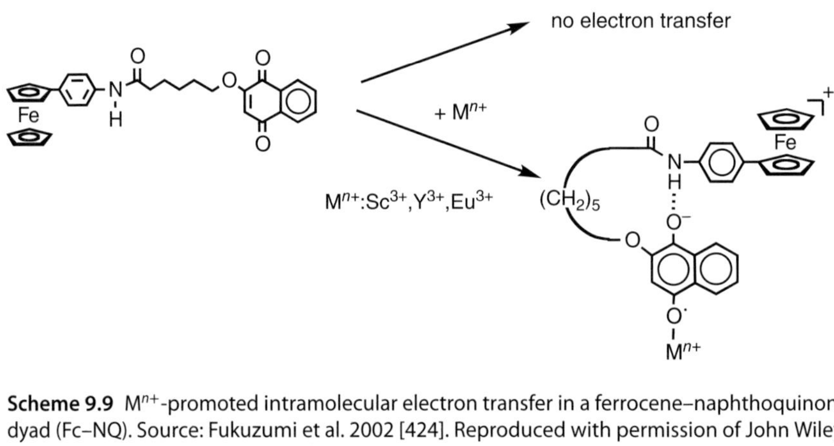

Scheme 9.9 M^{n+}-promoted intramolecular electron transfer in a ferrocene–naphthoquinone dyad (Fc–NQ). Source: Fukuzumi et al. 2002 [424]. Reproduced with permission of John Wiley & Sons.

The drastic change in the free energy change of electron transfer from Fc to NQ by the addition of Sc^{3+} to an MeCN solution of Fc–NQ is shown by the change in the CVs. The one-electron reduction potential of the NQ moiety is observed as a well-defined reversible wave at −0.81 V vs. SCE (Figure 9.24a) [425]. The one-electron oxidation potential for the Fc^+/Fc couple in Fc–NQ agrees with that of ferrocene (E_{ox} vs. SCE = 0.37 V) [386]. The free energy change of electron transfer from Fc to NQ is ΔG_{ET} = 1.18 eV, when the electron transfer is

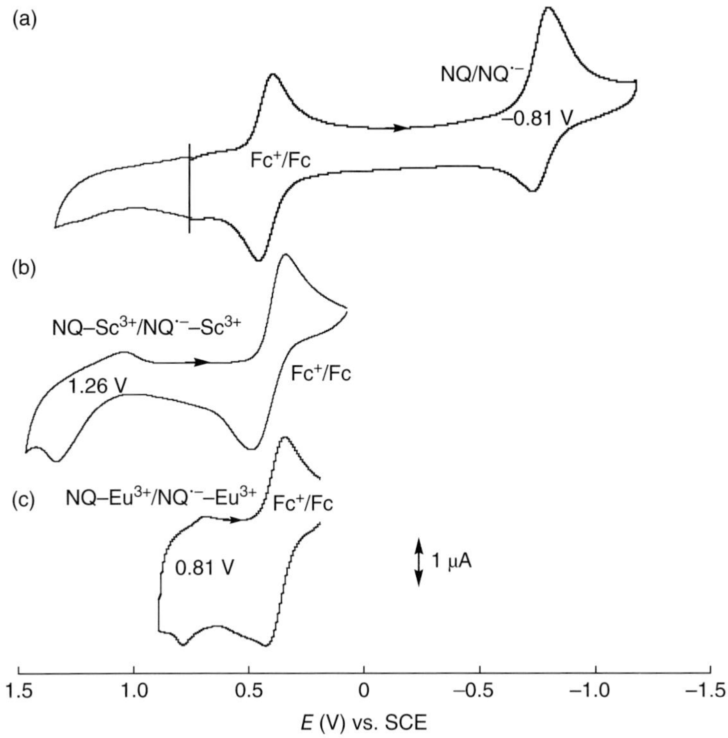

Figure 9.24 (a) CV of Fc–NQ (5.0×10^{-4} M) in deaerated MeCN containing 0.1 M Bu_4NPF_6 at 298 K with sweep rate of 50 mV s^{-1}. (b) CV of Fc–NQ (5.0×10^{-4} M) in the presence of Sc^{3+} (7.0×10^{-3} M) in deaerated MeCN containing 0.1 M Bu_4NPF_6 at 298 K with sweep rate of 50 mV s^{-1}. (c) CV of Fc–NQ (5.0×10^{-4} M) in the presence of Eu^{3+} (7.0×10^{-3} M) in deaerated MeCN containing 0.1 M Bu_4NPF_6 at 298 K with sweep rate of 100 mV s^{-1}. Source: Okamoto et al. 2003 [425]. Reproduced with permission of American Chemical Society.

thermodynamically impossible to occur. In the presence of 7.0×10^{-3} M Sc^{3+}, the E_{red} value exhibits a remarkable positive shift from −0.81 to 1.26 V (vs. SCE), whereas the one-electron oxidation potential of the Fc moiety remains the same irrespective of the absence or presence of Sc^{3+} (Figure 9.24b) [425]. In this case, the free energy change of electron transfer becomes negative: $\Delta G_{ET} = -0.89$ eV. Similar positive shifts of E_{red} are observed in the presence of Eu^{3+} (Figure 9.24c) and Y^{3+} [425]. Thus, electron transfer from Fc to NQ becomes thermodynamically possible by the addition of metal ions to an MeCN solution of Fc–NQ.

The positive shift of E_{red} in the presence of metal ion is ascribed to the binding of metal ion with $NQ^{\cdot-}$ (Eq. (9.10)). In such a case, E_{red} is given as a function of concentration of M^{n+},

$$\mathrm{Fc} - \mathrm{NQ}^{\bullet-} + \mathrm{M}^{n+} \underset{}{\overset{K_{red}}{\rightleftarrows}} \mathrm{F} - \mathrm{NQ}^{\bullet-}/\mathrm{M}^{n+} \tag{9.10}$$

in accordance with Nernst equation (Eq. (9.11)), where ΔE_{red} is the potential shift in the

$$\Delta E_{red} = (2.3RT/F)\ \log\ K_{red}[\mathrm{M}^{n+}] \tag{9.11}$$

presence of M^{n+} from the $E^0{}_{red}$ value in its absence and K_{red} is the formation constant of the Fc–$NQ^{\cdot-}/M^{n+}$ complex under the conditions such that $K_{red}[M^{n+}] \gg 1$ and $K_{ox}[M^{n+}] \ll 1$. The plots of ΔE_{red} vs. log $[M^{n+}]$ ($M^{n+} = Sc^{3+}$, Y^{3+}, and Eu^{3+}) are shown in Figure 9.25 [425]. The slope of each plot is determined as 0.059, which agrees with the expected slope (=2.3RT/F at 298 K) by the Nernst equation (Eq. (9.11)).

The intercepts of linear plots in Figure 9.25 afford the binding constants $K_{red}(Sc^{3+}) = 1.6 \times 10^{37}\ M^{-1}$, $K_{red}(Y^{3+}) = 1.1 \times 10^{31}\ M^{-1}$, $K_{red}(Eu^{3+}) = 3.3 \times 10^{29}\ M^{-1}$ [425]. The K_{red} values correspond to the free energy changes of the metal ion binding: 2.20 eV (Sc^{3+}), 1.83 eV (Y^{3+}), and 1.74 eV (Eu^{3+}) [425]. Such a large binding energy results in a remarkable change in the driving force of electron transfer in Fc–NQ from a highly negative value to a positive value in the presence of M^{n+}. The driving force of electron transfer in the presence of $M^{n+}(-\Delta G_{ET})$ is given by Eq. (9.12), where $-\Delta G^0{}_{ET}$ is the driving force in the absence of M^{n+} [425].

$$-\Delta G_{ET} = -\Delta G^0{}_{ET} + RT\ \ln(K_{red}[M^{n+}]) \tag{9.12}$$

The formation of the Fc^+–$NQ^{\cdot-}/Sc^{3+}$ complex in Scheme 9.9 was confirmed by the EPR spectrum as shown in Figure 9.26a together with the computer simulation spectrum [425]. The observation of superhyperfine splitting due to one Sc^{3+} (a(Sc) = 2.57 G) in Figure 9.26 clearly indicates the formation of the Fc^+–$NQ^{\cdot-}/Sc^{3+}$ complex. The remarkable stability of the Fc^+–$NQ^{\cdot-}/M^{n+}$ complex is ascribed to the hydrogen bonding of the amide proton with one carbonyl oxygen of semiquinone radical anion [425].

The rates of M^{n+}-promoted electron transfer were determined by monitoring the appearance of the absorption band at 420 and 800 nm [425]. The observed second-order rate constants (k_{ET}) increase linearly with increasing $[Sc^{3+}]$, $[Y^{3+}]$, and $[Eu^{3+}]$ [425]. The driving force dependence of log k_{ET} of M^{n+}-promoted electron transfer is shown in Figure 9.27, where the $-\Delta G_{ET}$ values are determined by Eq. (9.12) [425]. There are three separate linear correlations for the case of

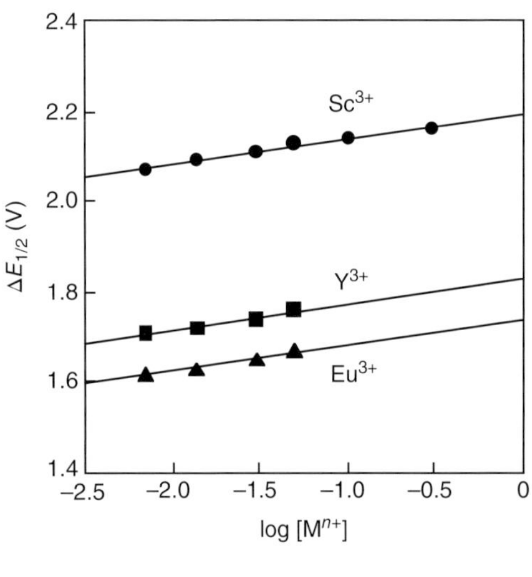

Figure 9.25 Plots of $\Delta E_{1/2}$ vs. log $[M^{n+}]$ for the one-electron reduction of Fc–NQ in the presence of Sc^{3+} (•), Y^{3+} (■), and Eu^{3+} (▲) in MeCN at 298 K. The plot of each $[M^{n+}]$ affords the same slope of 0.059. Source: Okamoto et al. 2003 [425]. Reproduced with permission of American Chemical Society.

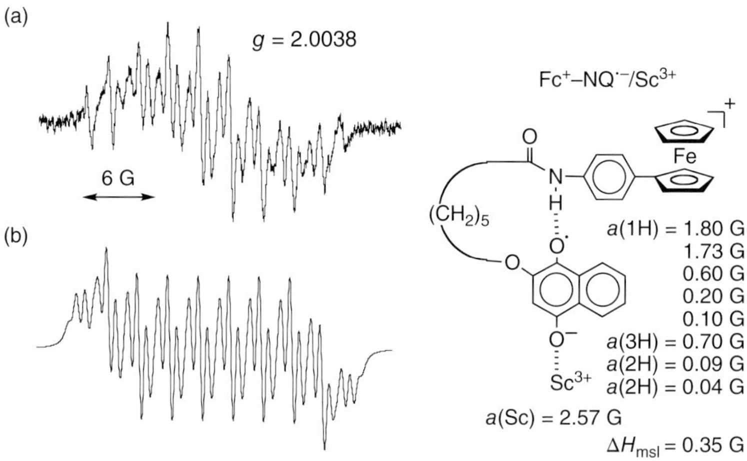

Figure 9.26 (a) EPR spectrum of Fc^+–$NQ^{\bullet -}$/Sc^{3+} (9.1 mM) produced in intramolecular electron transfer of Fc–NQ in the presence of $Sc(OTf)_3$ (20 mM) in deaerated MeCN at 298 K. (b) The computer simulation spectrum with the *hfc* values. Source: Okamoto et al. 2003 [425]. Reproduced with permission of American Chemical Society.

Sc^{3+}, Y^{3+}, and Eu^{3+} (Figure 9.27). The slope of each linear plot is determined to be 16.9 $(eV)^{-1}$, which corresponds to $1/2.3k_BT$ at 298 K, where k_B is the Boltzmann constant. This indicates that the change in driving force with concentration of M^{n+} is directly reflected on the change in the activation free energy, i.e. $\partial(\Delta G^{\neq}{}_{ET})/\partial(\Delta G_{ET}) = 1$. This is quite different from the slope expected from the Marcus equation (Eq. (9.8)): $\partial(\Delta G^{\neq}{}_{ET})/\partial(\Delta G_{ET}) = 0.5$ under the conditions that $-\Delta G_{ET} \ll \lambda_{DA}$. The driving force dependence of k_{ET} expected form the Marcus equation is shown as lines (a)–(c) in Figure 9.27, where three different λ_{DA} values (4.3, 4.4, and 4.5 eV) are used for the calculation using Eq. (9.8) [425]. The $\Delta G^{\neq}{}_{ET}$

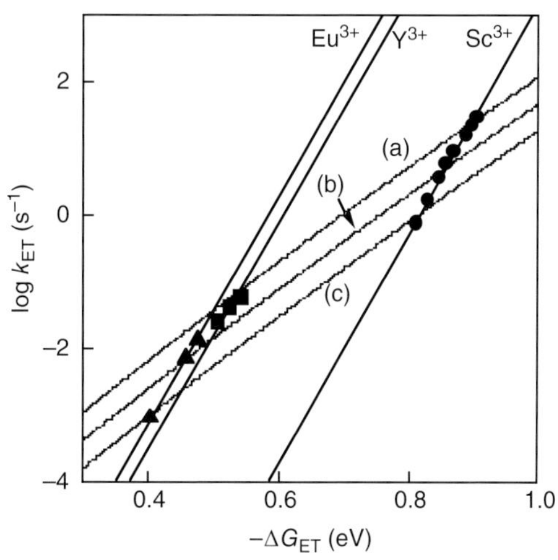

Figure 9.27 Plots of log k_{ET} vs. $-\Delta G_{ET}$ in M^{n+}-promoted intramolecular electron transfer in Fc–NQ in the addition of Sc^{3+} (•), Y^{3+} (■) and Eu^{3+} (▲) in deaerated MeCN at 298 K. The plot of k_{ET} on each $[M^{n+}]$ gives a straight line with the slope of 16.9. The lines (a), (b), and (c) represent the fit to Eq. (9.8) with (a) $\lambda = 4.3$ eV, (b) $\lambda = 4.4$ eV, (c) $\lambda = 4.5$ eV. Source: Okamoto et al. 2003 [425]. Reproduced with permission of American Chemical Society.

value is converted to the corresponding k_{ET} value using Eq. (9.13), by assuming that the M^{n+}-promoted

$$k_{ET} = (k_B T/h)\exp(-\Delta G^{\neq}{}_{ET}/k_B T) \tag{9.13}$$

intramolecular electron transfer in Fc–NQ is adiabatic. From the comparison of the calculated Marcus lines with the observed lines it is seen that the λ_{DA} value changes with concentration of M^{n+}.

Since the metal ion is involved only for the acceptor part, the dependence of λ_A on $[M^{n+}]$ should be responsible for the change in λ_{DA} with M^{n+} concentration. The electron self-exchange between Fc–NQ and the Fc–$NQ^{\cdot -}/M^{n+}$ complex occurs via formation of the Fc–NQ/M^{n+} complex as shown in Scheme 9.10 [425]. According to Scheme 9.10, the electron self-exchange rate constant (k_{ex}) is given by Eq. (9.14), where Z is the frequency factor for the

$$k_{ex} = ZK_{ox}[M^{n+}]\exp(-\lambda^0{}_A/4k_B T) \tag{9.14}$$

intermolecular ET reaction, and λ_{0A} is the reorganization energy for the electron self-exchange between Fc–NQ/M^{n+} and Fc–$NQ^{\cdot -}/M^{i+}$ [425]. Then, the reorganization energy between Fc–NQ and Fc–$NQ^{\cdot -}/M^{n+}$ (λ_A) is given by Eq. (9.15), by comparing Eq. (9.14) with $k_{ex} = Z\exp(-\lambda_A/4k_B T)$ [425]. From Eq. (9.9), the reorganization energy λ_{DA} is given by Eq. (9.16),

$$\lambda_A = \lambda^0{}_A - 4RT\ln(K_{ox}[M^{n+}]) \tag{9.15}$$

$$\lambda_{DA} = \lambda^0{}_{DA} - 2RT\ln(K_{ox}[M^{n+}]) \tag{9.16}$$

where $\lambda^0{}_{DA} = (\lambda_D + \lambda^0{}_A)/2$. This indicates that the λ_{DA} value decreases with increasing $[M^{n+}]$ [425].

When $\lambda_{DA} \gg -\Delta G_{ET}$, Eq. (9.8) is simplified to give Eq. (9.17). Then, the driving force dependence of k_{ET} is derived from Eqs. (9.13), (9.16), and (9.17) as Eq. (9.18),

$$\Delta G^{\neq}{}_{ET} = (\lambda_{DA}/4) + (\Delta G_{ET}/2) \tag{9.17}$$

where C is a constant that is independent of concentration of M^{n+} (Eq. (9.19)) [425]. Thus,

$$-RT\ln(k_{ET}/[M^{n+}]^{1/2}) = C + (\Delta G_{ET}/2) \tag{9.18}$$

Equation (9.18) provides the unified driving force dependence of k_{ET} including $[M^{n+}]$ [425].

$$C = (\lambda^0{}_{DA}/4) - RT\ln[k_B T(K_{ox})^{1/2}/h] \tag{9.19}$$

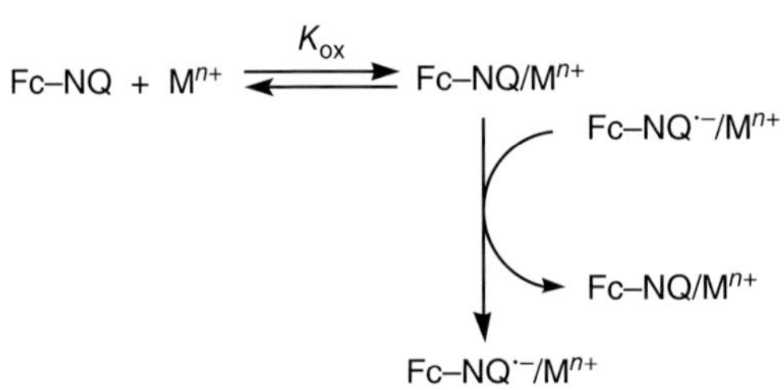

Scheme 9.10 M^{n+}-promoted intramolecular electron transfer in a ferrocene–naphthoquinone dyad (Fc–NQ). Source: Okamoto et al. 2003 [425]. Reproduced with permission of American Chemical Society.

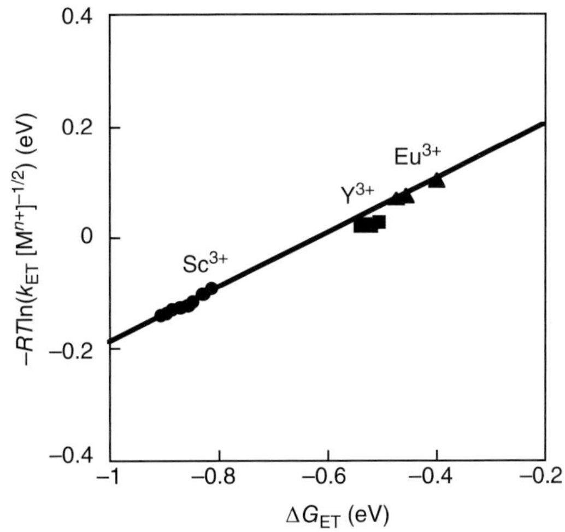

Figure 9.28 Plots of $-RT \ln(k_{ET}[M^{n+}]^{-1/2})$ vs. ΔG_{ET} in M^{n+}-promoted intramolecular electron transfer in Fc–NQ in the presence of Sc^{3+} (●), Y^{3+} (■), and Eu^{3+} (▲) in deaerated MeCN at 298 K. Plot of $-RT \ln(k_{ET}[M^{n+}]^{-1/2})$ vs. ΔG_{ET} on each $[M^{n+}]$ gives a straight line with the slope of 0.5. Source: Okamoto et al. 2003 [425]. Reproduced with permission of American Chemical Society.

A plot of $-RT \ln(k_{ET}/[M^{n+}]^{1/2})$ vs. ΔG_{ET} is shown in Figure 9.28, where a single linear correlation is obtained [425]. The slope is determined as 0.5, which agrees with prediction from Eq. (9.18). The single linear correlation in Figure 9.28 indicates that the C value is constant irrespective of the type of metal ions. The larger the binding between $NQ^{\bullet-}$ and M^{n+}, the larger is the $\lambda^0{}_{DA}$ value and also the larger is the K_{ox} value for the binding between NQ and M^{n+}. In such a case, the effects of different metal ions may be largely canceled in the C value in Eq. (9.19) [425].

As demonstrated above, the driving force dependence of k_{ET} of the metal ion-promoted intramolecular electron transfer can be well evaluated within the context of the Marcus theory of electron transfer in which the driving force of MCET increases with increasing concentrations of metal ions, whereas the reorganization energy of MCET decreases with concentrations of metal ions.

9.6 MCET Coupled with Hydrogen Bonding

The formation of a hydrogen bond between the amide proton and one carbonyl oxygen of $NQ^{\bullet-}$ was indicated in the Fc^+–$NQ^{\bullet-}/M^{n+}$ complex to stabilize the complex (vide supra). Electron-transfer reactions have been believed to be regulated through such non-covalent interactions that play an important role in biological ET systems, where electron donors and acceptors are usually bound to proteins at a fixed distance [426–430]. For example, in the bacterial photosynthetic reaction center (bRC) from *Rhodobacter* (*Rb*) *sphaeroides*, an electron is transferred from the singlet excited state of $(BChl)_2$ via bacteriochlorophyll (BChl) and bacteriopheophytin (Bphe) to the primary quinone (Q_A) and finally to the secondary quinone (Q_B). Although Q_A and Q_B are virtually identical, the difference in the hydrogen bonds with the amino acids makes it possible to transfer an electron from Q_A to Q_B [431]. In this case, the hydrogen bonds can not only provide a structural scaffold but also control the direction of electron transfer. Thus, the hydrogen bond has emerged as the paramount synthon for self-assembly to investigate the ET dynamics in supramolecular electron donor (D) and acceptor (A) ensembles [432, 433]. In general, the effects of hydrogen bonding on thermal intramolecular electron transfer in donor–acceptor linked systems with inert rigid spacers cannot be studied, because the electron transfer would be already over when donor and acceptor molecules are connected if the electron transfer occurred thermally. However, the thermal ET reaction can be started by adding metal ions to the intramolecular ET system, which would otherwise show no reactivity (vide infra). Thus, the quantitative effect of hydrogen bonding on the ET reactivity in metal ion-promoted electron transfer has been examined using a ferrocene–quinone dyad (Fc–Q) linked with an amide group and Fc–(Me)Q dyad, in which the amide proton acting as a hydrogen-bond acceptor, is replaced by the methyl group (vide infra) [434].

The crystal structure of Fc–Q is shown in Figure 9.29a, where the closest distance between the quinone oxygen atom and the amide hydrogen is 2.17 Å and the C—O bond lengths of two carbonyl groups of quinone are eventually the same (1.22 Å) [434]. These results indicate that there is no hydrogen bonding between the quinone oxygen atom and the amide proton in the neutral form in Fc–Q. In the reduced form, however, the hydrogen bond may be formed as suggested by the calculated structure. The optimized geometry of Ph–$Q^{\cdot -}$ in which Fc is omitted using the Amsterdam density function (ADF) calculation with the V (small) basis set is shown in Figure 9.29b [434]. The O–H distance between the quinone oxygen atom of $Q^{\cdot -}$ and the amide hydrogen in Figure 9.29b is 1.56 Å, which is much shorter than the distance in the X-ray structure of neutral Fc–Q (Figure 9.29a) [434]. This value is even shorter than the hydrogen bonding distance between the semiquinone radical anion and water (1.78 Å) [435]. The C—O bond length of the hydrogen-bonded carbonyl group (1.28 Å) becomes longer than the bond length of the other carbonyl group (1.26 Å) due to the weakening of the C—O bond by the hydrogen bonding with the amide proton. For comparison, the optimized geometry of Ph–(Me)$Q^{\cdot -}$ in which the N–H group is replaced by the N–Me group is shown in Figure 9.29c [434]. There is no significant structural difference between Ph–$Q^{\cdot -}$ and Ph–(Me)$Q^{\cdot -}$ except for the absence of the hydrogen bond in the latter case.

The formation of the hydrogen bond between the quinone oxygen atom and the amide proton in the reduced form (Fc–$Q^{\cdot -}$) is observed in photoinduced electron transfer [436]. Photoexcitation of the Q moiety in Fc–Q in deaerated PhCN with 388 nm femtosecond (150 fs width) laser light results in the appearance of a new absorption band (λ_{max} = 580 nm) at 1 ns after the laser excitation [436]. The absorption band at 580 nm is significantly red-shifted as compared to the diagnostic absorption band of semiquinone radical anion at 422 nm [437, 438], and this is assigned to $Q^{\cdot -}$, which is hydrogen bonded to the amide proton of the spacer [436]. Such a red-shift has been observed when semiquinone radical anion is bound to a hard acid such as Mg^{2+} [385]. No such absorption band was observed in the case of Fc–(Me)Q, which has no hydrogen bond acceptor [436]. The investigation of the photodynamics revealed that electron transfer from Fc to the singlet excited state of Q occurs rapidly to produce Fc–$Q^{\cdot -}$ without changing the conformation (<1 ps) and then $Q^{\cdot -}$ forms the hydrogen bonding with the

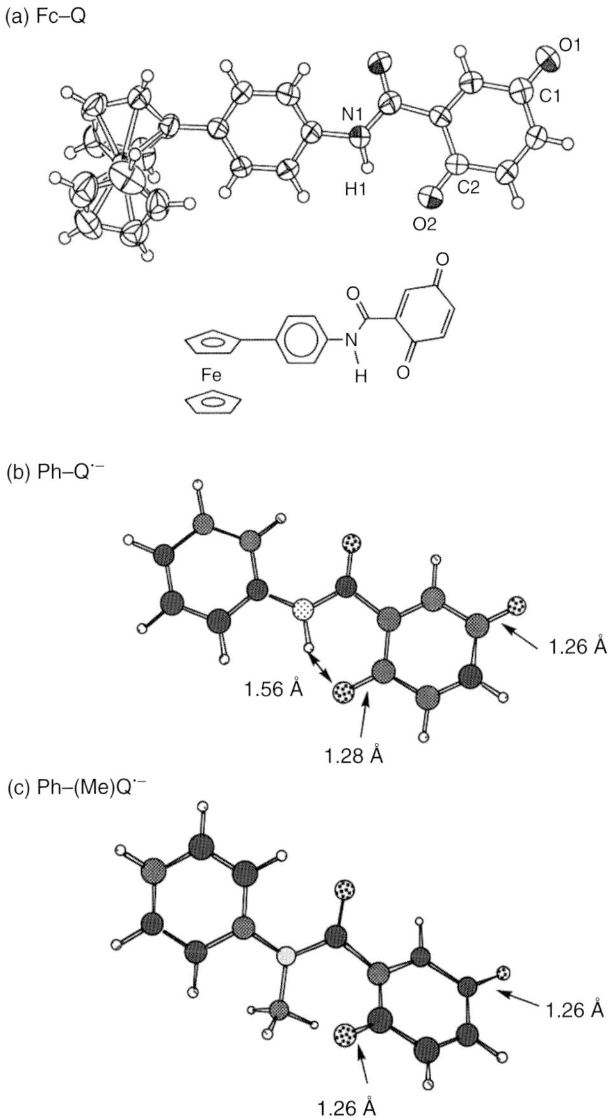

Figure 9.29 (a) ORTEP drawing of Fc–Q. (b) Optimized structure of Ph–Q$^{\bullet-}$. (c) Optimized structure of Ph–(Me)Q$^{\bullet-}$. The optimized structure is obtained by the ADF calculation with the V (small) basis set. Source: Fukuzumi et al. 2002 [436]. Reproduced with permission of American Chemical Society.

amide proton of the spacer ($\tau = \sim 5$ ps) and the resulting radical ion pair decays via a back electron transfer to the ground state as shown in Scheme 9.11 [436].

The effect of the hydrogen bonding on the one-electron reduction potential (E_{red}) of Q in Fc–Q was examined by the CVs of Fc–Q and Fc–(MeQ). The CV of Fc–Q (Figure 9.30a) exhibits two reversible one-electron redox couples at 0.39 V and −0.16 V (vs. SCE) due to the Fc^+/Fc couple and the

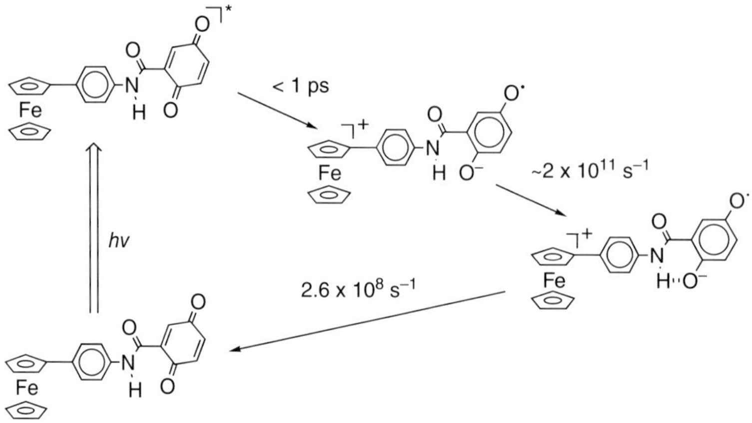

Scheme 9.11 Photoinduced electron transfer from Fc to Q in the Fc–Q dyad, followed by hydrogen bond formation between $Q^{\bullet-}$ and the amide proton in Fc–$Q^{\bullet-}$. Source: Fukuzumi et al. 2002 [436]. Reproduced with permission of American Chemical Society.

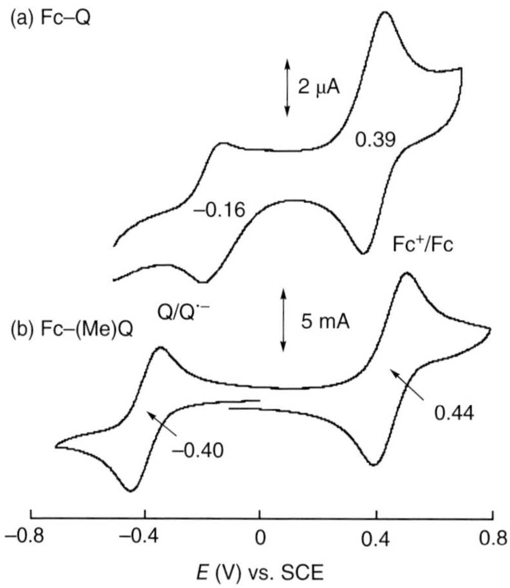

Figure 9.30 CVs of (a) Fc–Q (0.5 mM) and (b) Fc–(Me)Q (0.5 mM) in MeCN containing 0.1 M Bu_4NPF_6. Source: Fukuzumi et al. 2003 [434]. Reproduced with permission of American Chemical Society.

$Q/Q^{\bullet-}$ couple, respectively [434]. The one-electron reduction potential of Q (E_{red} = −0.16 V) is significantly shifted to a positive direction as compared to *p*-benzoquinone (−0.50 V) and this is more positive than the E_{red} value of *p*-benzoquinone possessing an electron-withdrawing substituent (−0.38 V for chloro-*p*-benzoquinone) [400]. The E_{red} value of Q (−0.40 V) in Fc–(Me)Q (Figure 9.30b) is significantly more negative as compared to Fc–Q (−0.16 V) [434]. Such a negative shift in the E_{red} value of Q in Fc–(Me)Q as compared to Fc–Q indicates that the radical anion ($Q^{\bullet-}$) is stabilized by the hydrogen bonding with the amide proton of the spacer.

The effect of hydrogen bonding on metal ion-promoted electron transfer from Fc to Q in Fc–Q can be quantitatively evaluated by comparing the rate constants of Fc–Q with those of Fc–(Me)Q) in which no hydrogen bond is formed [434]. Without metal ions no thermal electron transfer from Fc to Q and MeQ occurs in Fc–Q and Fc–(MeQ), respectively [434]. However, addition of Mg^{2+} (0.10 M) results in the formation of Fc^+ as indicated by the appearance of the absorption band due to Fc^+ at 800 nm together with the absorption band at 420 nm [434]. The absorption band at 420 nm, which may be assigned to $Q^{\cdot -}$ bound to Mg^{2+}, is blueshifted as compared to non-hydrogen bonded $Q^{\cdot -}$ (450 nm) and hydrogen bonded $Q^{\cdot -}$ (580 nm) [434]. Such a significant blueshift has been observed when $Q^{\cdot -}$ is bound to two Mg^{2+} ions: the absorption maximum of $Q^{\cdot -}$–Mg^{2+} complex (590 nm) is blueshifted to 410 nm for $Q^{\cdot -}$–$(Mg^{2+})_2$ [385].

The ET rates were determined by monitoring the appearance of the absorption band at 800 nm, obeying first-order kinetics [434]. The first-order rate constant increases linearly with increasing concentration of Mg^{2+}. The second-order rate constant (k_{et}) is determined from the slope of the linear plot of k_{obs} vs. $[Mg^{2+}]$ to be $1.4 \times 10^3\ M^{-1}\ s^{-1}$. When Fc–Q is replaced by Fc–(Me)Q, which contains no hydrogen-bond acceptor, the k_{et} value ($0.4\ M^{-1}\ s^{-1}$) is much smaller than the k_{et} value ($1.4 \times 10^3\ M^{-1}\ s^{-1}$) of the electron transfer in Fc–Q, which is promoted not only by Mg^{2+} but also by the hydrogen bonding between the $Q^{\cdot -}$ moiety and the amide proton. A variety of metal ions (M^{n+}: triflates salts) can also promote electron transfer from Fc to Q in Fc–(Me)Q (Scheme 9.12) [434].

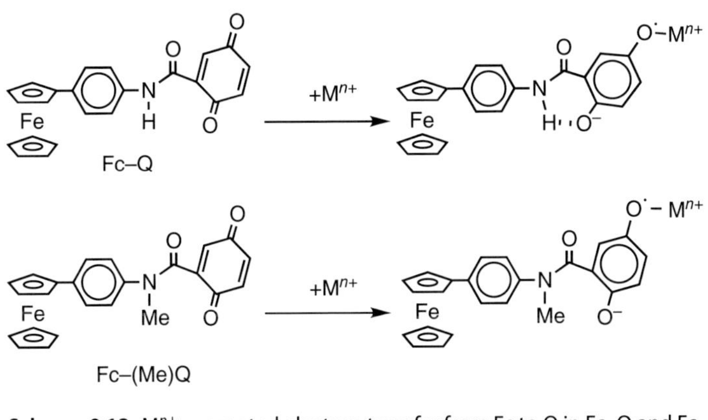

Scheme 9.12 M^{n+}-promoted electron transfer from Fc to Q in Fc–Q and Fc–(Me)Q. Source: Okamoto et al. 2003 [434]. Reproduced with permission of American Chemical Society.

The promoting effects of metal ions vary significantly depending on the Lewis acidity of metal ions. The Lewis acidity of metal ions has been determined as the ΔE values of $O_2^{\cdot -}/M^{n+}$ derived from the g_{zz}-values [201]. Figure 9.31 shows plots of $\log k_{et}$ vs. ΔE for the M^{n+}-promoted intramolecular electron transfer in Fc–Q and Fc–(Me)Q [434]. The k_{et} value of the Sc^{3+}-promoted electron transfer in Fc–Q is the largest among metal ions investigated and this is 10^4 times larger than the corresponding k_{et} value of Fc–(Me)Q [434]. The $\log k_{et}$ values of M^{n+}-promoted electron transfer in Fc–Q are linearly correlated with the ΔE values and each value is c. 10^4 times larger than the k_{et} values of Fc–(Me)Q [434]. The stabilization of the $Q^{\cdot -}$ moiety by hydrogen bonding with the amide proton in Fc–$Q^{\cdot -}$ results in the positive shift in E_{red} of Fc–Q as compared to Fc–(Me)Q in which the amide proton is replaced by the methyl group (Figure 9.30) [434].

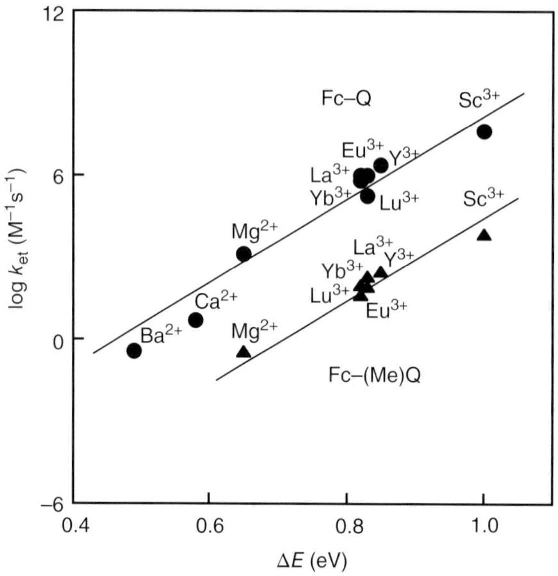

Figure 9.31 Plots of log k_{et} vs. ΔE in M^{n+}-promoted electron transfer in Fc–Q and Fc–(Me)Q in MeCN 298 K. Source: Okamoto et al. 2003 [434]. Reproduced with permission of American Chemical Society.

The 10^4 times difference in the k_{et} values corresponds to the difference in the one-electron reduction potential between Fc–Q (E_{red} vs. SCE = −0.16 V) and Fc–(Me)Q (E_{red} vs. SCE = −0.40 V), since the ratio of the rate constant is given by $\exp(0.24\,\text{eV}/k_BT)$, which is equal to 1.1×10^4 at 298 K [434]. Thus, the accelerating effect of hydrogen bonding on the rate of M^{n+}-promoted electron transfer in Fc–Q has been quantitatively clarified. In contrast to the photoinduced electron transfer in Fc–Q in Scheme 9.11, M^{n+}-promoted electron transfer in Fc–Q is coupled with the formation of the hydrogen bonding.

9.7 MCET Catalysis

9.7.1 Hydride Transfer vs. Cycloaddition

There are a variety of redox reactions in which electron transfer is involved as the rate-determining step. As demonstrated above, the ET step can be promoted by binding of metal ions to radical anions of electron acceptors. Thus, thermal and photochemical redox reactions, which would otherwise be unlikely to occur, are made possible to proceed efficiently by the catalysis of metal ions in the ET steps. Among such examples, the catalytic effects of metal ions on hydride transfer reactions from NADH analogs to substrates have particularly attracted considerable interest in relation to the essential role of metal ions in the redox reactions of dihydronicotinamide coenzymes in the native enzymatic system [438–441]. Hydride transfer from NADH analogs to substrates proceeds efficiently via metal ion-promoted electron transfer from NADH analogs to substrates [440–443]. On the other hand, cycloaddition reactions such as Diels–Alder reaction of anthracenes with p-benzoquinone derivatives are reported to proceed efficiently via metal ion-promoted electron transfer from anthracenes to p-benzoquinone derivatives [385, 444]. If both the hydride transfer and cycloaddition reactions can proceed via the same metal ion-promoted ET process, there may be the case in which the metal ion-catalyzed hydride transfer pathway competes with the cycloaddition pathway. Such an example has been shown in metal ion-catalyzed reactions of NADH analogs with *p*-benzoquinone (Q) (vide infra) [445].

Novel and efficient [2+3] cycloaddition reactions of NADH analogs with *p*-benzoquinone derivatives rather than the hydride transfer reactions occur in the presence of scandium triflate ($Sc(OTf)_3$) in MeCN to afford the cycloadducts [445]. When 1-benzyl-4-*t*-butyl-1,4-dihydronicotinamide (*t*-BuBNAH) is used as an NADH analog in the Sc^{3+}-catalyzed reaction with Q, the crystal structure of the cycloadduct has been determined successfully as shown in Figure 9.32 [445].

The dependence of the observed second-order rate constant (k_{obs}) on [Sc^{3+}] was examined for the cycloaddition reactions of *t*-BuBNAH with Q at various concentrations of Sc^{3+} [445]. The k_{obs} value increases with an increase in [Sc^{3+}] to exhibit a first-order dependence on [Sc^{3+}] at low concentrations, changing to a second-order dependence at high concentrations as shown in Figure 9.33a [445]. Virtually the same dependence of the rate constant (a mixture of first-order

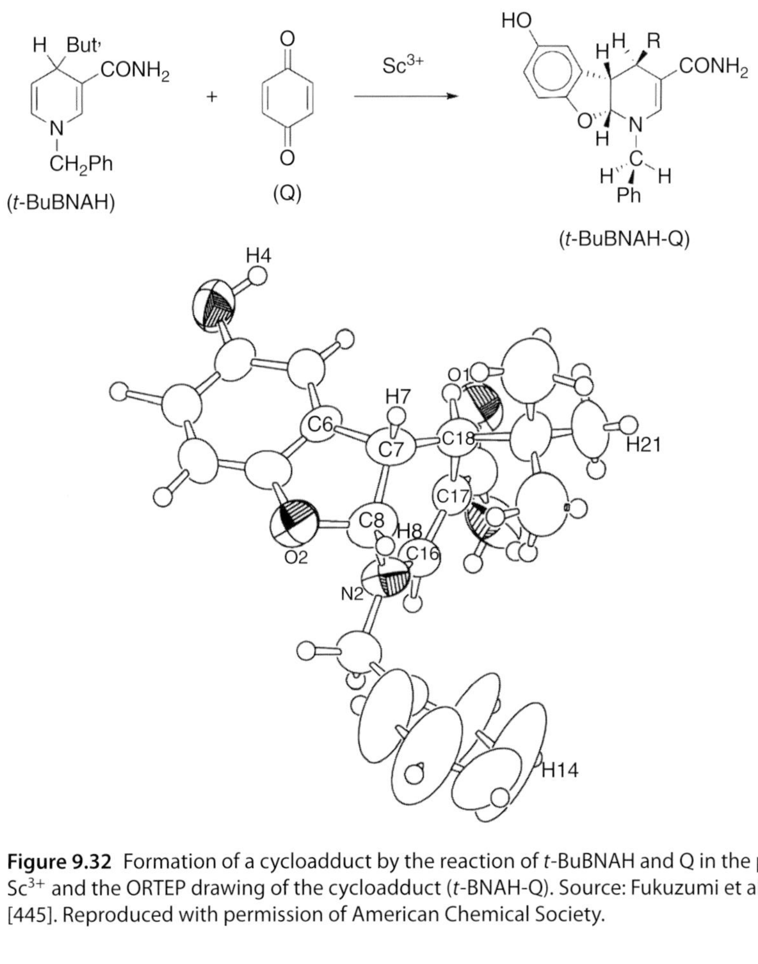

Figure 9.32 Formation of a cycloadduct by the reaction of *t*-BuBNAH and Q in the presence of Sc^{3+} and the ORTEP drawing of the cycloadduct (*t*-BNAH-Q). Source: Fukuzumi et al. 2001 [445]. Reproduced with permission of American Chemical Society.

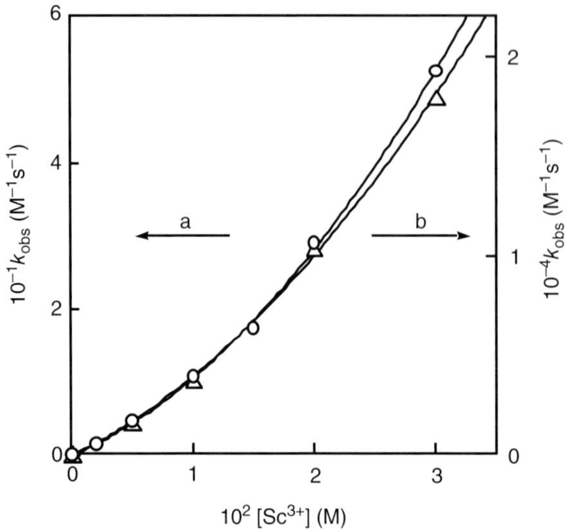

Figure 9.33 Dependence of k_{obs} on [Sc^{3+}] for (a) the cycloaddition reaction of *t*-BuBNAH (1.0×10^{-4} M) with Q (2.0×10^{-3} M) and (b) electron transfer from (TPP)Co (8.0×10^{-6} M) to Q (2.6×10^{-4} M) in the presence of $Sc(OTf)_3$ in deaerated MeCN at 298 K. Source: Fukuzumi et al. 2001 [445]. Reproduced with permission of American Chemical Society.

and second-order dependence) on [Sc^{3+}] is observed in electron transfer from (TPP)Co to Q (Figure 9.33b) [445]. This indicates that the cycloaddition of the NADH analog with Q proceeds via a rate-determining electron transfer from the NADH analog to Q, which is accelerated by the formation of the 1 : 1 and 1 : 2 complex between $Q^{\cdot-}$ and Sc^{3+} as in the case of an electron transfer from (TPP)Co to Q [445].

When Sc^{3+} is replaced by other metal ions such as Lu^{3+} [$Lu(OTf)_3$], Y^{3+} [$Y(OTf)_3$], or Mg^{2+} [$Mg(ClO_4)_2$] in the reaction of *t*-BuBNAH with Q, BNA^+ is formed together with the cycloadduct [445]. In the case of Mg^{2+}, the cycloaddition is no longer the main reaction but a hydride transfer from *t*-BuBNAH to Q occurs to yield *t*-BNA^+ as the major product (60%) [445]. When *t*-BuBNAH is replaced by BNAH in the reaction with Q catalyzed by metal ions other than Sc^{3+}, a hydride transfer from BNAH to Q becomes the main reaction and the yield of BNA^+ increases in the order Lu^{3+} (67%) < Y^{3+} (79%) < Mg^{2+} (93%) with a decrease in the Lewis acidity of the metal ion [201].

The Mg^{2+}-catalyzed hydride transfer from BNAH to Q is known to proceed via Mg^{2+}-promoted electron transfer from BNAH to Q, followed by a proton transfer from the resulting $BNAH^{\cdot+}$ to the $Q^{\cdot-}/(Mg^{2+})^2$ complex and the subsequent fast electron transfer from $BNA^{\cdot}$ to QH^-/Mg^{2+} [422, 446]. The change in the type of reaction depending on the Lewis acidity of the metal ion can be well explained by the ET mechanism in Scheme 9.13 [445]. The initial rate-determining electron transfer from BNAH to Q results in the formation of the radical ion pair ($BNAH^{\cdot+}$ and $Q^{\cdot-}$), where $Q^{\cdot-}$ forms 1 : 1 and 1 : 2 complexes with Sc^{3+}. This is followed by fast radical coupling between $Q^{\cdot-}$ and $BNAH^{\cdot+}$ to give a zwitterionic intermediate, which is eventually converted to the cycloadduct. Because there is a large spin density ($\rho = 0.466$) at C(5) of $BNAH^{\cdot+}$, the radical coupling of C(5) of $BNAH^{\cdot+}$ with $Q^{\cdot-}/(M^{n+})_2$ may occur efficiently (pathway (a)) in competition with the proton transfer from $BNAH^{\cdot+}$ to $Q^{\cdot-}/(M^{n+})_2$, which leads to the overall hydride transfer from BNAH to Q (pathway (b)) [445]. With decreasing Lewis acidity of the metal ion, the metal ion-promoted electron transfer from BNAH to Q is decelerated because of the weaker binding of the metal ion with $Q^{•-}$, while the proton transfer from $BNAH^{•+}$ to $Q^{•-}/(M^{n+})_2$ is accelerated due to the stronger basicity of $Q^{•-}/(M^{n+})_2$ [445]. This is the reason why the hydride transfer pathway from BNAH to Q becomes dominant in the presence of a weaker Lewis acid (e.g. Mg^{2+}) as compared to the selective cycloaddition reaction in the presence of Sc^{3+}. Thus, the hydride transfer pathway is changed to the cycloaddition pathway through the metal ion-promoted electron transfer from NADH analogs to *p*-benzoquinone derivatives depending on the Lewis acidity of the metal ion (Scheme 9.13) [445].

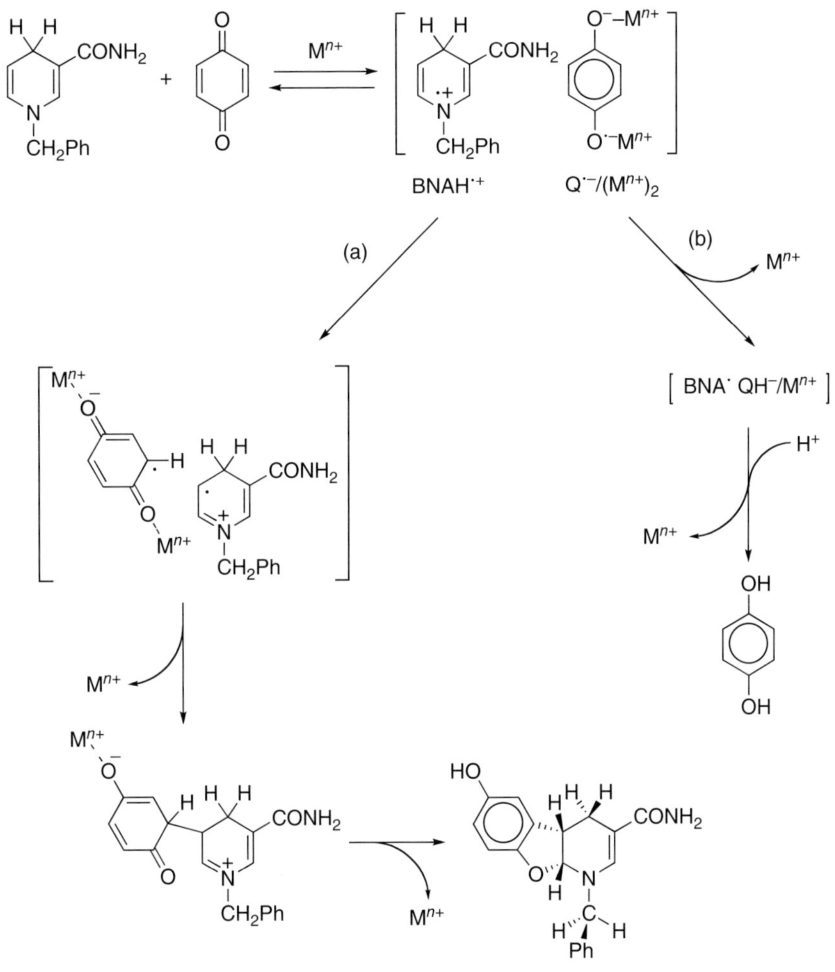

Scheme 9.13 (a) Cycloaddition and (b) hydride transfer pathways via M^{n+}-promoted electron transfer from BNAH to Q. Source: Fukuzumi et al. 2001 [445]. Reproduced with permission of American Chemical Society.

9.7.2 Suproxode Disumutase (SOD) Models

In the case of metal ion-promoted hydride transfer and cycloaddition reactions described above, binding of two metal ions to radical anions of electron acceptors can accelerate electron transfer from electron donors to acceptors, leading to more efficient catalysis of metal ions. In such cases, metal ions are not redox active, acting as Lewis acids to bind with radical anions that act as bases. On the other hand, a redox-active metal center is often associated with another metal center that can accelerate the redox processes in biological systems [447]. An important example of such systems is copper–zinc superoxide dismutase (Cu,Zn-SOD), which contains an imidazolate-bridged Cu^{II}–Zn^{II} bimetallic center in its active site [448–451]. This enzyme protects cells against oxidative damage by catalyzing the disproportionation (dismutation) of toxic superoxide ion ($O_2^{\bullet -}$) to dioxygen (O_2) and hydrogen peroxide (H_2O_2) [452]. Under physiological circumstances the disproportionation of $O_2^{\bullet -}$ must proceed very rapidly, i.e. at the diffusion-limited rate, in order to prevent the uncontrolled oxidations of cells by $O_2^{\bullet -}$ and/or its conjugated acid $HO_2^{\bullet}$ [453, 454]. The copper ion is coordinated to four imidazole N atoms of histidine residues in a distorted square-pyramidal geometry while the zinc ion located at a distance of 6.2 Å from the copper ion is coordinated to the carboxylato O atom of an aspartic acid residue and three imidazole N atoms of histidine residues in a distorted tetrahedral structure [448, 449]. A number of suproxode disumutase (SOD) model complexes have been designed and synthesized to provide valuable insights into the structure and catalytic function of the Cu,Zn-SOD active site [455–458]. The most important question to be answered is how the Cu,Zn-SOD active site can accelerate both the oxidation and reduction of $O_2^{\bullet -}$, required for the rapid disproportionation. Such a dual role would be made possible by the bimetallic system, since any monometallic system can only activate either the oxidation or the reduction of $O_2^{\bullet -}$. One metal site can act as a redox-active site and the other acts as a Lewis acid to control the redox reaction.

The important role of Zn^{II} ion in the bimetallic system to activate both the oxidation and reduction of $O_2^{\bullet -}$ has been implicated by a well-characterized SOD model, which is an imidazolate-bridged Cu^{II}–Zn^{II} heterodinuclear complex containing a dinucleating ligand, Hbdpi (Hbdpi = 4,5-bis(di(2-pyridylmethyl)aminomethyl)imidazole) [458]. The crystal structure of a Cu^{II}–Zn^{II} heterodinuclear complex $[CuZn(bdpi)(MeCN)_2]^{3+}$ and a Cu^{II} homodinuclear complex $[Cu_2(bdpi)(MeCN)_2]^{3+}$ are shown in Figure 9.34 [458]. The Cu^{II}–Zn^{II} distance of 6.197(2) Å in $[CuZn(bdpi)(MeCN)_2]^{3+}$ agrees well with that of native Cu,Zn-SOD (6.2 Å). Each metal has a pentacoordinate geometry with the imidazolate nitrogen, two pyridine nitrogens, the tertiary amine nitrogen, and a solvent (MeCN or H_2O). The coordination site occupied by a solvent can be susceptible for ligand substitution, providing a binding site for substrate superoxide. The Cu^{II}–Zn^{II} heterodinuclear complex has a structure that is essentially the same as the corresponding Cu^{II}–Cu^{II} homonuclear complex. The τ value of the Cu center of $[CuZn(bdpi)(MeCN)_2]^{3+}$ is obtained as $\tau_1 = 0.84$ in the Cu^{II} site and as $\tau_2 = 0.91$ in the Zn^{II} site, where α_1 = N(1)–Cu(1)–N(5) (122.6°), β_1 = N(3)–Cu(1)–N(9) (172.9°), α_2 = N(2)–Zn(1)–N(6) (120.5°), β_2 = N(4)–Zn(1)–N(10) (175.1°), respectively [458]. Thus, the coordination

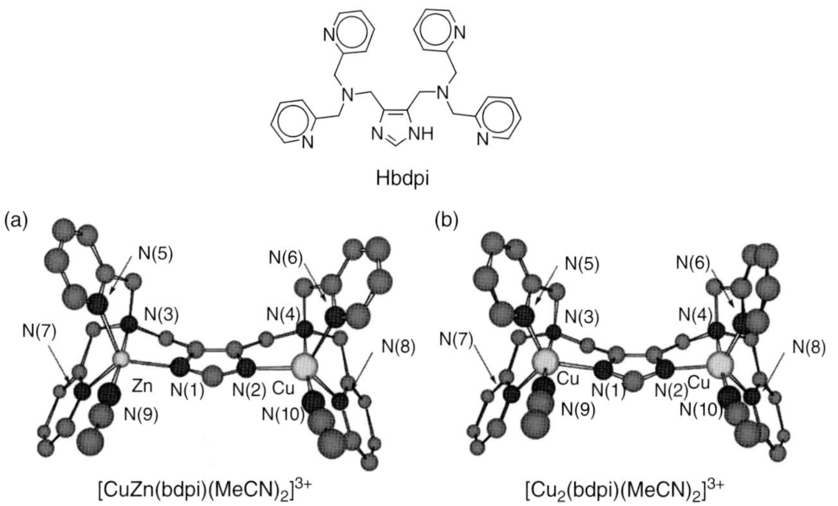

Figure 9.34 Crystal structures of (a) $[Cu^{II}Zn^{II}(bdpi)(MeCN)_2](ClO_4)_3$·2MeCN and (b) $[Cu_2(bdpi)(MeCN)_2](ClO_4)_3$·MeCN·$3H_2O$. The hydrogen atoms are omitted for clarity. Source: Ohtsu et al. 2000 [458]. Reproduced with permission of American Chemical Society.

geometry of Cu^{II} and Zn^{II} in the Cu^{II}–Zn^{II} heterodinuclear complex can be described as slightly distorted trigonal bipyramidal [458].

The EPR spectrum of the Cu^{II}–Zn^{II} heterodinuclear complex gave the well-defined EPR parameters ($g_{\parallel} = 2.10$, $g_{\perp} = 2.24$, $|A_{\parallel}| = 117$ G, and $|A_{\perp}| = 124$ G), which indicate that the Cu^{II} ion in the complex has a trigonal bipyramidal environment and a $d_z{}^2$ ground state [458]. The observation of well-defined EPR spectrum also confirms that the complex retains its imidazolate-bridged Cu^{II}–Zn^{II} heterodinuclear structure and that it is not a 1 : 1 mixture of Cu^{II}–Cu^{II} and Zn^{II}–Zn^{II} homodinuclear complexes.

A large positive shift (0.21 V) in the E_{red} value of the Cu^{II}–Zn^{II} complex is observed as compared to the corresponding Cu^{II} mononuclear complexes as shown in Figure 9.35 [458]. Such a positive shift of the E_{red} value of the Cu^{II}–Zn^{II} complex can be ascribed to the electronic effect of the imidazolate-bound Zn^{II} ion, which leads to a decrease in the electron density on the copper ion. This indicates that an important role of Zn^{II} ion in the imidazolate-bridged Cu^{II}–Zn^{II} complex is to accelerate outer-sphere electron transfer from $O_2{}^{\bullet-}$ to produce the Cu^{I}–Zn^{II} complex, when the free energy change of electron transfer becomes thermodynamically more favorable as compared to that without Zn^{II} ion [458].

The presence of Zn^{II} ion, which can act as a Lewis acid, also enables to accelerate electron transfer from the Cu^{I}–Zn^{II} complex to $O_2{}^{\bullet-}$, since $O_2{}^{\bullet-}$ can form a complex with metal ions acting as a Lewis acid to accelerate the ET reduction of $O_2{}^{\bullet-}$ (vide supra) [201]. The formation of $O_2{}^{\bullet-}$–Zn^{II} complex is confirmed using Zn^{II} complexes, $[Zn^{II}(MeIm(Py)_2)\text{–}(MeCN)](ClO_4)_2$ and $[Zn^{II}(MeIm(Me)_2)(MeCN)](ClO_4)_2$ ($MeIm(Py)_2$ = (1-methyl-4-imidazolyl-methyl)bis(2-pyridylmethyl)amine, $MeIm(Me)_2$ = (1-methyl-4-imidazolyl-methyl)bis(6-methyl-2-pyridylmethyl)amine), as shown in Scheme 9.14 [459]. The EPR spectra of the $O_2{}^{\bullet-}$–Zn^{II} complexes measured at 133 K (Figure 9.36)

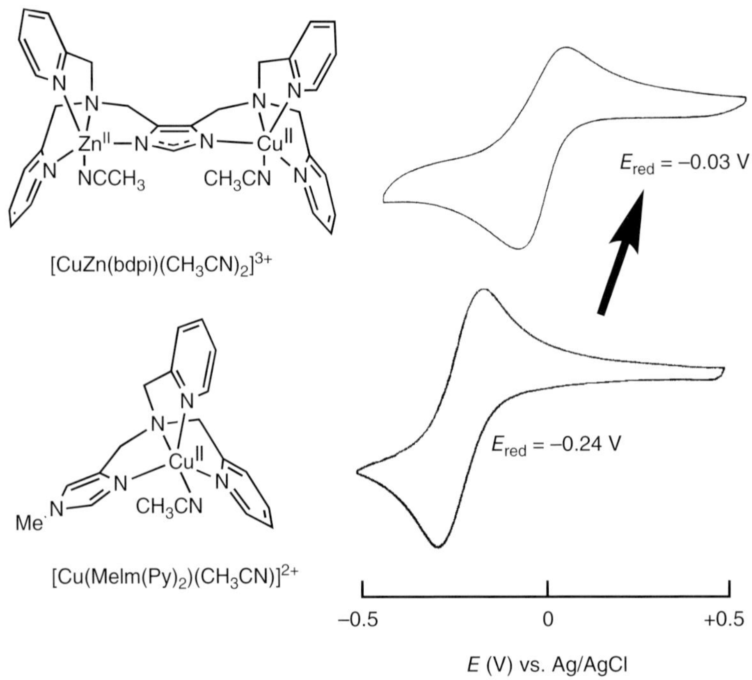

Figure 9.35 CVs of (a) $[CuZn(bdpi)(MeCN)_2]^{3+}$ (1.0 mM) and (b) $[Cu(MeIm(Py)_2(MeCN))^{2+}$ (1.0 mM) in MeCN containing 0.1 M TBAP; working electrode: grassy-carbon, counter electrode: Pt wire, reference electrode: Ag/AgCl, scan rate 50 mV s^{-1}. Source: Kaim and Schwederski 1994 [447]. Reproduced with permission of John Wiley & Sons.

gave the g_{zz}-values, which are significantly smaller than the value of free $O_2^{\bullet-}$ due to the complexation of Zn^{II} ion with $O_2^{\bullet-}$ [459].

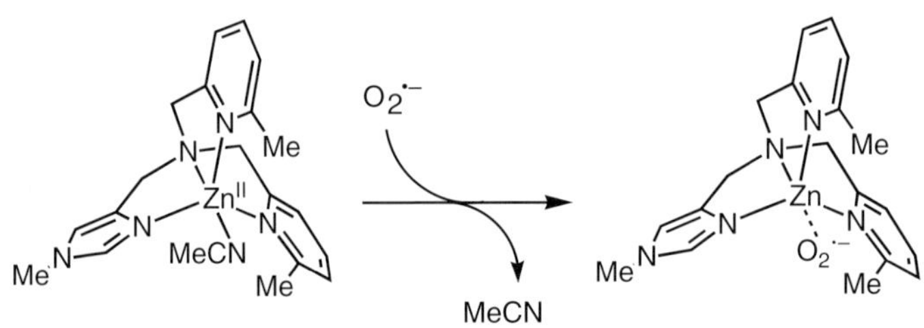

$[Zn^{II}(MeIm(Me)_2)(MeCN)]^{2+}$

Scheme 9.14 Formation of the $O_2^{\bullet-}$ complex with $[Zn^{II}(MeIm(Me)_2)]^{2+}$. Source: Ohtsu et al. 2001 [459]. Reproduced with permission of John Wiley & Sons.

The ΔE values of the $O_2^{\bullet-}$–Zn^{II} complexes have been evaluated from deviation of the g_{zz}-values from the free spin value [201]. The ΔE values of the $O_2^{\bullet-}$ complex with $[Zn^{II}(MeIm(Me)_2)]^{2+}$ (0.87 eV) and $[Zn^{II}(MeIm(Py)_2)]^{2+}$ (0.85 eV) are significantly larger than those of $O_2^{\bullet-}$ complexes with other divalent metal ions (Mg^{II} ion: 0.65 eV, Ca^{II} ion: 0.58 eV, Sr^{II} ion: 0.52 eV, and Ba^{II} ion: 0.49 eV) [201], reflecting the strong Lewis acidity of Zn^{II} ion as compared to other divalent metal ions [459]. The larger ΔE value of the $O_2^{\bullet-}$ complex with $[Zn^{II}(MeIm(Me)_2)]^{2+}$

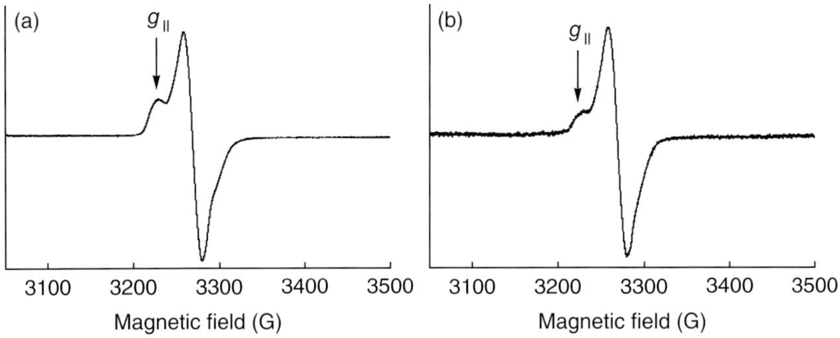

Figure 9.36 EPR spectra of (a) $[Zn^{II}(O_2^{\bullet -})(MeIm(Py)_2)]^+$ and (b) $[Zn^{II}(O_2^{\bullet -})\text{–}(MeIm(Me)_2)]^+$ in frozen EtCN at 133 K. Source: Ohtsu et al. 2001 [459]. Reproduced with permission of John Wiley & Sons.

as compared to $[Zn^{II}(MeIm(Py)_2)]^{2+}$ (0.85 eV) may be ascribed to the increased Lewis acidity of the Zn^{II} ion in $[Zn^{II}(MeIm(Me)_2)]^{2+}$ due to the steric effect of the *o*-methyl group of pyridine moiety of $[Zn^{II}(MeIm(Me)_2)]^{2+}$ [459]. Thus, the essential role of Zn^{II} ion in SODs is to accelerate both the oxidation and reduction of superoxide ion by controlling the redox potentials of Cu^{II} ion and superoxide ion in the catalytic cycle of SOD as shown in Scheme 9.15 [458]. A similar effect is also expected for another copper ion in the $Cu^{II}\text{–}Cu^{II}$ homodinuclear complex containing the same Hbdpi ligand [458].

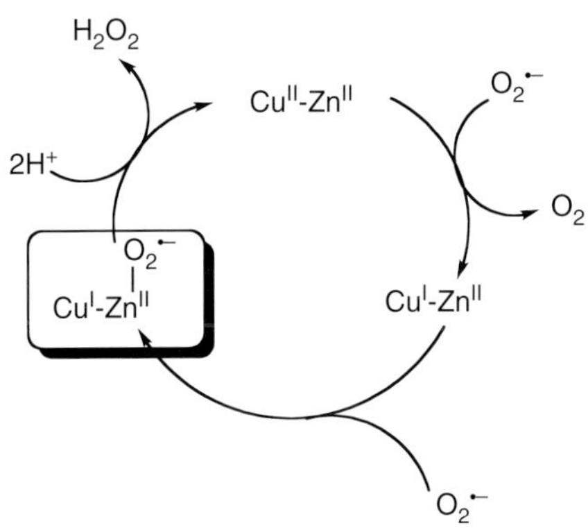

Scheme 9.15 The catalytic cycle of the imidazolate-bridged $Cu^{II}\text{–}Zn^{II}$ heterodinuclear SOD model complex. Source: Ohtsu et al. 2000 [458]. Reproduced with permission of American Chemical Society.

The $Cu^{II}\text{–}Cu^{II}$ homodinuclear, $Cu^{II}\text{–}Zn^{II}$ heterodinuclear, and Cu^{II} mononuclear complexes exhibit catalytic activity toward the dismutation of superoxide anions. The SOD activities were investigated by cytochrome *c* assay [460] using the xanthine oxidase reaction as the source of superoxide. The concentrations of complexes required to attain 50% inhibition of the reduction (defined as IC_{50}) were determined for these complexes. The highest SOD activity ($IC_{50} = 0.24\,\mu M$) was observed for the $Cu^{II}\text{–}Zn^{II}$ complex, $[CuZn(bdpi)(MeCN)_2]^{3+}$ [458]. Although the IC_{50} value is still larger than the value reported for native SOD ($0.04\,\mu M$), this is smaller than the best reported value. Thus, the imidazolate-bridged Cu^{II}–Zn^{II} heterodinuclear SOD model complex, which has a coordination site available for the binding of superoxide as shown by the X-ray structure in Figure 9.34a, exhibits *the highest activity* among the structurally established SOD models reported so far [458].

The same mechanism as in Scheme 9.15 can be applied to the disproportionation of semiquinone radical anion ($Q^{•-}$) by the imidazolate-bridged Cu^{II}–Zn^{II} complex as shown in Scheme 9.16 [459]. Since the one-electron oxidation

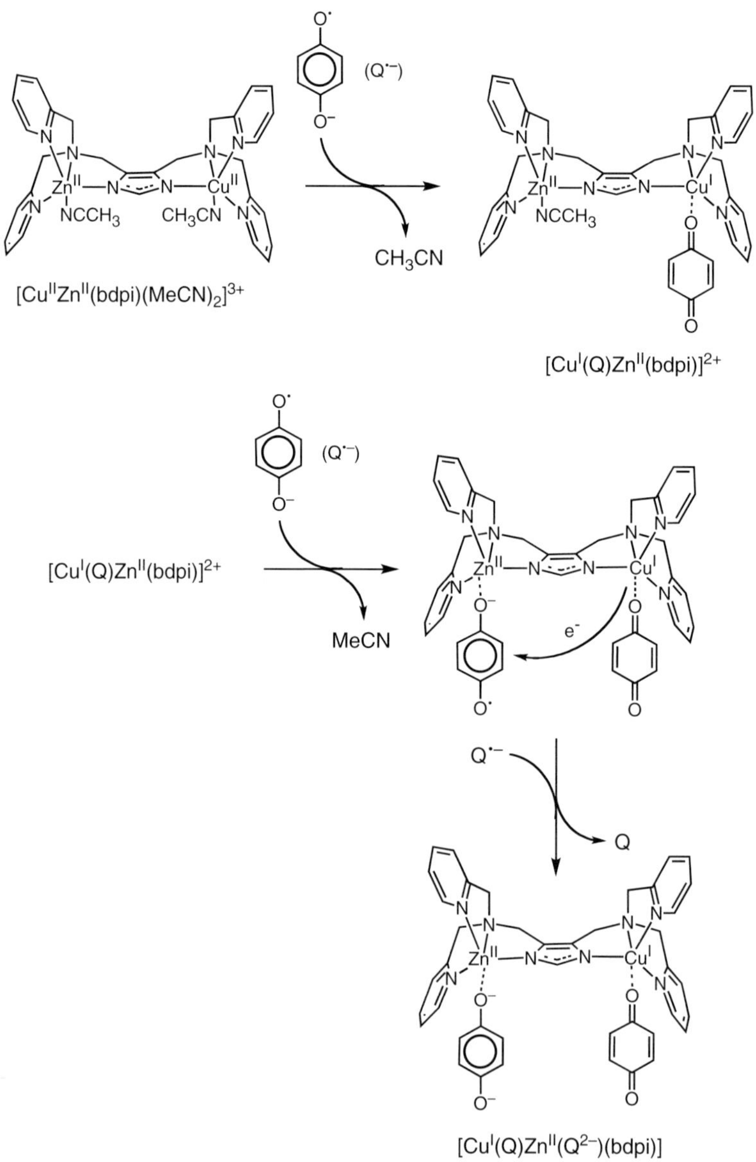

Scheme 9.16 Disproportionation of $Q^{•-}$ with $[CuZn(bdpi)(MeCN)_2]^{3+}$. Source: Ohtsu et al. 2001 [459]. Reproduced with permission of John Wiley & Sons.

potential (E_{ox} vs. SCE) of $Q^{\cdot-}$ (−0.51 V), which is equal to the one-electron reduction potential of Q [438], is more negative than the one-electron reduction potential (E_{red} vs. SCE) of the Cu^{II} center of $[CuZn(bdpi)(MeCN)_2]^{3+}$ (−0.03 V), electron transfer from $Q^{\cdot-}$ to the Cu^{II} center should occur to yield the Cu^{I}–Q complex. The complexation of $Q^{\cdot-}$ with Zn^{II} ion in $[Cu^{I}(Q)Zn^{II}(bdpi)]^{2+}$ may also facilitate an intramolecular electron transfer from the Cu^{I} center to $Q^{\cdot-}$ bound to the Zn^{II} center (Scheme 9.16) [459]. Thus, the Zn^{II} ion also plays an essential role in facilitating the ET reduction $Q^{\cdot-}$.

9.8 MCET of Metal-Oxo Complexes

The electron-transfer reduction of an iron(IV)-oxo complex, $[(N4Py)FeIV(O)]^{2+}$ (N4Py = *N*,*N*-bis(2-pyridylmethyl)-*N*-bis(2-pyridyl)methylamine), is accelerated by binding of redox-inactive metal ions (M^{n+}) to the corresponding ion(III)-oxo complex, as shown in the case of the electron-transfer reduction of O_2 in Figure 9.3 [461]. The one-electron reduction potential of $[(N4Py)Fe^{IV}(O)]^{2+}$ in MeCN (E_{red} = 0.49 V vs. SCE) [462] was shifted to a positive direction by binding of $Sc(OTf)_3$ to $[(N4Py)Fe^{III}(O)]^{+}$ [461]. The E_{red} values in the presence of $Sc(OTf)_3$ were determined from the equilibrium constants of electron transfer with various electron donors [461]. The E_{red} values increase linearly with increasing logarithm of concentration of $Sc(OTf)_3$ (log $[Sc^{3+}]$) as shown in Figure 9.37, where the slope is 118 mV/log $[Sc^{3+}]$ [461]. This indicates that the one-electron reduction of $[(N4Py)Fe^{IV}(O)]^{2+}$ is accompanied by binding of one or two Sc^{3+} ions to $[(N4Py)Fe^{III}(O)]^{+}$, in accordance with the Nernst equation (Eq. (9.20)),

$$E_{red} = E^{0}{}_{red} + (2.3RT/F)\log(1 + K_1[Sc^{3+}] + K_1K_2[Sc^{3+}]^2) \tag{9.20}$$

where $E^{0}{}_{red}$ is the one-electron reduction potential without Sc^{3+}, and K_1 and K_2 are the formation constants of 1 : 1 and 1 : 2 complexes of $[(N4Py)Fe^{III}(O)]^{+}$ with one Sc^{3+} and two Sc^{3+} ions, respectively (Scheme 9.17) [461].

The observed second-order rate constant (k_{obs}) of electron transfer from electron donors to $[(N4Py)Fe^{IV}(O)]^{2+}$ in the presence of $Sc(OTf)_3$ in MeCN increased

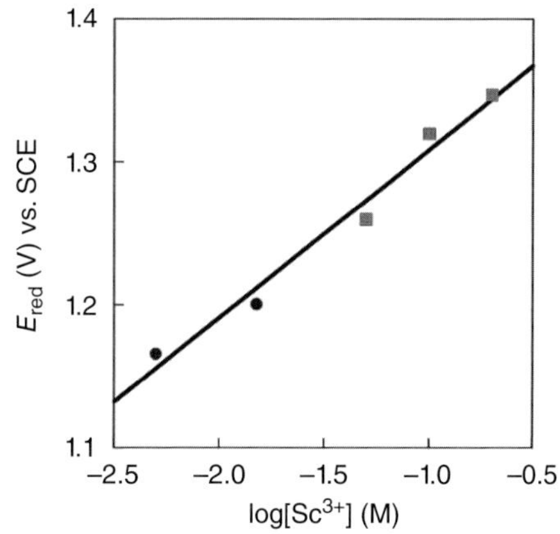

Figure 9.37 Dependence of E_{red} of $[(N4Py)Fe^{IV}(O)]^{2+}$ in MeCN at 298 K on log $[Sc^{3+}]$. The E_{red} values were derived from equilibrium constants (K_{et}) of the electron transfer from electron donors to $[(N4Py)Fe^{IV}(O)]^{2+}$ in the presence of $Sc(OTf)_3$ in MeCN at 298 K. Source: Morimoto et al. 2011 [461]. Reproduced with permission of American Chemical Society.

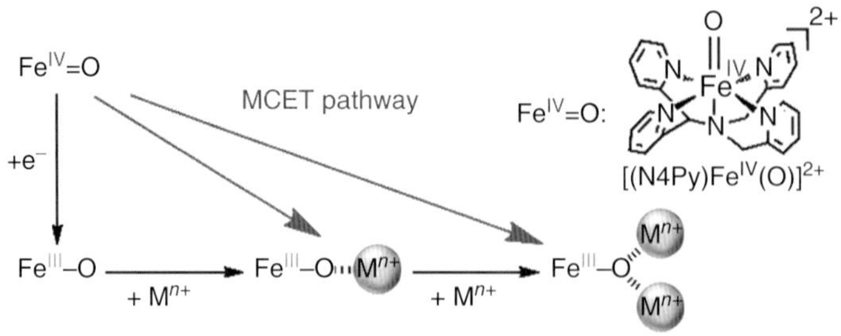

Scheme 9.17 Metal ion-coupled electron-transfer reduction of $[(N4Py)Fe^{IV}(O)]^{2+}$. Source: Morimoto et al. 2011 [461]. Reproduced with permission of American Chemical Society.

with a first-order dependence on $[Sc^{3+}]$ at low concentrations, changing to a second-order dependence at high concentrations, as given by Eq. (9.21) [461]. The k_1 and k_2 values were determined for

$$k_{obs} = k_1[Sc^{3+}] + k_2[Sc^{3+}]^2 \tag{9.21}$$

MCET from ferrocene to $[(N4Py)Fe^{IV}(O)]^{2+}$ in the presence of various metal ions [461]. Both log k_1 and log k_2 exhibit good linear correlations with the ΔE values of metal ions (Eq. (9.1)) as shown in Figure 9.38 [461]. Thus, the stronger the Lewis acidity of metal ions is, the larger the rate constants (k_1 and k_2) of MCET from ferrocene to $[(N4Py)Fe^{IV}(O)]^{2+}$ become.

The first crystallographic evidence for the binding of a redox-inactive metal ion to a high-valent metal-oxo complex was reported as the X-ray crystal structure of Sc^{3+}-bound $[(TMC)Fe^{IV}(O)]^{2+}$ (TMC = 1,4,8,11-tetramethyl-1,4,8,11-tetraazacyclotetradecane) [463], which was formulated as the neutral $[(TMC)(Fe^{IV}–O–Sc^{III})(OTf)_4(OH)]$ (Figure 9.39) [464]. Later, density functional theory (DFT) calculations suggested that the formulation could be $[(TMC)Fe^{III}–O–Sc^{III}(OTf)_4(OH_2)]$ [465], which was supported by Mössbauer and X-ray absorption spectroscopy of the crystals as well as ESI mass spectrometry and EPR spectroscopy [466]. It may be suggested that the $Fe^{IV}(O)$ complex is reduced to the Sc^{3+}-binding $Fe^{III}(O)$ species during the crystallization in the presence of ether; however, the source of one electron for the reduction of the

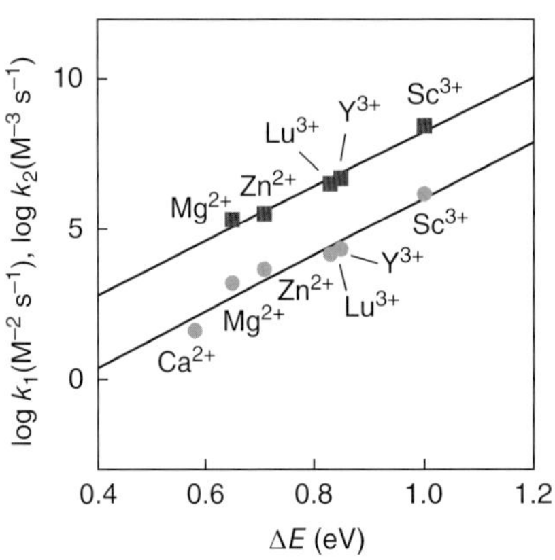

Figure 9.38 Plots of log k_1 (circles) and log k_2 (squares) of MCET from ferrocene to $[(N4Py)Fe^{IV}(O)]^{2+}$ in the presence of various metal ions in MeCN at 298 K vs. ΔE (quantitative measure of Lewis acidity of metal ions). Source: Morimoto et al. 2011 [461]. Reproduced with permission of American Chemical Society.

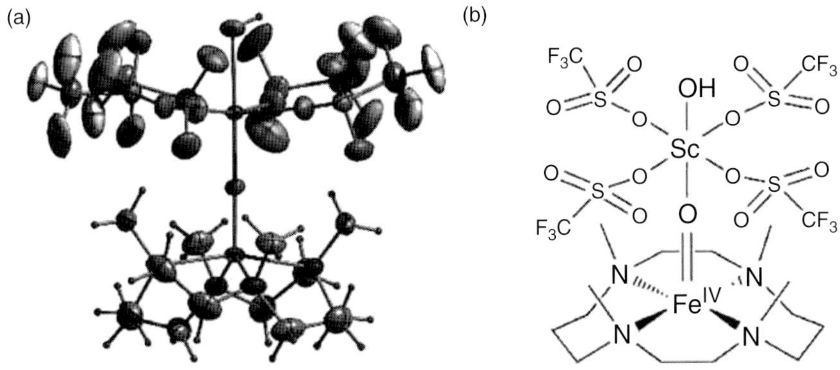

Figure 9.39 (a) ORTEP-style and (b) chemical structure of the Sc^{3+}-bound nonheme iron(IV)-oxo complex, $[(TMC)Fe^{IV}(O)–Sc(OTf)_4(OH)]$. Source: Fukuzumi et al. 2010 [464]. Reproduced with permission of Springer Nature.

$Fe^{IV}(O)$ complex to the $Fe^{III}(O)$ complex has yet to be clarified. In any case, the strong binding of Sc^{3+} ion to the iron-oxo moiety of the $Fe^{III}(O)$ complex results in enhancement of the two-electron reduction of the $Fe^{IV}(O)$ complex to the Fe^{II} complex [464].

$[(TMC)Fe^{III}(O_2)]^+–Sc^{3+}$ was suggested as an intermediate for Sc^{3+} ion-promoted O_2 activation by $[(TMC)Fe^{II}]^{2+}$ with BPh_4^- to produce $[(TMC)Fe^{IV}(O)]^{2+}$ [467]. It was proposed that electron transfer from $[(TMC)Fe^{II}]^{2+}$ to O_2 was followed by Sc^{3+} ion-promoted electron transfer from BPh_4^- to the superoxo complex ($[(TMC)Fe^{III}(O_2)]^{2+}$) to produce $[(TMC)Fe^{III}(O_2)]^+–Sc^{3+}$, which was further reduced to yield $[(TMC)Fe^{IV}(O)]^{2+}$ via the O—O bond cleavage [467]. Sc^{3+} ion could be replaced by Brønsted acids for the formation of $[(TMC)Fe^{IV}(O)]^{2+}$ from $[(TMC)Fe^{II}]^{2+}$ with O_2 and BPh_4^- [468]. However, the actual mechanism was later clarified to be autocatalytic radical chain reactions (Scheme 9.18) rather than direct O_2 activation by $[(TMC)Fe^{II}]^{2+}$ [469]. The initiation is started by Sc^{3+} ion-coupled electron transfer from BPh_4^- to $[(TMC)Fe^{IV}(O)]^{2+}$ to produce phenyl radical ($Ph^•$). The chain propagation step is composed of the addition of O_2 to $Ph^•$ and Sc^{3+} ion-coupled electron transfer from BPh_4^- to the resulting phenylperoxyl radical ($PhOO^•$) to produce the $PhOO^-/Sc^{3+}$ complex, accompanied by regeneration of $Ph^•$ (propagation step). The $PhOO^-/Sc^{3+}$ complex reacts with $[(TMC)Fe^{II}]^{2+}$ to produce $[(TMC)Fe^{IV}(O)]^{2+}$ and PhOH after the reaction with residual water [469]. Such radical chain autoxidation reactions of ion(II) complexes by O_2 have also been reported for formation of ion(IV)-oxo complexes from the reactions of ion(II) complexes with hydrogen sources and O_2 [470, 471].

Sc^{3+} ion has also been reported to bind to copper(II)-peroxo complexes in the Sc^{3+} ion-promoted catalytic two-electron reduction of O_2 by decamethylferrocene (Fc*) [472]. The Sc^{3+}-bound peroxo complex, $[(tmpa)Cu^{II}(O_2)–Sc(OTf)_3]$, (tmpa = tris(2-pyridyl-methyl)amine), was formed by addition of 1 equiv of $Sc(OTf)_3$ to an acetone solution of a *trans*-μ-1,2-peroxo-dicopper complex, $[(tmpa)Cu^{II}(O_2)Cu^{II}(tmpa)]^{2+}$ [472]. When $Sc(OTf)_3$ was replaced by HOTf,

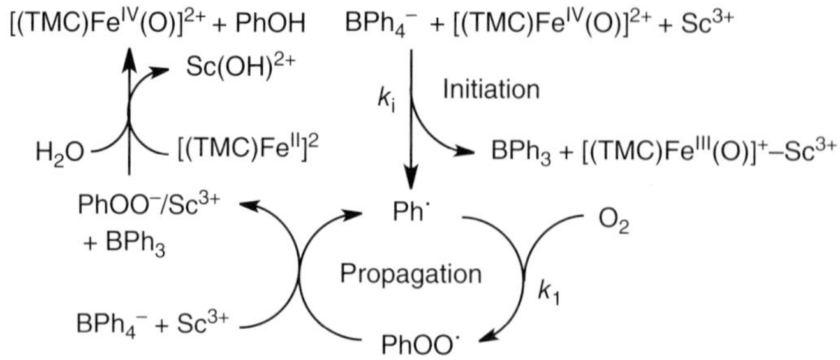

Scheme 9.18 Sc^{3+}-coupled electron-transfer chain reactions for formation of $[(TMC)Fe^{IV}(O)]^{2+}$ in reaction of $[(TMC)Fe^{II}]^{2+}$ with O_2 and BPh_4^-. Source: Nishida et al. 2014 [469]. Reproduced with permission of American Chemical Society.

$[(tmpa)Cu^{II}(OOH)]^+$ was further reduced by the PCET reduction, acting as an intermediate for the catalytic four-electron reduction of O_2 by Fc^* [473, 474].

In contrast to the case of ion(IV)-oxo complexes, $Sc(OTf)_3$ binds to manganese(IV)-oxo complexes, such as $[(N4Py)Mn^{IV}(O)]^{2+}$ and $[(Bn\text{-}TPEN)Mn^{IV}(O)]^{2+}$ (Bn-TPEN = *N*-benzyl-*N*,*N*′,*N*′-tris(2-pyridylmethyl)-1,2-diaminoethane) [475, 476]. Addition of up to 2 equiv of $Sc(OTf)_3$ to a solution of $[(Bn\text{-}TPEN)Mn^{IV}(O)]^{2+}$ in a solvent mixture of CF_3CH_2OH/CH_3CN resulted in a blueshift of the absorption band at $\lambda_{max} = 1020$ nm to that at $\lambda_{max} = 740$ nm with an isosbestic point at 900 nm (Figure 9.40a) [476]. Further addition of $Sc(OTf)_3$ resulted in more blueshift to give the absorption band at $\lambda_{max} = 690$ nm (Figure 9.40b) [476]. Such stepwise spectral changes indicate the binding of one and two Sc^{3+} ions to $[(Bn\text{-}TPEN)Mn^{IV}(O)]^{2+}$ to produce $Mn^{IV}(O)–(Sc^{3+})_1$ and $Mn^{IV}(O)–(Sc^{3+})_2$ complexes, respectively [476]. The formation constants of the $[(Bn\text{-}TPEN)Mn^{IV}(O)]^{2+}–(Sc^{3+})_1$ and $[(Bn\text{-}TPEN)–Mn^{IV}(O)]^{2+}–(Sc^{3+})_2$ complexes were determined from the spectral titrations to be 4.0×10^3 and $1.2 \times 10^3\ M^{-1}$, respectively [476]. The formation

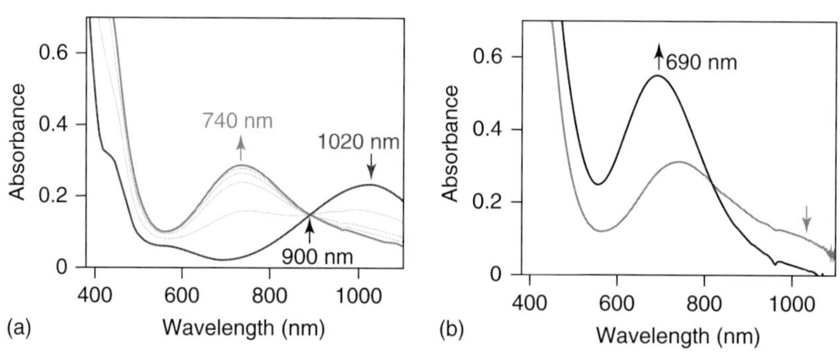

Figure 9.40 (a) UV–vis spectral changes showing the conversion from $[(Bn\text{-}TPEN)Mn^{IV}(O)]^{2+}$ to $[(Bn\text{-}TPEN)Mn^{IV}(O)]^{2+}–(Sc^{3+})_1$ upon addition of incremental amounts of Sc^{3+} (from 0.0 to 2.0 mM) in the titration experiment. (b) UV–vis spectral changes showing the conversion from $[(Bn\text{-}TPEN)Mn^{IV}(O)]^{2+}–(Sc^{3+})_1$ to $[(Bn\text{-}TPEN)Mn^{IV}(O)]^{2+}–(Sc^{3+})_2$ observed upon addition of Sc^{3+} ions (from 0 to 12 mM). Source: Yoon et al. 2013 [476]. Reproduced with permission of American Chemical Society.

constant of $[(N4Py)Mn^{IV}(O)]^{2+}$–$(Sc^{3+})_2$ complex was also determined to be $6.1 \times 10^3\ M^{-1}$ [476].

The extended X-ray absorption fine structure (EXAFS) results indicated that on going from no Sc^{3+} binding in $[(Bn\text{-}TPEN)Mn^{IV}(O)]^{2+}$ to one and two Sc^{3+} binding, the Mn=O bond elongates from 1.69 to 1.74(2) Å, which is consistent with a weakening of the Mn=O bond. The EXAFS data also revealed a short Mn–Sc distance (3.45(10) Å), which clearly indicates that Sc^{3+} ions bind to the $Mn^{IV}(O)$ moiety in both $[(Bn\text{-}TPEN)Mn^{IV}(O)]^{2+}$–$(Sc^{3+})_1$ and $[(Bn\text{-}TPEN)Mn^{IV}(O)]^{2+}$–$(Sc^{3+})_2$. DFT calculations of the $Mn^{IV}(O)$–$(Sc^{3+})_2$ complex suggested that one Sc^{3+} ion binds directly to the Mn–O moiety of the $Mn^{IV}(O)$ complex but the second Sc^{3+} ion is located at the secondary coordination sphere as shown in Figure 9.41 [476].

The E_{red} values of $Mn^{IV}(O)$ complexes, $[(Bn\text{-}TPEN)Mn^{IV}(O)]^{2+}$ and $[(N4Py)Mn^{IV}(O)]^{2+}$, in the presence of various concentrations of $Sc(OTf)_3$ were determined from the K_{et} values obtained by the electron-transfer titrations with $[Ru^{II}(bpy)_3]^{2+}$ (E_{ox} = 1.24 V vs. SCE) and $[Ru^{II}(5\text{-}Cl\text{-}phen)_3]^{2+}$ (5-Cl-phen = 5-chloro-1,10-phenanthroline, E_{ox} = 1.36 V vs. SCE) [476]. The E_{red} values of $[(Bn\text{-}TPEN)Mn^{IV}(O)]^{2+}$ and $[(N4Py)Mn^{IV}(O)]^{2+}$ in the absence of $Sc(OTf)_3$were shifted from 0.78 and 0.80 V vs. SCE [477] to 1.36 and 1.42 V vs. SCE in the presence of large excess $Sc(OTf)_3$, which correspond to the E_{red} values of $[(Bn\text{-}TPEN)Mn^{IV}(O)]^{2+}$–$(Sc^{3+})_2$ and $[(N4Py)Mn^{IV}(O)]^{2+}$–$(Sc^{3+})_2$, respectively (Figure 9.42) [476].

Although high-valent cobalt-oxo complexes still remain elusive, binding of redox-inactive metal ions is shown to stabilize a cobalt(IV)-oxo complex as shown in Scheme 9.19 [478]. The cobalt(III) complex, $[(TAML)Co^{III}]^-$ (TAML = tetraamido macrocyclic ligand), is readily oxidized by cerium ammonium nitrate (CAN) in acetone to afford the $[(TAML)Co^{IV}(O)\text{–}Ce]^{2+}$ complex. [478] The reaction of $[(TAML)Co^{III}]^-$ with iodosylbenzene (PhIO) in the presence of $Sc(OTf)_3$ also generated the $[(TAML)Co^{IV}(O)\text{–}Sc]^{2+}$ complex, which is stabilized by binding of Sc^{3+} ion to the metal-oxo moiety [478]. Theoretical calculations predict that the binding of Lewis acidic metal ions

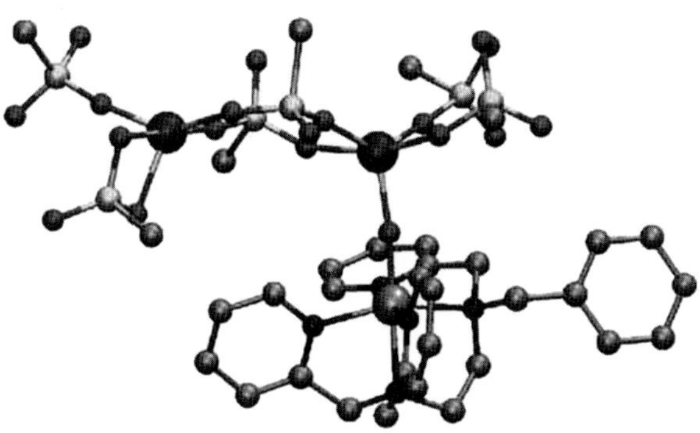

Figure 9.41 DFT-optimized structure of $[(Bn\text{-}TPEN)Mn^{IV}(O)\text{–}[Sc(OTf)_3]_2]^{2+}$, calculated at the B3LYP/LACVP level. The Mn—O bond lengths of $[Mn^{IV}(O)]^{2+}$, $[Mn^{IV}(O)]^{2+}$–Sc^{3+}, and $[Mn^{IV}(O)]^{2+}$–$(Sc^{3+})_2$ were calculated to be 1.68, 1.75, and 1.75 Å, respectively. Source: Yoon et al. 2013 [476]. Reproduced with permission of American Chemical Society.

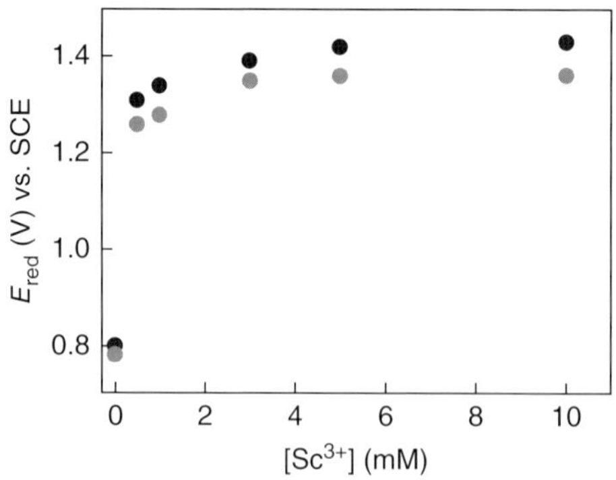

Figure 9.42 Dependence of E_{red} of $[(N4Py)Mn^{IV}(O)]^{2+}$ (black circles) and $[(Bn\text{-}TPEN)Mn^{IV}(O)]^{2+}$ (grey circles) on [Sc^{3+}] in CF_3CH_2OH/CH_3CN (v/v = 1 : 1) at 273 K. The E_{red} values were determined from the equilibrium constants (K_{et}) between one-electron donors and $[(N4Py)Mn^{IV}(O)]^{2+}$. Source: Yoon et al. 2013 [476]. Reproduced with permission of American Chemical Society.

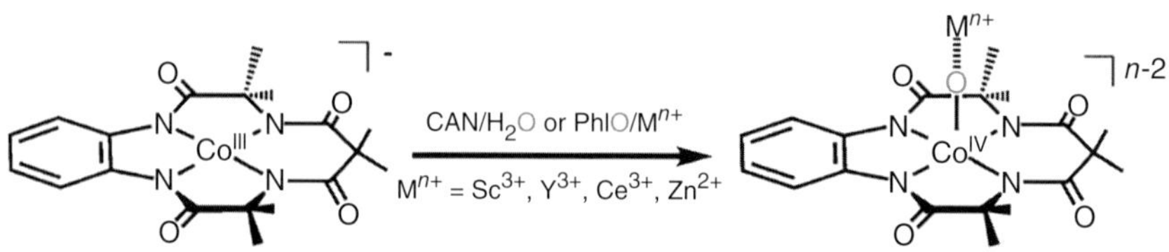

Scheme 9.19 Oxidation of a cobalt(III) complex to a cobalt(IV)-oxo-metal ion complexes by CAN/H_2O or PhIO/M^{n+}. Source: Hong et al. 2014 [478]. Reproduced with permission of John Wiley & Sons.

(M^{n+}) to the cobalt-oxo core increases the electrophilicity of the oxygen atom, resulting in the redox tautomerism of a highly unstable $[(TAML)Co^{III}(O^{\cdot})]^{2-}$ species to a more stable $[(TAML)Co^{IV}(O)(M^{n+})]$ core [478]. The Co^{IV} oxidation state of $[(TAML)Co^{IV}(O)–Sc]^{2+}$ was confirmed by EPR and XAS measurements. It should be noted that the determined Co–O distance of 1.67 Å for $[(TAML)Co^{IV}(O)–Sc]^{2+}$ in EXAFS is significantly shorter than the Co–O distances of 1.79 and 1.85 Å determined for $[(\alpha_2\text{-}P_2W_{17}O_{61}Co^{IV})_2O]^{14-}$ by X-ray diffraction [479] and $[(TMG_3tren)Co^{IV}(O)(Sc)(OTf)_3]]^{2+}$ (TMG_3tren = (tris[2-(*N*-tetramethylguanidyl)ethyl]-amine) by EXAFS [480]. The latter species was later suggested to contain a $Co^{III}(OH)$ unit rather than a $Co^{IV}(O)$ unit [481], and further investigation is needed to clarify the Co–O(H) moiety.

9.9 PCET of Metal-Oxo Complexes

As in the case of $Sc(OTf)_3$ (vide supra), the E_{red} value of $[(N4Py)Fe^{IV}(O)]^{2+}$ is shifted to a positive direction by the binding of HOTf by $[(N4Py)Fe^{III}(O)]^{+}$ [482]. The dependence of E_{red} of $[(N4Py)Fe^{IV}(O)]^{2+}$ on log [HOTf] and log [$Sc(OTf)_3$] is shown in Figure 9.43, where the slopes are determined to be 113 ± 6 mV for HOTf and 118 ± 8 mV for $Sc(OTf)_3$, which agree with Eq. (9.20), where $Sc(OTf)_3$ is replaced by HOTf [482]. Not only one but also two HOTf molecules are bound to $[(N4Py)Fe^{III}(O)]^{+}$. The binding of HOTf molecules to $[(N4Py)Fe^{III}(O)]^{+}$ are stronger than that of $Sc(OTf)_3$ molecules, resulting in the more positive E_{red}

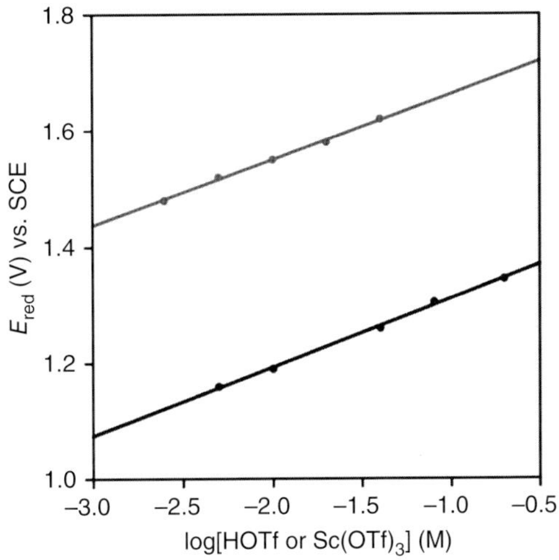

Figure 9.43 Dependence of E_{red} of $[(N4Py)Fe^{IV}(O)]^{2+}$ on log [HOTf] (grey circles) and log $[Sc(OTf)_3]$ (black circles) in deaerated MeCN at 298 K. Black and grey lines are fitted by Eq. (9.20), where Sc^{3+} is replaced by HOTf. Source: Park et al. 2014 [482]. Reproduced with permission of American Chemical Society.

values of $[(N4Py)Fe^{IV}(O)]^{2+}$ in the presence of HOTf than those in the presence of the same concentration of $Sc(OTf)_3$ (Figure 9.41) [482].

When HOTf was replaced by the deuterated acid (DOTf), the k_{et} value of PCET from $[Ru^{II}(bpy)_3]^{2+}$ to $[(N4Py)Fe^{IV}(O)]^{2+}$ with DOTf in MeCN at 298 K was larger than that with the same concentration of HOTf as shown in Figure 9.44 [482]. Both the k_{et} values with DOTf and HOTf increased with increasing concentration of DOTf and HOTf in accordance with Eq. (9.21), where $[Sc^{3+}]$ is replaced by [DOTf] and [HOTf] [482]. The kinetic isotope effect (KIE) value (k_{obs}(HOTf)/k_{obs}(DOTf)) was determined to be 0.25 ± 0.05 at HOTf (DOTf) = 2.5 mM, which increases to a constant value of 0.40 ± 0.04 at higher concentrations of HOTf (DOTf) [482]. Such inverse KIEs indicate that binding of protons to the Fe-oxo group to produce an O—H (or O—D)

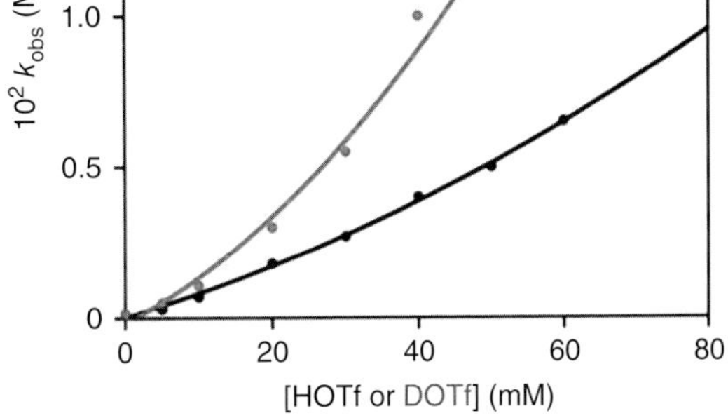

Figure 9.44 Plots of k_{obs} vs. [HOTf] (black circles) and [DOTf] (grey circles) for oxidation of toluene by $[(N4Py)Fe^{IV}(O)]^{2+}$ in the presence of HOTf and DOTf in MeCN at 298 K, respectively. Source: Park et al. 2014 [482]. Reproduced with permission of American Chemical Society.

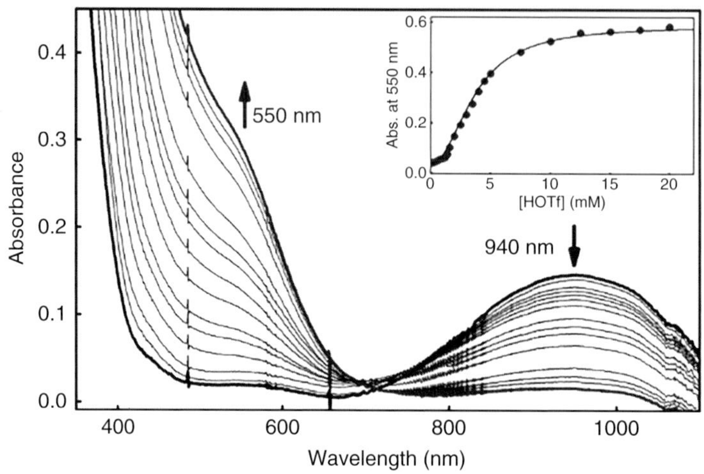

Figure 9.45 UV–vis spectral changes observed in the titration of $[(N4Py)Mn^{IV}(O)]^{2+}$ (black bold line) with HOTf. HOTf (0–20 mM) was added incrementally into the solution of 1 (0.50 mM) in $CF_3CH_2OH–CH_3CN$ (v/v = 1 : 1) at 273 K. Inset shows the spectral titration monitored at 550 nm (black circles) due to the formation of $[(N4Py)Mn^{IV}(O)]^{2+}–(HOTf)_2$. Source: Chen et al. 2015 [484]. Reproduced with permission of Royal Society of Chemistry.

bond is involved in the rate-determining step of PCET from $[Ru^{II}(bpy)_3]^{2+}$ to $[(N4Py)Fe^{IV}(O)]^{2+}$, because an inverse KIE results from a larger zero-point energy difference between the O—H and O—D bonds in the transition state relative to the ground state [265, 483]. Thus, the O—H bonds in the reduced species (e.g. $[(N4Py)Fe^{III}(OH)]^{2+}$ and $[(N4Py)Fe^{III}(OH_2)]^{3+}$) are much stronger than those in $[(N4Py)Fe^{IV}(OH)]^{3+}$ and$[(N4Py)Fe^{IV}(OH_2)]^{4+}$, respectively. The inverse KIEs clearly indicate that electron transfer is coupled with the binding of one and two HOTf molecules to $[(N4Py)Fe^{III}(O)]^{+}$ [482].

In contrast to the case of $[(N4Py)Fe^{IV}(O)]^{2+}$, two HOTf molecules are directly bound to $[(N4Py)Mn^{IV}(O)]^{2+}$ to produce $[(N4Py)Mn^{IV}(O)]^{2+}–(HOTf)_2$ as shown in Figure 9.45, where absorption change at 550 nm due to $[(N4Py)Mn^{IV}(O)]^{2+}–(HOTf)_2$ exhibits a sigmoidal curvature (inset of Figure 9.45) [484]. The EXAFS data of $[(N4Py)Mn^{IV}(O)]^{2+}–(HOTf)_2$ indicate an elongated Mn—O bond distance of 1.74 Å as compared with 1.67 Å for the Mn—O bond distance $[(N4Py)Mn^{IV}(O)]^{2+}$ [484]. This Mn—O bond distance in $[(N4Py)Mn^{IV}(O)]^{2+}–(HOTf)_2$ is nearly identical to that of the $[(N4Py)Mn^{IV}(O)]^{2+}$ complex binding two $Sc(OTf)_3$ molecules [475, 476].

The E_{red} value of $[(N4Py)Mn^{IV}(O)]^{2+}$ increased with increasing concentration of HOTf to reach the constant value of 1.65 V vs. SCE at more than 30 mM of HOTf at 273 K (Figure 9.46) [484]. The saturated E_{red} value of 1.65 V vs. SCE in the presence of a large excess of HOTf corresponds to that of $[(N4Py)Mn^{IV}(O)]^{2+}–(HOTf)_2$. The large difference ($\Delta E_{red}$ = 0.85 V) in the E_{red} values between $[Mn^{IV}(O)]^{2+}$ and $[Mn^{IV}(O)]^{2+}–(HOTf)_2$ indicates that the binding of HOTf to $[Mn^{III}(O)]^{+}$ is much stronger than that to $[Mn^{IV}(O)]^{2+}$, as expected from the increased basicity of the oxo moiety in $[Mn^{III}(O)]^{+}$ [484]. Thus, the HOTf-bound $Mn^{IV}(O)$ complex acts as the strongest oxidant among the nonheme Mn-oxo complexes bearing the same ligand system [484].

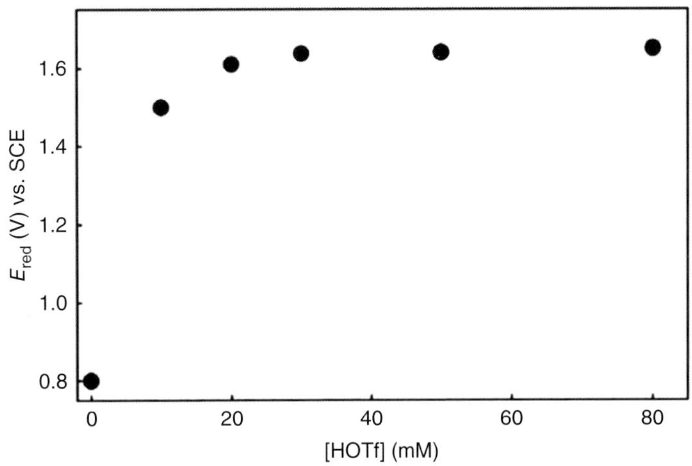

Figure 9.46 Dependence of E_{red} of $[(N4Py)Mn^{IV}(O)]^{2+}$ on the concentration of HOTf (0–80 mM) in $CF_3CH_2OH–CH_3CN$ (v/v = 1 : 1) at 273 K. Source: Chen et al. 2015 [484]. Reproduced with permission of Royal Society of Chemistry.

9.10 Unified Mechanism of MCET and PCET of Metal-Oxo Complexes

Not only the rates of electron-transfer reactions of $[(N4Py)Fe^{IV}(O)]^{2+}$ but also the rates of the oxidative C—H bond cleavage of toluene derivatives, benzyl alcohol derivatives, sulfoxidation of thioanisole derivatives, and epoxidation of styrene derivatives with $[(N4Py)Fe^{IV}(O)]^{2+}$ were remarkably enhanced by the presence of perchloric acid ($HClO_4$), triflic acid (HOTf), and $Sc(OTf)_3$ in acetonitrile at 298 K [482, 485–489]. The enhancement of electron-transfer reactions is ascribed to the positive shifts of E_{red} values of $[(N4Py)Fe^{IV}(O)]^{2+}$ in the presence of $Sc(OTf)_3$ and HOTf (Figure 9.43) [482]. The driving force dependence of $\log k_{et}$ of electron transfer from coordinatively saturated metal complexes (electron donors) to $[(N4Py)Fe^{IV}(O)]^{2+}$ in the absence and presence of 10 mM of HOTf (black line in Figure 9.47) is well fitted by the Marcus equation of outer-sphere electron transfer (Eq. (9.22)) [6],

$$k_{et} = Z_{exp}[-(\lambda/4)(1 + \Delta G_{et}/\lambda)^2/k_B T] \tag{9.22}$$

where Z is the frequency factor, which is $k_B TK/h$ (k_B is the Boltzmann constant, T is absolute temperature, K is the formation constant of the precursor complex of electron transfer, and h is the Planck constant), using the same value of reorganization energy of electron transfer ($\lambda = 2.74$ eV). The Z value of outer-sphere electron-transfer reactions is taken as 1.0×10^{11} M^{-1} s^{-1} [6].

The $\log k_{obs}$ values of oxidation of toluene and thioanisole derivatives by $[(N4Py)Fe^{IV}(O)]^{2+}$ in the absence of an acid (nos. 1–12, black) are much larger than those expected by Eq. (9.22) with $\lambda = 2.74$ eV (black line in Figure 9.47), because the oxidation of toluene and thioanisole derivatives by $[(N4Py)Fe^{IV}(O)]^{2+}$ in the absence of an acid proceeds via hydrogen atom transfer (HAT) and oxygen atom transfer (OAT), which require significant interactions in the transition state

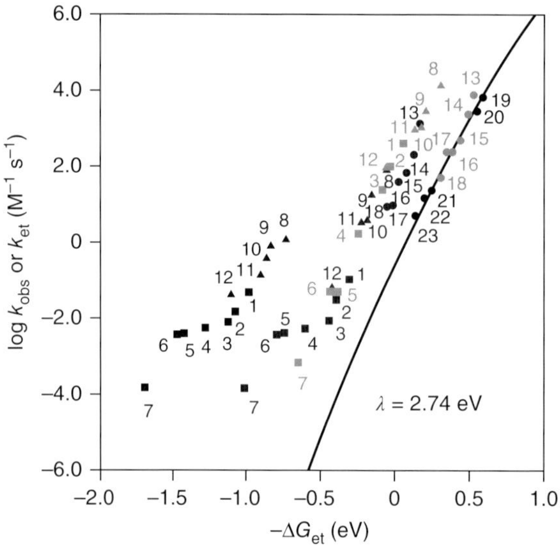

Figure 9.47 Plots of log k_{obs} for oxidation of toluene and thioanisole derivatives [(1) hexamethylbenzene, (2) 1,2,3,4,5-tetramethylbenzene, (3) 1,2,4,5-tetramethylbenzene, (4) 1,2,4-trimethylbenzene, (5) 1,4-dimethylbenzene, (6) 1,3,5-trimethylbenzene, (7) toluene, (8) *p*-Me-thioanisole, (9) thioanisole, (10) *p*-Cl-thioanisole, (11) *p*-Br-thioanisole, and (12) *p*-CN-thioanisole] by $[(N4Py)Fe^{IV}(O)]^{2+}$ in the absence (black) and presence of 10 mM of HOTF (black) and $Sc(OTf)_3$ (grey) in MeCN at 298 K vs. the driving force of electron transfer [$-\Delta G_{et} = e(E_{red} - E_{ox})$] from toluene derivatives (squares) and thioanisole derivatives (triangles) to $[(N4Py)Fe^{IV}(O)]^{2+}$ in the presence of HOTf (black) and $Sc(OTf)_3$ (grey). The black and grey circles show the driving force dependence of the rate constants (log k_{et}) of electron transfer from electron donors [(13) $[Fe^{II}(Ph_2phen)_3]^{2+}$, (14) $[Fe^{II}(bpy)_3]^{2+}$, (15) $[Ru^{II}(4,4'\text{-}Me_2phen)_3]^{2+}$, (16) $[Ru^{II}(5,5'\text{-}Me_2phen)_3]^{2+}$, (17) $[Fe^{II}(Clphen)_3]^{2+}$, and (18) $[Ru^{II}(bpy)_3]^{2+}$] to $[(N4Py)Fe^{IV}(O)]^{2+}$ in the presence of HOTf (10 mM) and $Sc(OTf)_3$ (10 mM) in MeCN at 298 K. The black line is drawn using Eq. (3.1) with $\lambda = 2.74$ eV. The black circles show the driving force dependence of the rate constants (log k_{et}) of electron transfer from electron donors [(19) decamethylferrocene, (20) octamethylferrocene, (21) 1.1′-dimethylferrocene, (22) *n*-amylferrocene, and (23) ferrocene] to $[(N4Py)Fe^{IV}(O)]^{2+}$ in the absence of acids in MeCN at 298 K. Source: Park et al. 2014 [482]. Reproduced with permission of American Chemical Society.

as compared with outer-sphere electron-transfer processes [482]. In contrast to the results without acid, log k_{obs} values of oxidation of toluene and thioanisole derivatives by $[(N4Py)Fe^{IV}(O)]^{2+}$ in the presence of HOTf and $Sc(OTf)_3$ are parallel with those of outer-sphere electron-transfer reactions of $[(N4Py)Fe^{IV}(O)]^{2+}$ [482]. Such parallel relations suggest that the oxidation of toluene and thioanisole derivatives by $[(N4Py)Fe^{IV}(O)]^{2+}$ in the presence of HOTf and $Sc(OTf)_3$ proceeds via the rate-determining electron-transfer steps [482]. The larger log k_{obs} values than the log k_{et} values of outer-sphere electron-transfer reactions result from the larger K values of precursor complexes formed prior to electron transfer, because the stronger interaction with toluene and thioanisole derivatives is expected as compared with that with coordinatively saturated metal complexes. This was confirmed by the driving force dependence of logarithm of the first-order rate constants (k_{ET}) of electron transfer in the precursor complexes, which were determined from the dependence of the pseudo-first-order rate constants on concentrations of substrates [482].

The unified correlations of driving force dependence of log k_{ET} of oxidation of all kinds of substrates by $[(N4Py)Fe^{IV}(O)]^{2+}$ in Figure 9.48 indicate that oxidation of toluene and thioanisole derivatives by $[(N4Py)Fe^{IV}(O)]^{2+}$ in the presence of HOTf and $Sc(OTf)_3$ proceeds via the rate-determining PCET and MCET from substrates to $[(N4Py)Fe^{IV}(O)]^{2+}$, following formation of the precursor complexes as shown in Scheme 9.20, provided that the free energy change of electron transfer is smaller than 0.5 eV [471]. When the free energy change of electron transfer is larger than 0.5 eV, HAT and OAT processes become predominant [482].

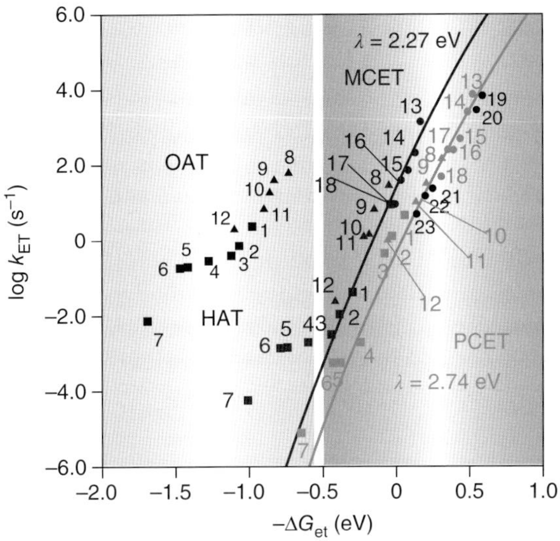

Figure 9.48 Plots of log k_{ET} for C—H bond cleavage of toluene derivatives and sulfoxidation of thioanisole derivatives [(1) hexamethylbenzene, (2) 1,2,3,4,5-tetramethylbenzene, (3) 1,2,4,5-tetramethylbenzene, (4) 1,2,4-trimethylbenzene, (5) 1,4-dimethylbenzene, (6) 1,3,5-trimethylbenzene, (7) toluene, (8) *p*-Me–thioanisole, (9) thioanisole, (10) *p*-Cl–thioanisole, (11) *p*-Br–thioanisole, and (12) *p*-CN–thioanisole] by $[(N4Py)Fe^{IV}(O)]^{2+}$ in the absence (black) and presence of acids (10 mM), HOTf (black), and $Sc(OTf)_3$ (grey), in MeCN at 298 K vs. the driving force of electron transfer $[-\Delta G_{et} = e(E_{red} - E_{ox})]$ from toluene derivatives (squares) and thioanisole derivatives (triangles) to $[(N4Py)Fe^{IV}(O)]^{2+}$ in the presence of HOTf (black) and $Sc(OTf)_3$ (grey). The black and grey circles show the driving force dependence of the rate constants (log k_{et}) of electron transfer from electron donors [(13) $[Fe^{II}(Ph_2phen)_3]^{2+}$, (14) $[Fe^{II}(bpy)_3]^{2+}$, (15) $[RuII(4,4'\text{-}Me_2phen)_3]^{2+}$, (16) $[Ru^{II}(5,5'\text{-}Me_2phen)_3]^{2+}$, (17) $[Fe^{II}(Clphen)_3]^{2+}$, and (18) $[Ru^{II}(bpy)_3]^{2+}$ to $[(N4Py)Fe^{IV}(O)]^{2+}$ in the presence of acids (10 mM), HOTf and $Sc(OTf)_3$, in MeCN at 298 K, respectively. The black circles show the driving force dependence of the rate constants (log k_{et}) of electron transfer from electron donors [(19) decamethylferrocene, (20) octamethylferrocene, (21) 1.1′-dimethylferrocene, (22) *n*-amylferrocene, and (23) ferrocene] to $[(N4Py)Fe^{IV}(O)]^{2+}$ in the absence of acids in MeCN at 298 K. Source: Park et al. 2014 [482]. Reproduced with permission of American Chemical Society.

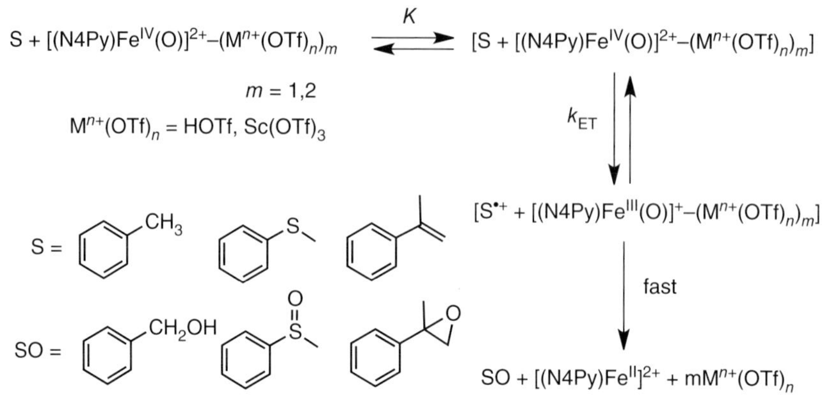

Scheme 9.20 Unified mechanism of oxidation of toluene and thioanisol derivatives by $[(N4Py)Fe^{IV}(O)]^{2+}$ in the presence of HOTf and $Sc(OTf)_3$. Source: Park et al. 2014 [482]. Reproduced with permission of American Chemical Society.

The rate-determining PCET and MCET of toluene derivatives are followed by rapid proton transfer and the oxygen rebound to produce the corresponding benzyl alcohol derivatives and $[(N4Py)Fe^{II}]^{2+}$, which is oxidized by $[(N4Py)Fe^{IV}(O)]^{2+}$ to $[(N4Py)Fe^{III}]^{3+}$ in the presence of HOTf and $Sc(OTf)_3$ [482]. In the case of oxidation of thioanisole derivatives and styrene derivatives by $[(N4Py)Fe^{IV}(O)]^{2+}$ in the presence of HOTf and $Sc(OTf)_3$, the PCET and MCET are followed by rapid $O^{\cdot-}$ transfer to produce the corresponding sulfoxide and epoxide derivatives and $[(N4Py)Fe^{II}]^{2+}$ (Scheme 9.19), which is also oxidized by $[(N4Py)Fe^{IV}(O)]^{2+}$ to $[(N4Py)Fe^{III}]^{2+}$ in the presence of HOTf and $Sc(OTf)_3$ [482, 489].

In the case of oxidation of *N*,*N*-dimethylaniline (DMA) by $[(N4Py)Fe^{IV}(O)]^{2+}$, competition between DMA and $[(N4Py)Fe^{IV}(O)]^{2+}$ for the binding of Sc^{3+} ion results in overall enhancement of electron transfer from DMA to $[(N4Py)Fe^{IV}(O)]^{2+}$, leading to the formation of DMA dimer radical cation ($TMB^{\cdot+}$, tetramethylbenzidine radical cation) [490], in contrast to the demethylation in the absence of an acid [491–493].

In contrast to the case of $[(N4Py)Fe^{IV}(O)]^{2+}$, two HOTf molecules can bind to $[(N4Py)Mn^{IV}(O)]^{2+}$ to produce $[(N4Py)Mn^{IV}(O)]^{2+}–(HOTf)_2$ (vide supra), when electron transfer from electron donors to $[(N4Py)Mn^{IV}(O)]^{2+}–(HOTf)_2$ proceeds via an outer-sphere pathway rather than an inner-sphere pathway because of the steric effects of two HOTf molecules bound to the oxo moiety of $[(N4Py)Mn^{IV}(O)]^{2+}$ [494]. In such a case, the logarithm of the rate constants of epoxidation of styrene derivatives with $[(N4Py)Mn^{IV}(O)]^{2+}–(HOTf)_2$, sulfoxidation of thioanisole derivatives, and electron transfer from coordinatively saturated metal complexes such as ferrocene derivatives to $[(N4Py)Mn^{IV}(O)]^{2+}–(HOTf)_2$ exhibit a remarkably unified correlation with the driving force of outer-sphere electron transfer in light of the Marcus theory of electron transfer (Figure 9.49) [494]. Thus, mechanisms of OAT (epoxidation and sulfoxidation) reactions of styrene and thioanisole derivatives and HAT (hydroxylation) reactions of toluene derivatives by $[(N4Py)Mn^{IV}(O)]^{2+}–(HOTf)_2$ have been unified as the same reaction pathway via outer-sphere electron transfer, followed by the fast bond-forming step, which exhibits the singly unified electron-transfer driving force dependence of the rate constants as outer-sphere electron-transfer reactions [494].

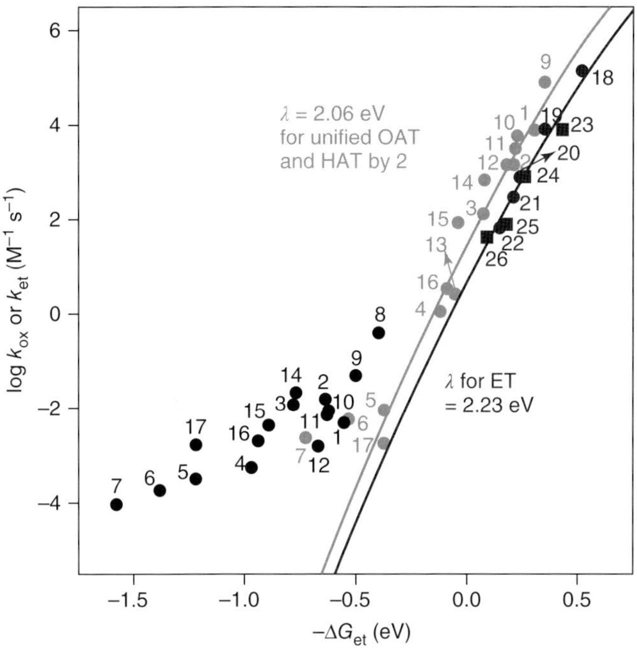

Figure 9.49 Electron-transfer driving force ($-\Delta G_{et}$) dependence of log k_{ox} for epoxidation of styrene derivatives [(1) 4-MeO–styrene, (2) *trans*-stilbene, (3) *cis*-stilbene, (4) 4-Me–styrene, (5) styrene, (6) 3-Cl–styrene, and (7) 2,6-di-Cl–styrene], sulfoxidation of thioanisole derivatives [(8) 4-MeO–thioanisole, (9) 4-Me–thioanisole, (10) thioanisole, (11) 4-F–thioanisole, (12) 4-Br–thioanisole, and (13) 4-CN–thioanisole] and hydroxylation of toluene derivatives [(14) hexamethylbenzene, (15) pentamethylbenzene, (16) 1,2,4,5-tetramethylbenzene, and (17) mesitylene] by $[(N4Py)Mn^{IV}(O)]^{2}$ (black circles) and $[(N4Py)Mn^{IV}(O)]^{2+}$–$(HOTf)_2$ (grey circles) and log k_{et} for electron transfer from various electron donors [coordinatively saturated metal complexes; (18) $[Ru^{II}(Me_2\text{-bpy})_3]^{2+}$, (19) $[Ru^{II}(bpy)_3]^{2+}$, (20) $[Ru^{II}(5\text{-Cl-phen})_3]^{2+}$, (21) $[Ru^{II}(5\text{-Br-bpy})_3]^{2+}$, and (22) $[Ru^{II}(5\text{-NO}_2\text{-phen})_3]^{2+}$] to $[(N4Py)Mn^{IV}(O)]^{2+}$–$(HOTf)_2$ (black circles) in TFE/MeCN (1 : 1 v/v) at 273 K. The black squares show the driving force dependence of the rate constants (log k_{et}) of electron transfer from electron donors [(23) ferrocene, (24) bromoferrocene, (25) acetylferrocene, and (26) dibromoferrocene] to $[(N4Py)Mn^{IV}(O)]^{2+}$ in TFE/MeCN (1 : 1 v/v) at 273 K. Source: Lee et al. 2019 [494]. Reproduced with permission of American Chemical Society.

In the case of the epoxidation of *cis*-stilbene by $[(N4Py)Mn^{IV}(O)]^{2+}$–$(HOTf)_2$, the isomerization of *cis*-stilbene radical cation to *trans*-stilbene radical cation occurs after outer-sphere electron transfer from *cis*-stilbene to $[(N4Py)Mn^{IV}(O)]^{2+}$–$(HOTf)_2$ to yield *trans*-stilbene oxide selectively. This is taken as evidence for the occurrence of electron transfer in the acid-catalyzed epoxidation, because *cis*-stilbene radical cation is rapidly isomerized to *trans*-stilbene radical cation, which affords *trans*-stilbene oxide selectively [494].

9.11 MCET of Metal-Peroxo Complexes

The reaction of $[(TMC)Fe^{II}]^{2+}$ with $O_2^{\cdot-}$–M^{n+} affords the iron(III)-peroxo-metal ion complex ($[(TMC)Fe^{III}(O_2)]^{+}$–M^{n+}) [495, 496]. The $[(TMC)Fe^{III}(\mu,\eta^2:\eta^2\text{-}O_2)]^{+}$–$M^{n+}$ complexes were also synthesized by reacting $[(TMC)Fe^{III}(O_2)]^{+}$

with redox-inactive metal ions [495, 496]. The absorption maxima due to the ligand to metal charge transfer (LMCT) transition of $[(TMC)Fe^{III}(O_2)]^{+}–M^{n+}$ vary depending on the Lewis acidity of metal ions (Figure 9.50) and the $h\nu_{max}$ value increased linearly with increasing the Lewis acidity of M^{n+} (ΔE), which was determined from the g_{zz} values of $O_2^{\cdot -}–M^{n+}$ complexes as shown in Figure 9.51 [498]. It is of interest to note that the $h\nu_{max}$ value of $[(TMC)Fe^{III}(O_2)]^{+}–Ca^{2+}$ is similar to that of $[(TMC)Fe^{III}(O_2)]^{+}–Sr^{2+}$ [496]. The EXAFS data together with DFT calculations indicate that the optimized $Fe–M^{2+}$ distances and the $Fe–O–M^{2+}$ angles in $[(TMC)Fe^{III}(O_2)]^{+}–Ca^{2+}$ and $[(TMC)Fe^{III}(O_2)]^{+}–Sr^{2+}$ are 4.11 Å and 4.34 Å and 140° and 142°, respectively [496].

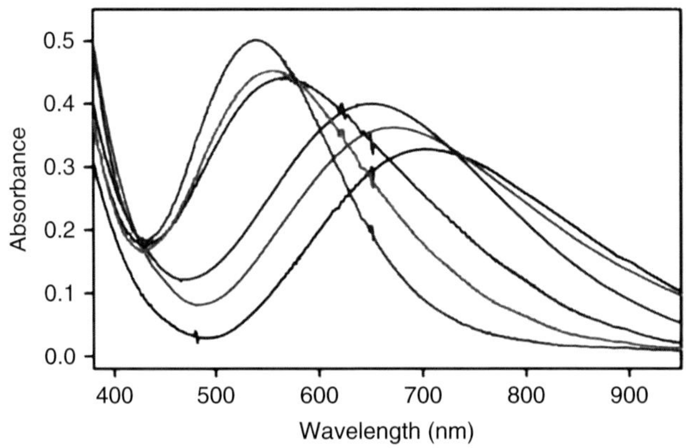

Figure 9.50 UV–vis spectra of $[(TMC)Fe^{III}(O_2)]–M^{n+}$ complexes ($M^{n+} = Sc^{3+}, Y^{3+}, Lu^{3+}, Zn^{2+}, Ca^{2+}$, and Sr^{2+}) obtained in the reaction of the isolated $[(TMC)Fe^{III}(O_2)](OTf)$ (black, 0.50 mM) with redox-inactive metal ions (M^{n+}) in MeCN at −20 °C: λ_{max} = 535 nm (Sc^{3+}), λ_{max} = 560 nm (Y^{3+}), λ_{max} = 570 nm (Lu^{3+}), λ_{max} = 650 nm (Zn^{2+}), λ_{max} = 670 nm (Ca^{2+}), and λ_{max} = 710 nm (Sr^{2+}). Source: Bang et al. 2014 [496]. Reproduced with permission of Springer Nature.

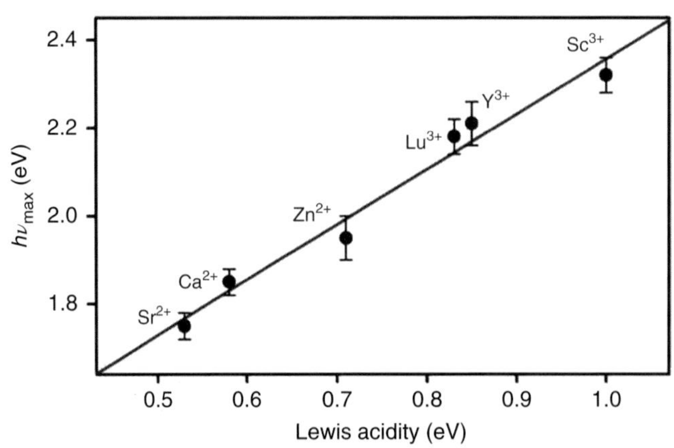

Figure 9.51 Plot of $h\nu_{max}$ of $[(TMC)Fe^{III}(O_2)]^{+}–M^{n+}$ complexes vs. Lewis acidity of metal ions (ΔE). Source: Bang et al. 2014 [496]. Reproduced with permission of Springer Nature.

It was found that the electron-transfer reduction of $[(TMC)Fe^{III}(O_2)]^+–M^{n+}$ to undergo the O—O bond cleavage was accelerated by increasing the Lewis acidity of M^{n+}, but the electron-transfer oxidation of $[(TMC)Fe^{III}(O_2)]^+–M^{n+}$ to release O_2 was decelerated by increasing the Lewis acidity of M^{n+} [496]. Thus, it was proposed that in addition to the role(s) of the Ca^{2+} (or Sr^{2+}) ion in facilitating the O—O bond making process in water oxidation reaction, another possible role of the Ca^{2+} (or Sr^{2+}) ion is to release O_2 by controlling the redox potential of the manganese-peroxo (or –hydroperoxo) species in oxygen evolving complex (OEC) [496]. The stronger Lewis acid that binds to the $Mn^V(O)$ complex in OEC may facilitate the O—O bond formation with water but decelerate the oxidation of the resulting Mn^{III}-peroxo (or Mn^{III}-hydroperoxo) complex. This may be a reason why nature does not choose a stronger Lewis acidic redox-inactive metal ion (e.g. Zn^{2+}) but instead chooses Ca^{2+} ion in the oxidation of water to evolve O_2 and why Sr^{2+} ion, which has virtually the same Lewis acidity as Ca^{2+}, is the only surrogate that replaces Ca^{2+} ion for the reactivity of the OEC in PSII [497].

10

Catalytic Reduction of O_2

Bimetallic catalysis also plays a vital role in the four-electron reduction of O_2 to water, which is the reverse process of photosynthesis, to maintain the life of an aerobic organism by respiration [4, 5, 498, 499]. Cytochrome *c* oxidases (C*c*Os), the last protein in the electron transport chain, are responsible for catalyzing the reduction of O_2 to water by the soluble electron carrier cytochrome *c* [498, 499]. The X-ray structures of C*c*Os have revealed that the catalytic site of C*c*Os consists of the bimetallic complex of heme *a* and Cu (Fe_{a3}/Cu_B) where the two metal ions in this binuclear center are 4.5 Å apart and coordinate a hydroxide ion in the fully oxidized state (159,160) [500, 501]. The catalytic four-electron reduction of O_2 has merited not only great biological interest but also technological significance such as fuel cells [502–507]. The most important question is how the C*c*O enzyme catalyzes the four-electron reduction of O_2 to water without releasing the two-electron reduced species (H_2O_2) [508]. This has been clarified by studying the catalytic four-electron reduction of O_2 vs. two-electron reduction of O_2 using a one-electron reductant in a homogeneous solution (vide infra).

The catalytic reduction of O_2 by ferrocene derivatives (one-electron reductants such as cytochrome *c*) with a monomer cobalt porphyrin and cofacial dicobalt porphyrins was examined in the presence of perchloric acid ($HClO_4$) in PhCN [509]. This has allowed for the difference between the single and bimetallic systems to be clarified.

No oxidation of ferrocene derivatives occurs by O_2 in the presence of $HClO_4$ in PhCN at 298 K. The addition of cobalt porphyrin catalysts and $HClO_4$ to an air-saturated MeCN solution of ferrocene derivatives results in efficient oxidation of ferrocene derivatives by O_2 [386, 509, 510]. The formation of ferrocenium ion was monitored by the rise in absorbance at 620–700 nm (ε(650 nm) = 290 M^{-1} cm^{-1} for $[Fe(C_5H_4Me)_2]^+$) [386]. The concentration of $[Fe(C_5H_4Me)_2]+$ (3.4×10^{-3} M) formed in the Co(OEP)-catalyzed reduction of O_2 by $Fe(C_5H_4Me)_2$ is twice that of the O_2 concentration (1.7×10^{-3} M) (Figure 10.1a) [509]. Thus, only two-electron reduction of O_2 occurs and no further reduction occurs to produce more than 2 equiv of $[Fe(C_5H_4Me)_2]^+$ (Eq. (10.1)). It was confirmed that H_2O_2 (1.6×10^{-3} M) is formed in the two-electron reduction of O_2 by iodometric measurements [509].

$$2Fe(C_5H_4Me)_2 + O_2 + 2H^+ \xrightarrow[\text{Co(OEP)}]{} 2[Fe(C_5H_4Me)_2]^+ + H_2O_2 \qquad (10.1)$$

Electron Transfer: Mechanisms and Applications, First Edition. Shunichi Fukuzumi.

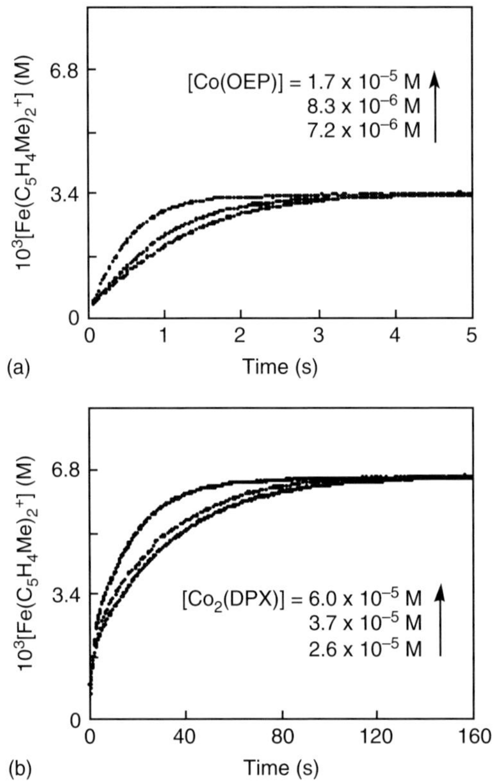

Figure 10.1 Time profiles of formation of $[Fe(C_5H_4Me)_2]^+$ monitored at 650 nm ($\varepsilon = 290\ M^{-1}\ cm^{-1}$) in the ET oxidation of $Fe(C_5H_4Me)_2$ (1.0×10^{-1} M) by O_2 (1.7×10^{-3} M), catalyzed by (a) Co(OEP) and (b) Co_2(DPX) in the presence of $HClO_4$ (2.0×10^{-2} M) in PhCN at 298 K. Source: Fukuzumi et al. 2004 [509]. Reproduced with permission of American Chemical Society.

When a cofacial dicobalt porphyrin [Co_2(DPX)] (DPX = bisporphyrin with dimethylxanthene spacer) is used as a catalyst instead of a single cobalt porphyrin Co(OEP), the concentration of $[Fe(C_5H_4Me)_2]^+$ formed in the Co_2(DPX)-catalyzed reduction of O_2 by $Fe(C_5H_4Me)_2$ (6.8×10^{-3} M in the case of Co_2(DPX) compared to 3.4×10^{-3} M in the case of Co(OEP)) is four times the O_2 concentration (1.7×10^{-3} M) as shown in Figure 10.1b [509]. Thus, the four-electron reduction of O_2 by $Fe(C_5H_4Me)_2$ occurs efficiently in the presence of a catalytic amount of Co_2(DPX) and $HClO_4$ (2.0×10^{-2} M) in PhCN (Eq. (10.2)). It was

$$4Fe(C_5H_4Me)_2 + O_2 + 4H^+ \xrightarrow[Co_2(DPX)]{} 4[Fe(C_5H_4Me)_2]^+ + 2H_2O \qquad (10.2)$$

confirmed that no H_2O_2 was formed in the catalytic reduction of O_2 by $Fe(C_5H_4Me)_2$ [509]. The rates of two-electron and four-electron reduction of O_2 increase with increasing the catalyst concentration (Figure 10.1) [509].

The rate of formation of $[Fe(C_5H_5)_2]^+$ in the Co(OEP)-catalyzed ET oxidation of $Fe(C_5H_5)_2$ by O_2 in the presence of $HClO_4$ in O_2-saturated PhCN at 298 K obeyed pseudo-first-order kinetics [509]. The pseudo-first-order rate constant (k_{obs}) increases linearly with an increase in the catalyst concentration [509]. From the slope the second-order rate constant of the catalytic electron transfer (k_{cat}) is determined. The k_{cat} values remain constant with the change in O_2 and $HClO_4$ concentrations [509]. The k_{cat} values determined from the formation of

$[Fe(C_5H_5)_2]^+$ and $[Fe(C_5H_4Me)_2]^+$ in the Co(OEP)-catalyzed ET oxidation of $Fe(C_5H_5)_2$ and $Fe(C_5H_4Me)_2$ by O_2 are found to be two times of the k_{et} values of electron transfer from $Fe(C_5H_5)_2$ and $Fe(C_5H_4Me)_2$ to $[Co^{III}(OEP)]^+$ in the absence of O_2, respectively [509]. This indicates that the turnover-determining step (t.d.s.) for the catalytic two-electron reduction of O_2 is the ET step from ferrocene derivatives to $[Co^{III}(OEP)]^+$ as shown in Scheme 10.1 [509]. The initial electron transfer from $Fe(C_5H_5)_2$ to $[Co^{III}(OEP)]^+$ is followed by the fast electron transfer from $Co^{II}(OEP)$ to O_2 in the presence of an acid to produce the $[Co^{III}(OEP)(O_2H)]^+$, which is further reduced by $Fe(C_5H_5)_2$ in the presence of an acid to produce H_2O_2, accompanied by regeneration of $[Co^{III}(OEP)]^+$. The catalytic mechanism of two-electron reduction of O_2 in Scheme 9.20 is virtually the same as that reported for (TPP)Co-catalyzed two-electron reduction of O_2 by ferrocene derivatives [386].

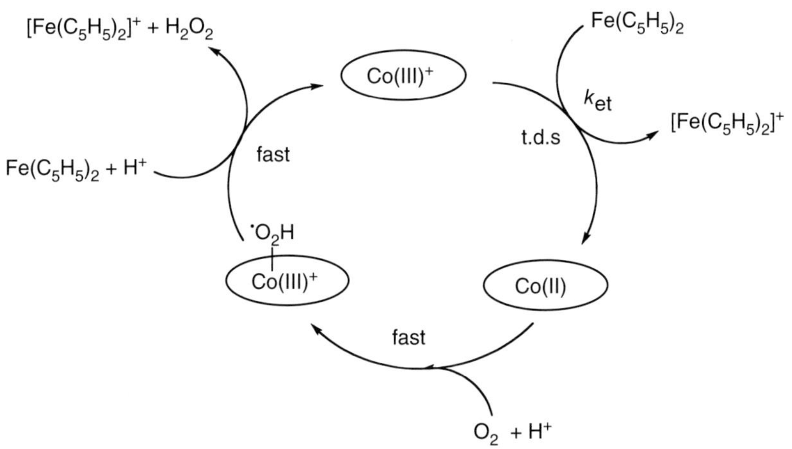

Scheme 10.1 The catalytic cycle for the two-electron reduction of O_2 by $Fe(C_5H_5)_2$ with Co(OEP) in the presence of $HClO_4$ in PhCN. Source: Fukuzumi et al. 2004 [509]. Reproduced with permission of American Chemical Society.

According to Scheme 9.20, the rate of formation of ferrocenium ions is given by Eq. (10.3), where the catalytic rate constant (k_{cat}) corresponds to $2k_{et}$.

$$d[Fe(C_5H_5)_2{}^+]/dt = 2k_{et}[Fe(C_5H_5)_2][Co(OEP)] \text{ in the presence of } O_2 \quad (10.3)$$

The other cofacial dicobalt porphyrins [Co_2(DPA), Co_2(DPB), and Co_2(DPD)] (DPA = bisporphyrin with anthracene spacer, DPB = bisporphyrin with biphenylene spacer, DPD = bisporphyrin with dibenzofuran spacer) also catalyze the reduction of O_2 by $Fe(C_5H_4Me)_2$, but the amount of $[Fe(C_5H_4Me)_2]^+$ formed is less than 4 equiv of O_2 [509]. Thus, the clean four-electron reduction of O_2 by $Fe(C_5H_4Me)_2$ occurs only in the case of Co_2(DPX) used as a catalyst. The reported crystal structures of Co_2(DPB) [511], Co_2(DPA) [512], Co_2(DPX) [513], and Co_2(DPD)(2MeOH) [513] indicate that the metal–metal separations in Co_2(DPA) (4.53 Å) and Co_2(DPX) (4.58 Å) are suitable for the oxygen binding

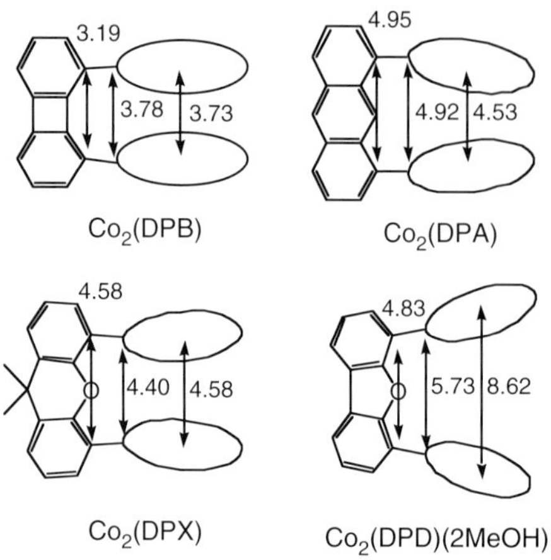

Figure 10.2 Selected distance (Å) in Co_2(DPB), Co_2(DPA), Co_2(DPX), and Co_2(DPD)(2MeOH). Source: Fukuzumi et al. 2004 [509]. Reproduced with permission of American Chemical Society.

between two metal centers, whereas the metal–metal separation in Co_2(DPB) (3.73 Å) is too short and the separation in Co_2(DPD)(2MeOH) (8.62 Å) is too long as shown in Figure 10.2 [509]. In fact, the largest superhyperfine coupling constant is observed for the μ-superoxo species of Co_2(DPX), which acts as the most efficient catalyst for the selective four-electron reduction of O_2 by ferrocene derivatives [509].

The mechanism of the selective catalytic four-electron reduction of O_2 by ferrocene derivatives is proposed as shown in Scheme 10.2 by combining with the catalytic mechanism of the two-electron reduction of O_2 in Scheme 10.1 [509]. The initial two-electron reduction of the $Co^{III}{}_2$ complex by ferrocene derivatives gives the $Co^{II}{}_2$ complex, which reacts with O_2 to produce the μ-peroxo $Co^{III}–O_2–Co^{III}$ complex. The heterolytic O—O bond cleavage of the $Co^{III}–O_2–Co^{III}$ complex affords the high valent Co^{IV}-oxo porphyrin π-radical cation, which is readily reduced by ferrocene derivatives in the presence of proton to yield H_2O. Such high valent metal-oxo species of both heme and non-heme types are known to act as strong electron acceptors [462, 514, 515]. Alternatively, the homolytic O—O bond cleavage affords two Co^{III}-oxyl species, which are reduced by ferrocene derivatives to H_2O in the presence of proton. The critical point to distinguish between the two-electron and four-electron reduction pathways is the formation of the μ-peroxo $Co^{III}–O_2–Co^{III}$ complex, which requires an appropriate Co–Co distance in the cofacial dicobalt complex. The Co–Co distance in Co_2(DPX) is best suited for the formation of the μ-peroxo $Co^{III}–O_2–Co^{III}$ complex, resulting in the catalytic four-electron reduction of O_2.

A kinetic study has indicated that the proton-coupled electron transfer from $[Co^{III}Co^{II}(DPX)]^+$ to O_2 is the turnover-determining step (t.d.s.) in the catalytic four-electron reduction of O_2 in Scheme 10.2 [509]. When $Fe(C_5H_5)_2$ is replaced by a much stronger reductant such as $Fe(C_5Me_5)_2$, the kinetics of formation of $[Fe(C_5Me_5)_2]^+$ changes drastically from normal first-order kinetics in the case of $Fe(C_5H_5)_2$ to zero-order kinetics as shown in Figure 10.3a, where the rate remains constant irrespective of the concentration of $Fe(C_5Me_5)_2$ [509]. Furthermore the zero-order rate constant remains constant with variation of concentrations of $HClO_4$ and O_2 as shown in Figure 10.3c,d, respectively [498]. In contrast, the k_{cat}

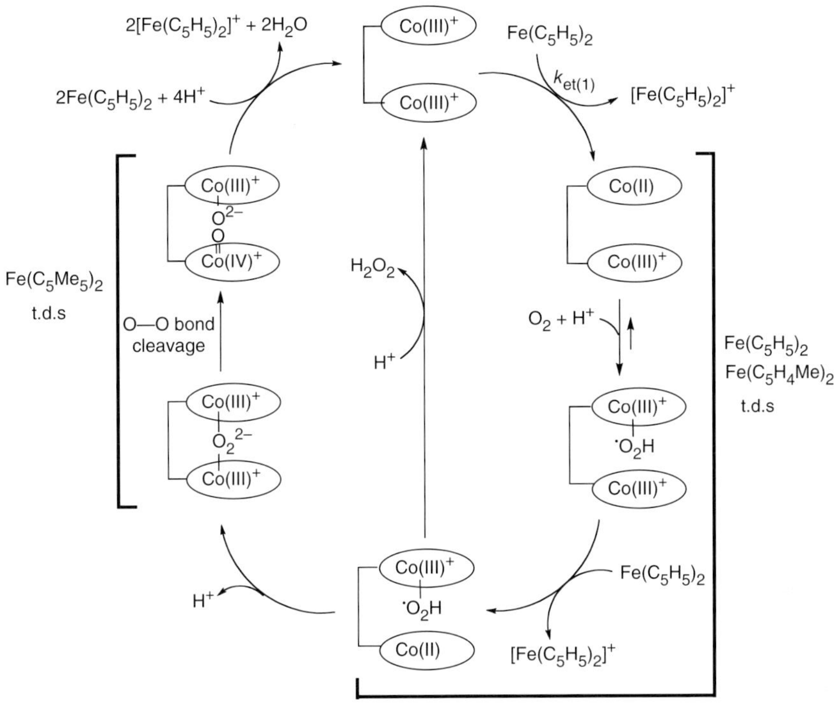

Scheme 10.2 The catalytic mechanism of the four-electron reduction of O_2 by ferrocene with a cofacial cobalt porphyrin. Source: Fukuzumi et al. 2004 [509]. Reproduced with permission of American Chemical Society.

values in the case of $Fe(C_5H_5)_2$ increase linearly with increasing concentrations of $HClO_4$ and O_2 [509]. This indicates that the turnover-determining step changes from the proton-coupled electron transfer from $[Co^{III}Co^{II}(DPX)]^+$ to O_2 in the case of $Fe(C_5H_5)_2$ to the reaction step which has nothing to do with the ET step involving $Fe(C_5Me_5)_2$, $HClO_4$, and O_2. Such a process must be the O—O bond cleavage of the $Co^{III}–O_2–Co^{III}$ complex in Scheme 10.1. The O—O bond cleavage rate has thereby been determined as $320\ s^{-1}$ from the slope in Figure 10.3b [509].

The Co_2(DPA)-catalyzed reduction of O_2 also occurs efficiently by an NADH analog (a two-electron reductant), 9-alkyl-10-methyl-9,10-dihydroacridines (AcrHR: R = Me, Et, and CH_2COOEt), to yield 9-alkyl-10-methylacridinium ion ($AcrR^+$) and H_2O (Scheme 10.3) [516, 517]. In the case of R = Bu^t and CMe_2COOMe, however, the catalytic reduction of O_2 by AcrHR results in oxygenation of the alkyl group of AcrHR rather than dehydrogenation to yield 10-methylacridinium ion ($AcrH^+$) and ROH (Scheme 10.3) [517].

In the case of $AcrH_2$, the initial slow electron transfer from $AcrH_2$ to the $Co^{III}{}_2$ complex is followed by the C(9)-H cleavage in competition with the back electron transfer from the $Co^{III}Co^{II}$ complex to $AcrH_2^{\cdot+}$ to produce AcrH·. Since the catalytic rate-determining step is deprotonation of $AcrH_2^{\cdot+}$, the $Co^{III}Co^{II}$ complex reacts rapidly with O_2 and H^+ to give the $Co^{III}Co^{III}O_2H$ complex, followed by electron transfer from $AcrH^\cdot$ to the $Co^{III}Co^{III}O_2H$ complex to produce

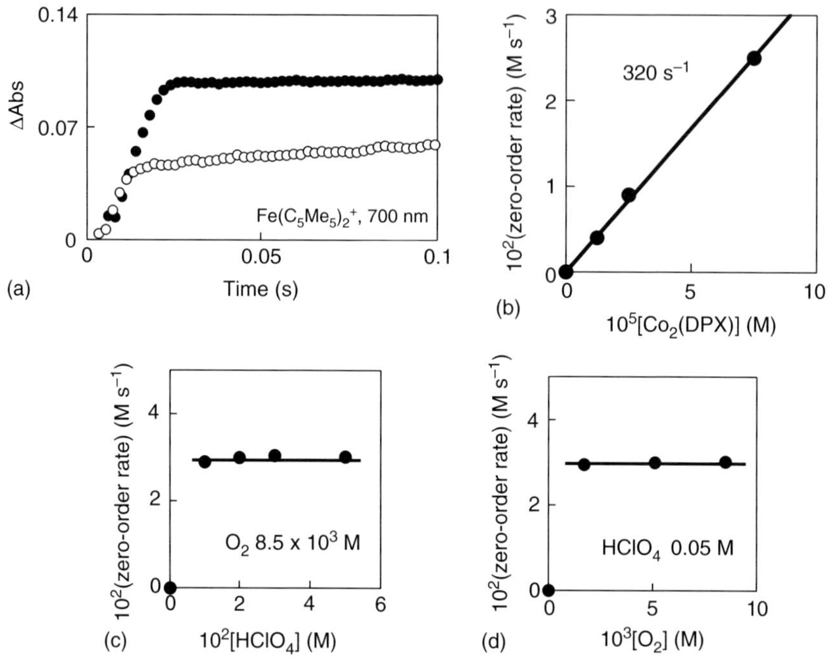

Figure 10.3 (a) Time profiles of formation of $[Fe(C_5Me_5)_2]^+$ monitored by absorbance at 700 nm ($\varepsilon = 240\ M^{-1}\ cm^{-1}$) in the ET oxidation of $Fe(C_5Me_5)_2$ [2.5×10^{-4} M (○), 4.0×10^{-4} M (•)] by O_2 (8.5×10^{-3} M), catalyzed by Co_2(DPX) (8.0×10^{-5} M) in the presence of 0.05 M $HClO_4$ in PhCN at 298 K. Source: (a) Babcock 1999 [498]. Reproduced with permission of PNAS. (b) Plot of the zero-order rate constant vs. [Co_2(DPX)]. (c) Plot of the zero-order rate vs. [$HClO_4$] in the presence of O_2 (8.5×10^{-3} M). (d) Plot of the zero-order rate vs. [O_2] in the presence of 0.05 M $HClO_4$. Source: (b–d)Fukuzumi et al. 2004 [509]. Reproduced with permission of American Chemical Society.

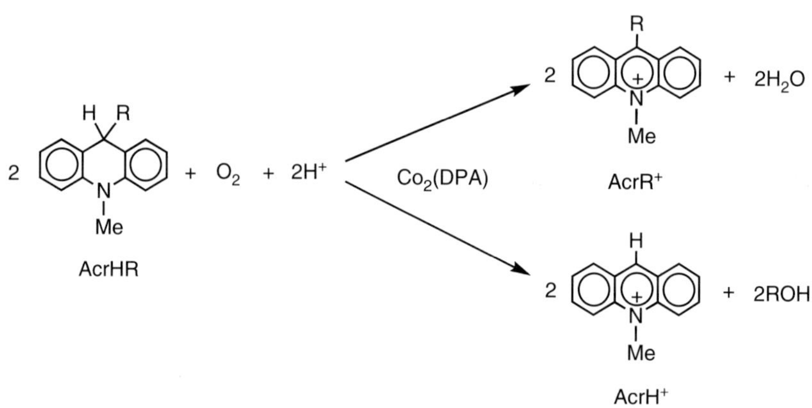

Scheme 10.3 Catalytic dehydration vs. oxygenation of the R group of AcrHR with Co_2(DPA). Source: Fukuzumi et al. 2004 [517]. Reproduced with permission of American Chemical Society.

AcrH$^+$ and the Co^{II}Co^{III}O$_2$H complex [517]. After deprotonation, the μ-peroxo Co^{III}–O$_2$–Co^{III} complex is formed as in the case of the catalytic four-electron reduction of O$_2$ by ferrocene derivatives. The heterolytic O—O-bond cleavage of the Co^{III}–O$_2$–Co^{III} complex affords the high valent Co^{IV}-oxo porphyrin π-radical cation, which is readily reduced by AcrH$_2$ in the presence of proton to yield H$_2$O, accompanied by the formation of AcrH$^+$ [517].

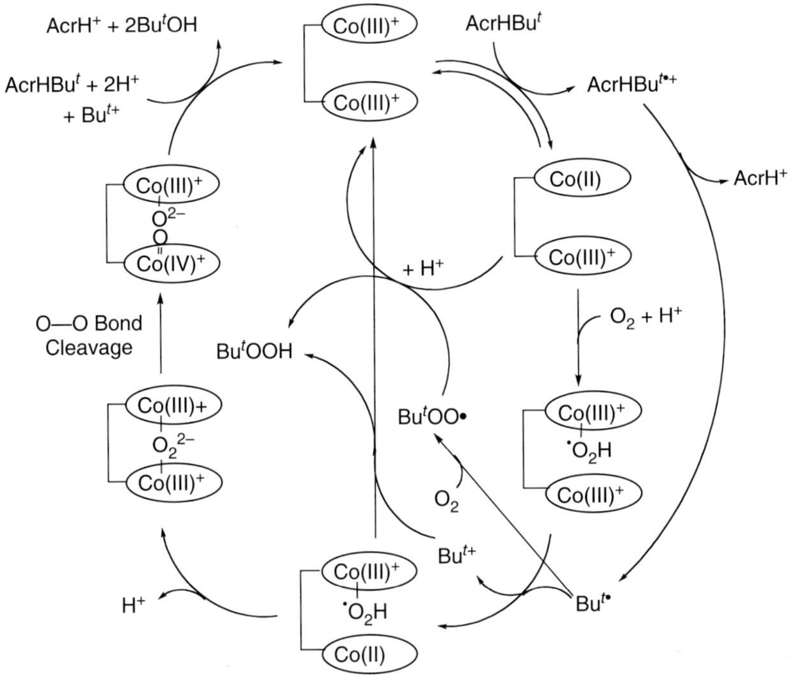

Scheme 10.4 Mechanism of the catalytic oxygenation of the R group of AcrHR by O$_2$ with a cofacial cobalt porphyrin. Source: Fukuzumi et al. 2004 [517]. Reproduced with permission of American Chemical Society.

In the case of AcrHBu^t, the mechanism of the catalytic four-electron reduction of O$_2$, accompanied by the oxygenation of Bu^t, is modified as shown in Scheme 10.4 [517]. The initial electron transfer from AcrHBu^t to the Co$^{III}{}_2$ complex results in the homolytic C(9)—C bond cleavage to produce Bu$^{t\cdot}$ and AcrH$^+$ [518, 519]. Since the homolytic C(9)—C bond cleavage is also the catalytic rate-determining step, the Co^{III}Co^{III}O$_2$H complex is formed by the reaction of O$_2$ and H$^+$, followed by electron transfer from Bu$^{t\cdot}$ (E_{ox} = 0.09 V vs. SCE) [520] to the Co^{III}Co^{III}O$_2$H complex to produce Bu^{t+} and the Co^{II}Co^{III}O$_2$H complex. The subsequent step may be the same as in the case of the four-electron reduction of O$_2$ by AcrH$_2$. The high valent Co^{IV}-oxo porphyrin π-radical cation may be formed by the heterolytic O—O bond cleavage of the Co^{III}–O$_2$–Co^{III} complex, being reduced by AcrHBu^t in the presence of proton to yield Bu^{t}OH, accompanied by the formation of AcrH$^+$ (Scheme 10.4) [517]. However, Bu$^{t\cdot}$ produced in the initial electron transfer from AcrHBu^t to the Co$^{III}{}_2$ complex is readily trapped by O_2 to give the peroxyl radical $Bu^tOO^{\cdot}$ [521]. Such alkylperoxyl radicals ($ROO^{\cdot}$) are regarded as rather strong one-electron oxidants judging from the highly positive one-electron reduction potentials [517]. Thus, the initial electron transfer from $AcrHBu^t$ to the $Co^{III}{}_2$ complex may also be followed by the subsequent electron transfer from the $Co^{III}Co^{II}$ complex to $Bu^tOO^{\cdot}$ to produce Bu^tOOH after protonation, accompanied by regeneration of the $Co^{III}{}_2$ complex (Scheme 10.4) [517]. Thus, the four-electron reduction of O_2 by AcrHR catalyzed by cofacial dicobalt porphyrins occurs efficiently via electron transfer from AcrHR to cofacial dicobalt porphyrins, followed by the C(9)—H and C(9)—C bond cleavage of $AcrHR^{\cdot+}$ depending on the type of R, leading to the dehydration and oxygenation of AcrHR, respectively [517]. In each case, the formation of the μ-peroxo $Co^{III}–O_2–Co^{III}$ complex plays a crucial role in the four-electron reduction of O_2 by AcrHR as in the case of the reduction by ferrocene derivatives [517].

11

Catalytic Oxidation of H_2O

The catalytic mechanism for water oxidation by ceric ammonium nitrate (CAN), $(NH_4)_2[Ce^{IV}(NO_3)_6]$, has been clarified using all-inorganic mononuclear ruthenium complexes bearing Keggin-type lacunary heteropolytungstate, $[Ru^{III}(H_2O)SiW_{11}O_{39}]^{5-}$ and $[Ru^{III}(H_2O)GeW_{11}O_{39}]^{5-}$, as water oxidation catalysts (WOCs) [522]. The Ru^V=O species was produced by the two-electron oxidation of the Ru^{III} complex with 2 equiv of CAN (Scheme 10.3) [522]. The formation of the Ru^V=O complex was detected by electron paramagnetic resonance (EPR) and resonance Raman spectroscopy [522]. A kinetic study suggested that the rate-determining step (r.d.s.) is the nucleophilic attack of H_2O on the Ru^V=O complex to produce the Ru^{III}OOH complex, which is the O—O bond formation step, where the reverse O—O bond cleavage process is in competition with the follow-up two-electron oxidation processes to generate O_2 (Scheme 11.1) [522].

Virtually, the same mechanism, as shown in Scheme 10.4, was proposed for the photocatalytic water oxidation by persulfate using a water-soluble cobalt porphyrin (CoFPS), which possesses fluorine groups at the 2- and 6-positions of the *meso*-aryl rings of the porphyrin. CoFPS exhibited significantly improved stability with a high TON (570), as compared with non-fluorinated Co porphyrin (cobalt tetraphenylporphine tetrasulfonic acid [CoTPPS]: TON = 122) [523]. The increased stability results from resistance of CoFPS against singlet oxygen, the formation of which was monitored in situ by using 9,10-diphenylanthracene as a chemical probe [523]. Dynamic light scattering (DLS) indicated that CoFPS remained homogeneous during the water oxidation catalysis [523].

In the case of the photocatalytic oxidation, 2 equiv of $[Ru(bpy)_3]^{3+}$ produced by the oxidative quenching of $[Ru(bpy)_3]^{2+*}$ by persulfate can oxidize Co^{III}FPS to produce $Co^{IV}(O)FPS^{\cdot+}$, which is equivalent to $Co^V(O)$FPS (Scheme 11.2) [523]. The nucleophilic attack of water on $Co^V(O)$FPS affords the hydroperoxo complex ($[Co^{III}(OOH)FPS]^-$), which is further oxidized by 2 equiv of $[Ru(bpy)_3]^{3+}$ to produce O_2 and regenerate Co^{III}FPS (Scheme 11.2) [523]. The nucleophilic attack of water on the high-valent cobalt-oxo species has also been proposed for electrocatalytic water oxidation with Co hangman corroles [524], non-heme Co complexes [525], and cationic Co porphyrins [526].

When a Ru^{III}-aqua complex, $Ru(k_3^{O,N,N}bda)(L)_2(OH_2)$ (L = 6-F-isoquinoline), was employed as a WOC for the water oxidation by CAN, a Ru^V=O species produced by the oxidation of Ru^{III}-OH_2 by 2 equiv of CAN was proposed to undergo

Electron Transfer: Mechanisms and Applications, First Edition. Shunichi Fukuzumi.

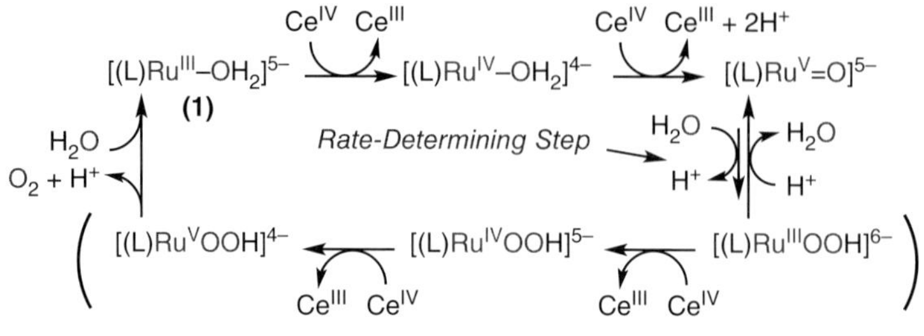

Scheme 11.1 Catalytic cycle for thermal water oxidation with CAN using $[(L)Ru^{III}–OH_2]$, $([Ru^{III}(H_2O)SiW_{11}O_{39}]^{5-}$, and $[Ru^{III}(H_2O)GeW_{11}O_{39}]^{5-})$, as WOCs. Source: Murakami et al. 2011 [522]. Reproduced with permission of American Chemical Society.

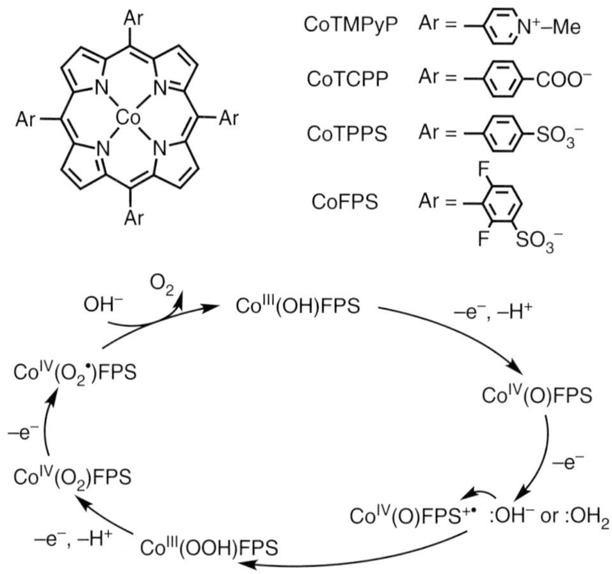

Scheme 11.2 Catalytic cycle for photodriven water oxidation by persulfate using CoFPS as a WOC. Source: Nakazono et al. 2015 [523]. Reproduced with permission of John Wiley & Sons.

radical coupling to produce the dinuclear peroxo complex (Ru^{IV}–O–O–Ru^{IV}), as shown in Scheme 11.3 [527]. O_2 may be evolved directly by the reaction of Ru^{IV}–O–O–Ru^{IV} with water or Ru^{IV}–O–O–Ru^{IV} is further oxidized to a superoxo complex that reacts with H_2O to evolve O_2, accompanied by regeneration of Ru^{III}-OH_2 and Ru^{IV}-OH (Scheme 11.3) [527]. The radical coupling process may be the rate-determining step in the catalytic water oxidation with CAN, because the rate of water oxidation was proportional to the square of the concentration of $Ru^{III}–OH_2$ [527]. The formation of the radical superoxo intermediate $[Ru^{IV}–O–O–Ru^{IV}]^{\cdot 3+}$ was indicated in the electrochemical experiments and supported by density functional theory (DFT) calculations [527]. However, the direct detection of the intermediates has yet to be made. $[Ru(bda)(6\text{-fluoroisoquinoline})_2]$ showed an extremely high catalytic activity with a time of flight technique (TOF) of $1000\ s^{-1}$, whereas $[Ru(bda)(6\text{-bromophthalazine})_2]$ exhibited a high TON (~100 000) [527].

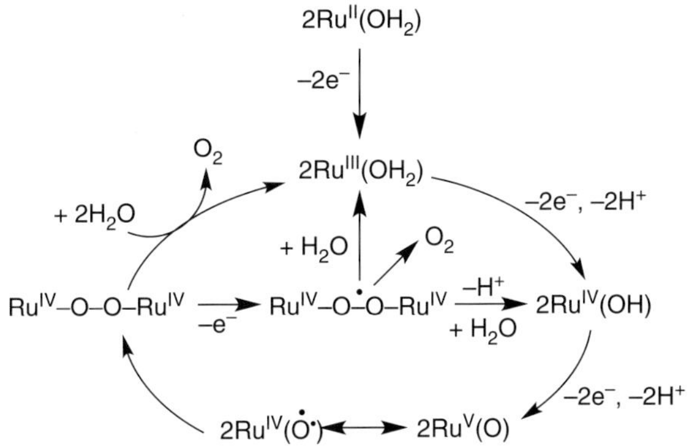

Scheme 11.3 Catalytic cycle for water oxidation by CAN with a Ru^{III}-aqua complex (Ru–OH_2: [Ru($k_3^{O,N,N}$bda)(L)$_2$(OH_2)] (L = 6-F isoquinoline)). Source: Wang et al. 2014 [527]. Reproduced with permission of Royal Society of Chemistry.

A large TON (5300) was obtained in the photocatalytic water oxidation by persulfate with a ruthenium tris(bipyridyl)-type dye (RuPS) and a pyrazolate-based dinuclear ruthenium water oxidation catalyst, {[Ru^{II}(py-SO_3)$_2$(H_2O)]$_2$(*m*-Mebbp)}$^-$, at pH 7.0 [528]. The mechanism of O—O bond formation (nucleophilic attack of water in Schemes 10.3 and 10.4 vs. radical coupling in Scheme 11.2) in the catalytic water oxidation can be controlled by the relative position of the Ru–O units, in addition to the electronic effects exerted by the auxiliary ligands [529–531].

A radical coupling mechanism was also proposed for the O—O bond formation in the photocatalytic water oxidation by persulfate with $[Ru(bpy)_3]^{2+}$ and CoTPPS, as shown in Scheme 11.4 [532]. The observation of a second-order dependence of the rate on concentration of CoTPPS in the photocatalytic water oxidation by persulfate indicates that the radical coupling of two Co-oxyl radicals is the rate-determining step in Scheme 11.4 [532]. The change in the O—O bond formation mechanism from the nucleophilic attack of water on the Co^V(O)

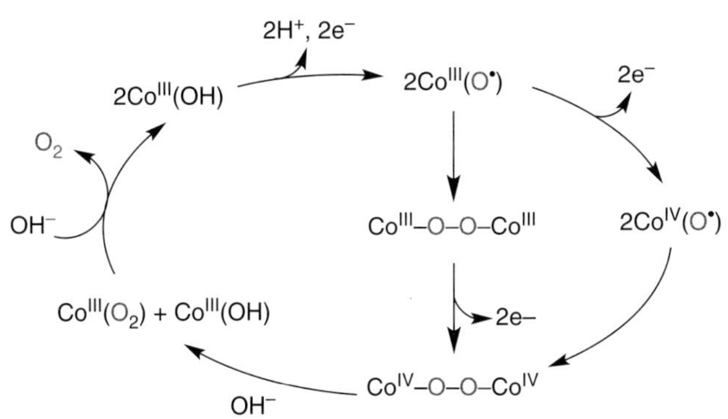

Scheme 11.4 Catalytic cycle for photodriven water oxidation by persulfate with $[Ru(bpy)_3]^{2+}$ as a photocatalyst and CoFPS as a WOC. Source: Nakazono et al. 2013 [532]. Reproduced with permission of Royal Society of Chemistry.

species (or $Co^{IV}(O)$ porphyrin radical cation) of CoFPS in Scheme 11.2 to the radical coupling of the $Co^{IV}(O)$ species of CoTPPS in Scheme 11.4 was suggested on the basis of the observation of less steric effect in CoTPPS relative to CoFPS owing to the shift of the sulfate groups from the 3- to the 4-positions on the aryl rings for the bimolecular radical coupling reaction [532]. The calculated Mulliken spin densities at the oxo ligand for the formally Co^V complexes show that CoFPS has more radical character on the oxyl oxygen (spin = 1.42) than CoTPPS does (spin = 1.09), thus suggesting that CoFPS may be more efficient for the bimolecular radical coupling reaction [532]. However, the electronic effect seems to be less important in the catalytic water oxidation with cobalt porphyrins. Any intermediates in Scheme 11.4 have yet to be detected or identified.

A bis-hydroxo-bridged dinuclear Co^{III}-pyridylmethylamine complex, $[\{Co(TPA)\}_2(\mu\text{-}OH)_2]^{4+}$ (**1**), acts as a homogeneous catalyst for photocatalytic water oxidation by persulfate with $[Ru(bpy)_3]^{2+}$, affording a high quantum yield (88%) and chemical yield (72%) with a high turnover number (TON = 742) for O_2 formation without forming catalytically active Co-oxide (CoO_x) nanoparticles [533]. The pH dependence of the catalytic activity indicates that the active species of water oxidation is a species (**2**) derived from the two-electron proton-coupled electron-transfer (PCET) oxidation of **1** and the PCET formation of **2** is the rate-determining step of the water oxidation, as shown in Scheme 11.5 [533].

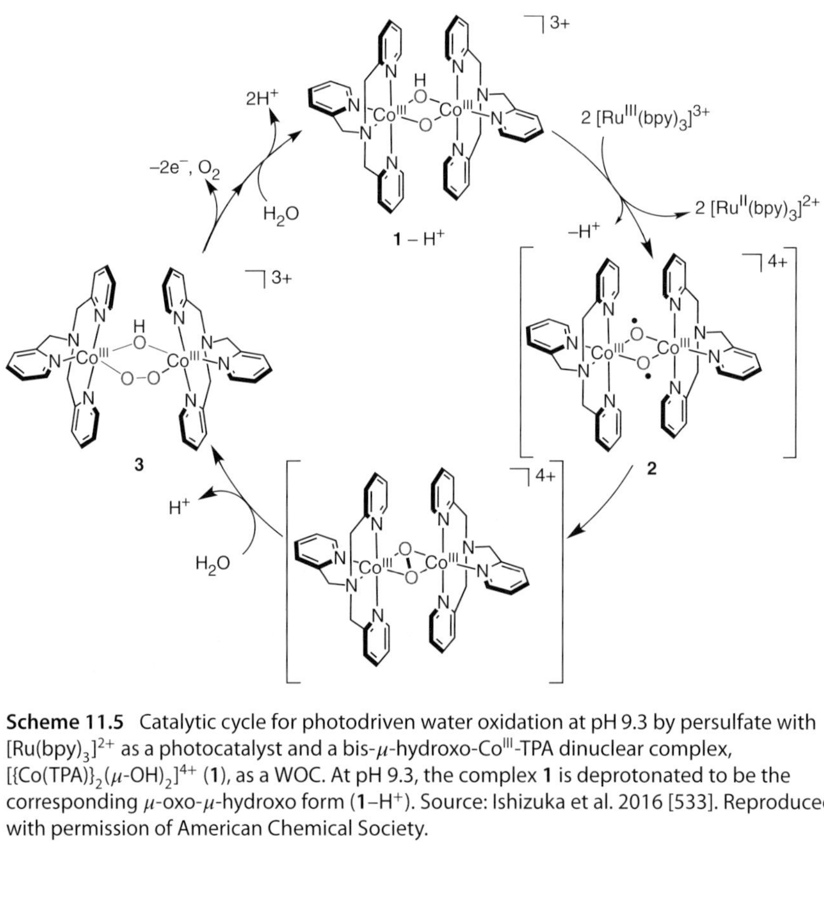

Scheme 11.5 Catalytic cycle for photodriven water oxidation at pH 9.3 by persulfate with $[Ru(bpy)_3]^{2+}$ as a photocatalyst and a bis-μ-hydroxo-Co^{III}-TPA dinuclear complex, $[\{Co(TPA)\}_2(\mu\text{-}OH)_2]^{4+}$ (**1**), as a WOC. At pH 9.3, the complex **1** is deprotonated to be the corresponding μ-oxo-μ-hydroxo form (**1**–H^+). Source: Ishizuka et al. 2016 [533]. Reproduced with permission of American Chemical Society.

The DFT calculations suggest that the most stable state of **2** is the triplet state of the bis-μ-oxyl Co^{III}-dinuclear complex, $[Co^{III}{}_2(\mu\text{-}O^{\cdot-})_2(TPA)_2]^{4+}$, rather than a bis-$\mu$-oxo-$Co^{IV}$ dinuclear complex [533]. The bis-μ-oxyl-$Co^{III}{}_2$ complex in the singlet state is thermally accessible on the basis of the result, suggesting that the singlet species is only 0.5 kcal mol^{-1} higher than the triplet state, being converted thermally to a putative μ:η^2,η^2-peroxo-Co^{III} dinuclear complex, the energy of which is only higher by 0.7 kcal mol^{-1} than that of the singlet bis-μ-oxyl-$Co^{III}{}_2$ complex [533]. Based on isotope-labeling experiments, the O—O bond formation has been proposed to occur via intramolecular radical coupling between the two oxyl ligands located diagonally in the singlet $Co^{III}{}_2(\mu\text{-}O^{\cdot-})_2$ rectangular core to afford the putative μ:η^2,η^2-peroxo-Co^{III} intermediate, which reacts with water to produce a proposed μ-peroxo-μ-hydroxo dinuclear Co^{III} complex (**3**) [533]. The complex **3** is further oxidized to evolve O_2, accompanied by regeneration of **1** (Scheme 11.5) [533].

12

Production of Hydrogen Peroxide from Water and Oxygen as a Solar Fuel

H_2O_2 was produced by the electrocatalytic O_2 reduction by using an H-type cell with an agar salt bridge [534]. Cobalt porphyrins in Figure 10.1 were used as catalysts, which were mounted on a glassy carbon electrode by a drop-casting method, because cobalt porphyrins are well known to act as selective catalysts for the two-electron reduction of O_2 to produce H_2O_2 [535–542]. Among the Co porphyrins in Figure 12.1, [Co (TCPP)] showed the highest onset potential of 0.3 V for the catalytic O_2 reduction [534]. Based on these results, [Co(TCPP)] and Pt wire electrode were connected to a conventional Si photovoltaic solar cell, which generates electrical power of 0.5 V at 2.5 mA [534]. After 11 hours running, the amount of H_2O_2 produced reached 1.46×10^{-5} mol [534]. Although the catalytic behavior as well as cell and electrode structures should be improved more for better performance, H_2O_2 was produced by the reduction of O_2 using a conventional photovoltaic solar cell [534].

It has been reported that H_2O_2 can be produced by a combination of photoinduced electron transfer from Ru^{II} complexes ($[Ru^{II}L]^{2+}$) to O_2 and the catalytic water oxidation by $[Ru^{III}L]^{3+}$ with a water oxidation catalyst (WOC) as shown in Scheme 12.1 [543]. Photoinduced electron transfer from the excited state of $[Ru^{II}L]^{2+}$ to O_2 produces $[Ru^{III}L]^{3+}$ and $O_2^{\cdot-}$, the latter of which is protonated in the presence of acids to produce $HO_2^{\cdot}$, and the disproportionation of $HO_2^{\cdot}$ yields H_2O_2 and O_2 [543, 544]. Various metal complexes and metal nanoparticles are known to catalyze the water oxidation by $[Ru^{III}L]^{3+}$ to produce O_2 [544–555]. Thus, the overall reaction of the photocatalytic two-electron reduction of O_2 and the catalytic four-electron oxidation of H_2O is the photocatalytic oxidation of H_2O by O_2 to produce H_2O_2 [543]. When $[Ru^{II}(Me_2phen)_3]^{2+}$ (Me_2phen = 4,7-dimethyl-1,10-phenanthroline) and $[Co^{III}(Cp^*)(bpy)(H_2O)]^{2+}$ ($Cp^* = \eta^5$-pentamethylcyclo-pentadienyl, bpy = 2,2′-bipyridine) were used as a photocatalyst and a WOC, respectively, in the presence of $Sc(NO_3)_3$ in water, the optimized quantum yield of the photocatalytic H_2O_2 production at $\lambda = 450$ nm was determined to be 37% [543]. The value of conversion efficiency from solar energy to chemical energy was also determined to be 0.25% [532]. The photocatalytic H_2O_2 production proceeds via photoinduced electron transfer from the excited state of $[Ru^{II}(Me_2phen)_3]^{2+}$ to O_2 to produce $[Ru^{III}(Me_2phen)_3]^{3+}$ and $O_2^{\cdot-}$ that binds to Sc^{3+} [543]. The disproportionation of $O_2^{\cdot-}$–Sc^{3+} in water yields H_2O_2, while 4 equiv of $[Ru^{III}(Me_2phen)_3]^{3+}$ oxidize H_2O to O_2 in the presence of a WOC [543].

Electron Transfer: Mechanisms and Applications, First Edition. Shunichi Fukuzumi.

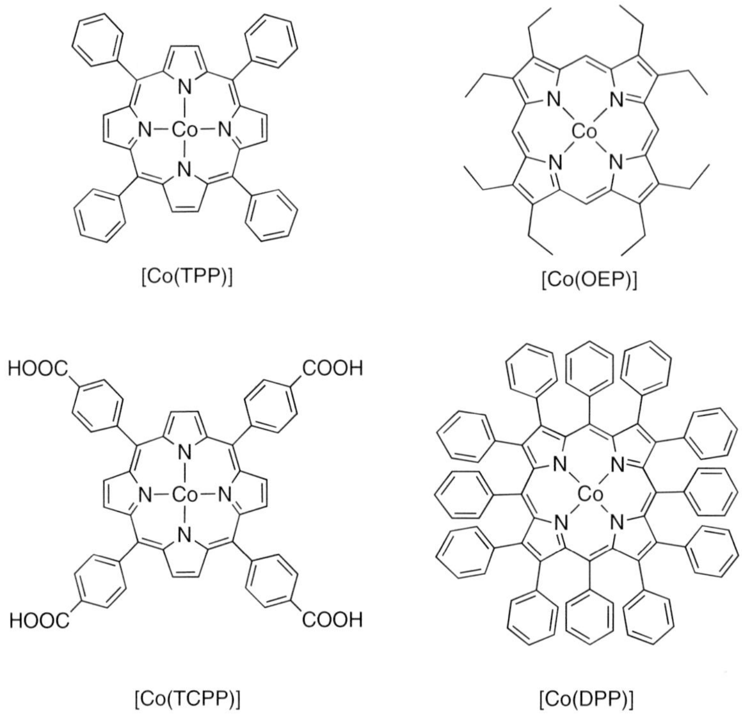

Figure 12.1 Structures of Co porphyrin catalysts for two-electron reduction of O_2. Source: Wua et al. 2015 [545]. Reproduced with permission of Elsevier.

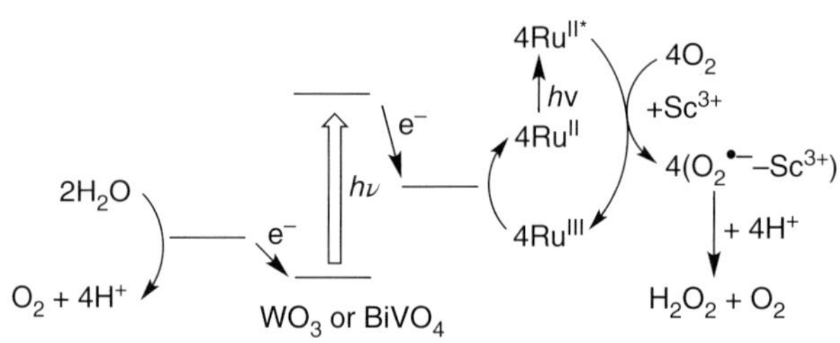

Scheme 12.1 Photocatalytic water oxidation by O_2 to produce H_2O_2. Source: Kato et al. 2013 [543]. https://pubs.rsc.org/en/content/articlelanding/2013/ee/c3ee42815j#!divAbstract; https://creativecommons.org/licenses/by/3.0/. Licensed under CCBY 3.0.

Graphitic carbon nitride containing electron-deficient aromatic diimide units (g-C_3N_4/PDI; PDI = pyromellitic diimide) was reported to act as a photocatalyst for the photocatalytic production of H_2O_2 from H_2O and O_2 [556]. However, the amount of H_2O_2 formed by 48 hours photoirradiation was only 30 μmol [556].

The photocatalytic production of H_2O_2 form H_2O and O_2 was also performed using $NiFe_2O_4$ nanoparticles composed of earth-abundant metals as a WOC [557] in the presence of $[Ru^{II}(Me_2phen)_3]^{2+}$ and $Sc(NO_3)_3$ under visible light irradiation ($\lambda > 420$ nm) [558]. The $NiFe_2O_4$ catalyst was reused after H_2O_2 production ceased due to the decomposition of the photocatalyst, by adding an

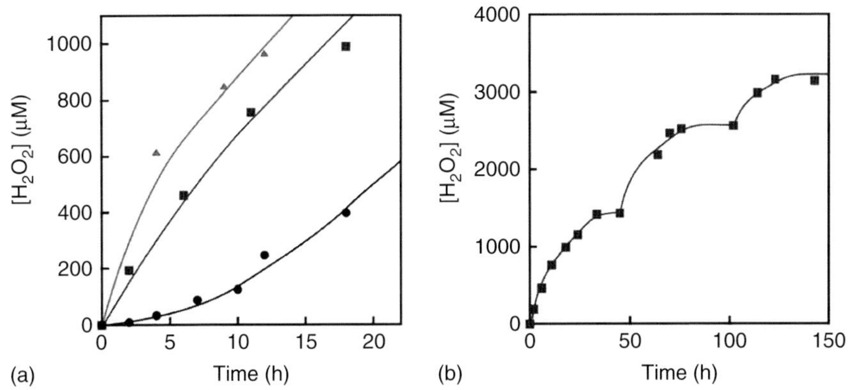

Figure 12.2 (a) Time courses of H_2O_2 production under visible light irradiation ($\lambda > 420$ nm) of $[Ru^{II}(Me_2phen)_3]^{2+}$ (200 μM) in the presence of $Sc(NO_3)_3$ (100 mM) and $NiFe_2O_4$ (0.17 g L^{-1}) with diameters of 1300 nm (black circles), 120 nm (black squares), and 91 nm (grey triangles) in O_2-saturated H_2O (3.0 mL, $[O_2] = 1.2$ mM). (b) Time course of H_2O_2 production in the presence of $NiFe_2O_4$ (0.17 g L^{-1}) and $Sc(NO_3)_3$ (100 mM) under visible light irradiation ($\lambda > 420$ nm) of $[Ru^{II}(Me_2phen)_3]^{2+}$ (200 μM) in O_2-saturated H_2O (3.0 mL, $[O_2] = 1.2$ mM). $[Ru^{II}(Me_2phen)_3]^{2+}$ was added twice to the reaction suspension at 50 and 100 hours during photoirradiation. The amount of $[Ru^{II}(Me_2phen)_3]^{2+}$ added each time at 50 and 100 hours to the reaction suspension was calculated in terms of the concentration increase of 200 μM. Source: Isaka et al. 2015 [558]. Reproduced with permission of Royal Society of Chemistry.

aqueous solution of $[Ru^{II}(Me_2phen)_3]^{2+}$ to the reaction suspension repeatedly. The concentration of H_2O_2 in the resulting suspension increased to be as high as 3.3 mM, ensuring the high stability of the nanoparticles as a WOC (Figure 12.2b) [558]. The initial rate of H_2O_2 production was accelerated 22 and 33 times when using $NiFe_2O_4$ nanoparticles with diameters of 120 and 91 nm, respectively, as compared to the as-prepared $NiFe_2O_4$ with a diameter of 1300 nm (Figure 12.2a) [558]. This increase in reactivity results from an increase in the surface area of the catalyst [558].

A series of heteropolynuclear cyanide metal complexes containing different metal ions, $Co_3[Fe(CN)_6]_2$, $Co_3[Co(CN)_6]_2$, $Cu_3[Co(CN)_6]_2$, $Co[Ni(CN)_4]$, $Fe_3[Cr(CN)_6]_2$, $Mn_3[Fe(CN)_6]_2$, $Co_3[Mn(CN)_6]_2$, $Co_3[Fe(CN)_6]_2$, $Co[Pd(CN)_4]$, and $Co[Pt(CN)_4]$, were also employed as WOCs in the photocatalytic production of H_2O_2 from H_2O and O_2 in an aqueous solution containing $[Ru^{II}(Me_2phen)]_3]^{2+}$ and $Sc(NO_3)_3$ under visible light irradiation with a xenon lamp using a UV light cut filter ($\lambda > 420$ nm) [559]. Among the various heteropolynuclear cyanide complexes, $Fe_3[Co(CN)_6]_2$ exhibited the highest catalytic reactivity [559]. The catalytic reactivity was further enhanced by using heteropolynuclear cyanide complexes $(Fe_xCo_{1-x})_3[Co(CN)_6]_2$ ($x = 0, 0.10, 0.50, 0.75, 0.90$, and 1) as shown in Figure 12.3 [559]. The catalytic activity was maximized when the Fe-to-Co ratio in the (Fe_xCo_{1-x}) moiety of $(Fe_xCo_{1-x})_3[Co(CN)_6]_2$ was 0.75 [559]. A complex with a larger Fe ratio would have more active sites for water oxidation, whereas a complex with a smaller Fe ratio stabilizes high-valence metal ions formed during water oxidation [559]. The optimized ratio of Fe in $(Fe_xCo_{1-x})_3[Co(CN)_6]_2$ ($x = 0.75$) is obtained by the balance of the two opposite effects of Fe on the water oxidation reaction [559].

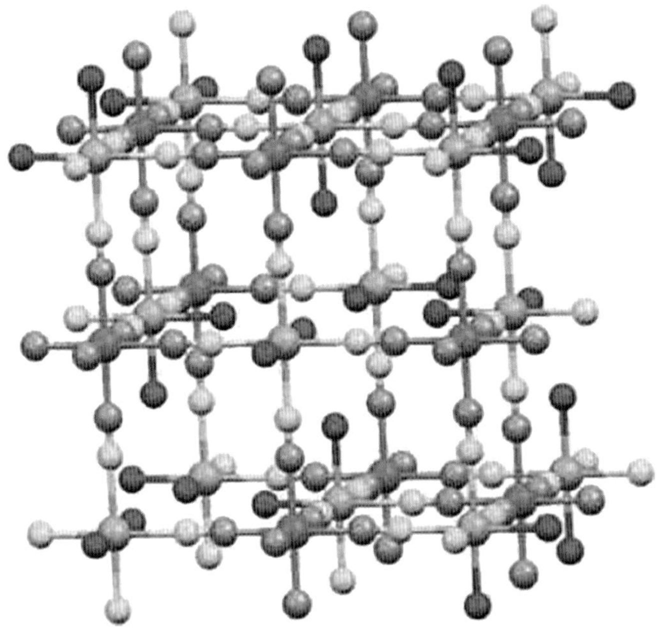

Figure 12.3 A schematic drawing of $(Fe_xCo_{1-x})_3[Co(CN)_6]_2$ where $x = 0, 0.10, 0.50, 0.75, 0.90$, and 1. Source: Isaka et al. 2016 [559]. https://pubs.rsc.org/en/content/articlelanding/2016/cy/c5cy01845e#!divAbstract; https://creativecommons.org/licenses/by/3.0/. Licensed under CCBY 3.0.

When the water oxidation and O_2 reduction were performed in a one-compartment cell as described above, the oxidation of H_2O_2 inevitably competed with the water oxidation in the presence of a WOC to preclude the production of H_2O_2 at higher concentrations. Thus, the photocatalytic production of H_2O_2 from H_2O and O_2 was performed using a two-compartment cell composed of a semiconductor photocatalyst anode in one cell and a carbon cloth cathode in the presence of Ru complexes and $Sc(NO_3)_3$ in an O_2-saturated aqueous solution in the other cell to achieve higher concentrations of H_2O_2 as compared with those with a one-compartment cell [560]. A schematic representation of the two-compartment cell is shown in Figure 12.4, where WO_3 or $BiVO_4$ was used as a photoanode [561, 562] for the photooxidation of water and a carbon mesh cathode was used for photoreduction of O_2 by Ru complexes [560]. The two compartments were separated by a Nafion membrane [561]. When $[Ru^{II}((MeO)_2bpy)_3]^{2+}$ ($(MeO)_2bpy$ = 4,4′-dimethoxy-2,2′-bipyridine) was employed as a photocatalyst for the two-electron reduction of O_2 in the presence of $Sc(NO_3)_3$ (100 mM) in an O_2-saturated aqueous solution at the carbon mesh cathode and the $BiVO_4$ photoanode, the best performance was obtained in terms of the photocatalytic production of H_2O_2 from H_2O and O_2 as shown in Figure 12.5 [560].

The photocatalytic production of H_2O_2 from H_2O and O_2 in the two-compartment cell is shown in Scheme 12.2 [560]. Electron transfer from $[Ru^{II}((MeO)_2bpy)_3]^{2+*}$ to O_2 occurs to produce $[Ru^{III}((MeO)_2bpy)_3]^{3+}$ and $O_2^{\cdot-}$.

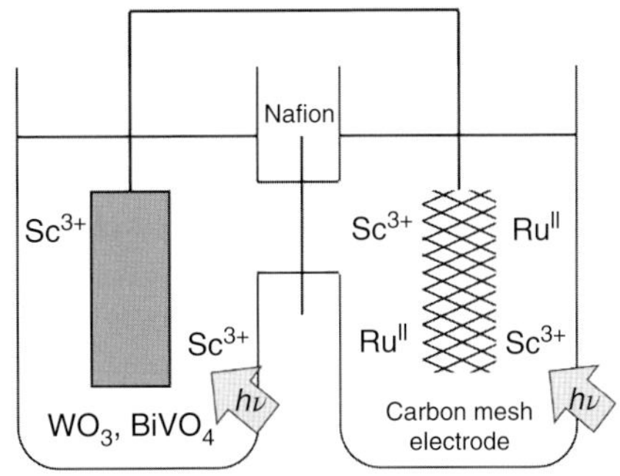

Figure 12.4 Schematic representation of a two-compartment cell employed in this study for photocatalytic production of H_2O_2 from H_2O and O_2 under visible light irradiation. Source: Isaka et al. 2016 [560]. Reproduced with permission of Royal Society of Chemistry.

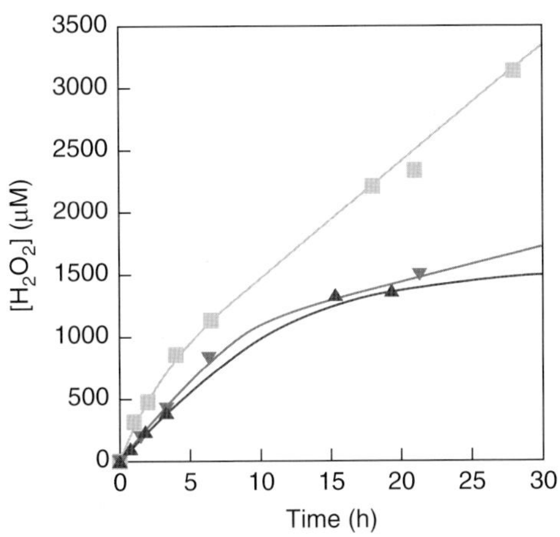

Figure 12.5 Production of H_2O_2 under photoirradiation of a two-compartment cell composed of a semiconductor photocatalyst anode ($BiVO_4$ (grey square and grey inverted triangle) or WO_3 (black triangle)) in one cell and a carbon cloth cathode in the presence of $[Ru^{II}((MeO)_2bpy)_3]^{2+}$ (160 μM, black triangle) or $[Ru^{II}(Me_2phen)_3]^{2+}$ (160 μM, grey inverted triangle and grey square) in the other cell. Both cells were filled with aqueous solution of $Sc(NO_3)_3$ (100 mM, 8.0 mL for each cell). A Xe lamp ($\lambda > 420$ nm) and a solar simulator were used to irradiate the Ru complex and the semiconductor photocatalyst, respectively. Source: Isaka et al. 2016 [560]. Reproduced with permission of Royal Society of Chemistry.

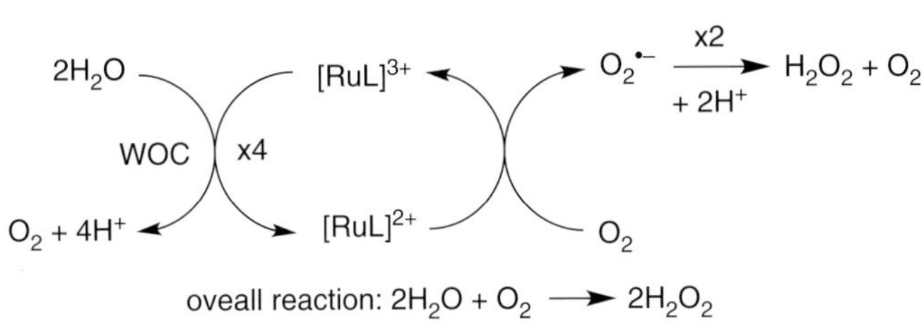

Scheme 12.2 Photocatalytic water oxidation by O_2 to H_2O by double photoexcitation. Source: Isaka et al. 2016 [560]. Reproduced with permission of Royal Society of Chemistry.

Strong binding of Sc^{3+} to $O_2^{\cdot-}$ to give $O_2^{\cdot-}$–Sc^{3+} complex [376, 563] prohibits back electron transfer from the $O_2^{\cdot-}$–Sc^{3+} complex to $[Ru^{III}((MeO)_2bpy)_3]^{3+}$. The $O_2^{\cdot-}$–Sc^{3+} complex disproportionates in the presence of H^+ to produce H_2O_2. On the other hand, the photoexcitation of $BiVO_4$ results in the formation of holes that oxidize water to O_2. Electrons in the conduction band of $BiVO_4$ reduce $[Ru^{III}((MeO)_2bpy)_3]^{3+}$ to regenerate $[Ru^{II}((MeO)_2bpy)_3]^{2+}$.

13

Production and Usage of Hydrogen Peroxide as a Solar Fuel in Seawater

It is highly desired to utilize the most earth-abundant seawater instead of pure water for the practical use of H_2O_2 as a solar fuel because of global shortage of clean water. Efficient photocatalytic production of H_2O_2 from seawater instead of pure water and O_2 in the air was performed in a two-compartment photoelectrochemical cell using WO_3 as a photocatalyst for water oxidation and a cobalt chlorin complex (Co^{II}(Ch)) supported on a glassy-carbon substrate for the selective two-electron reduction of O_2 as shown in Figure 13.1 [563]. The simulated 1 sun illumination of m-WO_3/FTO in the anode cell afforded the efficient photocatalytic production of H_2O_2 in the cathode cell in the two-compartment photoelectrochemical configuration without any external bias potential. The time courses of photocatalytic H_2O_2 production are shown in Figure 13.2 [563]. Very little amount of H_2O_2 was obtained in the absence of Co^{II}(Ch) on CP electrode, indicating that Co^{II}(Ch) adsorbed on CP efficiently catalyzes the two-electron reduction of O_2 to produce H_2O_2 [536, 537], before the charge recombination of the photoexcited electron in the conduction band and h^+ in the VB of WO_3. The rate of photocatalytic production of H_2O_2 in seawater was markedly enhanced compared with that in pure water. After the illumination for 24 hours, the amount of H_2O_2 produced in seawater reached c. 48 mM, which was high enough to operate an H_2O_2 fuel cell (vide infra) [563].

A similar enhancement in photocatalytic activity was observed in a NaCl solution, in which the concentration of Cl^- was the same as that of Cl^- in the seawater. The enhancement effect of Cl^- on the photocatalytic activity for water oxidation can be interpreted by the following Cl^--catalyzed water oxidation mechanism [564–566]. Upon photoexcitation of WO_3, the oxidation of Cl^- by photogenerated hole to form chlorine (Cl_2) occurs prior to the oxidation of water as given by Eq. (13.1) [564]. Cl_2 is converted to HClO depending on the pH of the solution (Eq. (13.2)) [564]. HClO is decomposed to O_2 and Cl^- under solar irradiation

$$2Cl^- + 2h^+ \xrightarrow{h\nu} Cl_2 \tag{13.1}$$

$$Cl_2 + H_2O \rightleftarrows HClO + H^+ + Cl^- \tag{13.2}$$

Electron Transfer: Mechanisms and Applications, First Edition. Shunichi Fukuzumi.

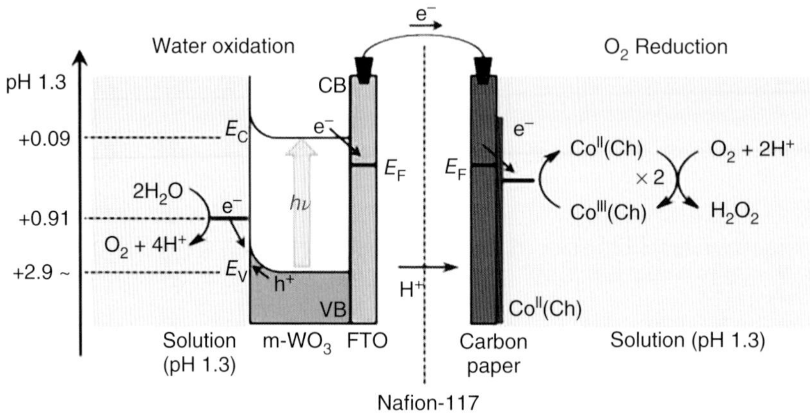

Figure 13.1 Photocatalytic production of H_2O_2 from water and O_2 using m-WO_3/FTO photoanode and $Co^{II}(Ch)$/CP cathode in water or seawater under simulated 1 sun (AM 1.5G) illumination. Source: Mase et al. 2016 [563]. https://www.nature.com/articles/ncomms11470#rightslink; http://creativecommons.org/licenses/by/4.0/. Licensed under CCBY 4.0.

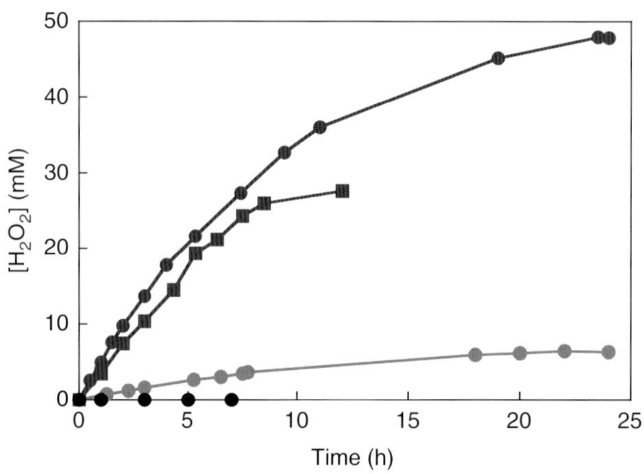

Figure 13.2 Time courses of H_2O_2 production with m-WO_3/FTO photoanode and $Co^{II}(Ch)$/CP cathode in pH 1.3 water (light grey circle), in pH 1.3 seawater (dark grey circle), and in an NaCl aqueous solution (pH 1.3) (black square) under simulated 1 sun (AM 1.5G) illumination. Time course of H_2O_2 production in the absence of $Co^{II}(Ch)$ on carbon paper under simulated 1 sun (AM 1.5G) illumination in pH 1.3 water is shown as black circle. Source: Mase et al. 2016 [563]. https://www.nature.com/articles/ncomms11470#rightslink; http://creativecommons.org/licenses/by/4.0/. Licensed under CCBY 4.0.

(Eq. (13.3)) [565]. Thus, the overall reaction of the water oxidation is catalyzed by Cl^- (Eq. (13.4)). The formation of $HClO/Cl_2$ was confirmed during the photocatalytic water oxidation in the presence of Cl^- [563].

$$2HClO \xrightarrow{h\nu} O_2 + 2H^+ + 2Cl^- \tag{13.3}$$

$$2H_2O + 4h^+ \xrightarrow{h\nu} O_2 + 4H^+ \tag{13.4}$$

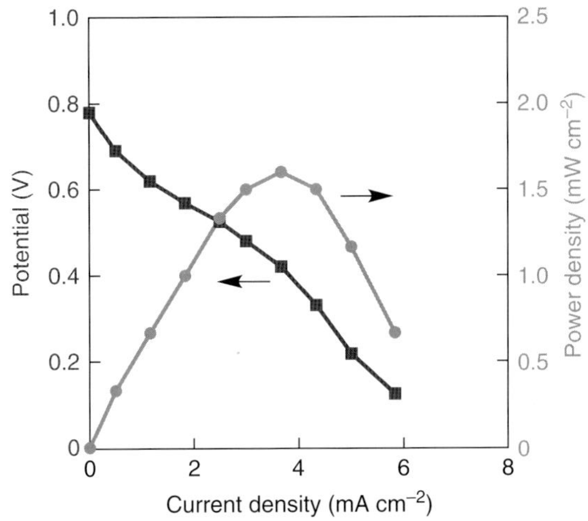

Figure 13.3 *I–V* (light grey) and *I–P* (dark grey) curves of the one-compartment H_2O_2 fuel cell with a Ni mesh anode and $Fe^{II}{}_3[Co^{III}(CN)_6]_2$/carbon cloth cathode in the reaction solution containing H_2O_2 (47.9 mM) produced by photocatalytic reaction in seawater as shown in Figure 5.5 (light grey circle) [563]. Source: Mase et al. 2016 [563]. https://www.nature.com/articles/ncomms11470#rightslink; http://creativecommons.org/licenses/by/4.0/. Licensed under CCBY 4.0.

The amount of O_2 evolved in seawater in the anode cell after one hour (12.7 μmol) was more than three times larger than that in water (3.7 μmol). Thus, the enhancement of photocatalytic production of H_2O_2 in seawater (Figure 13.2) results from the Cl^--catalyzed photooxidation of water [563].

The solar energy conversion efficiency for the photocatalytic production of H_2O_2 in seawater was determined to be 0.55% under simulated 1 sun illumination [563]. The best solar energy conversion efficiency was determined to be 0.94% when the illumination intensity was reduced to 0.1 sun. This efficiency exceeds that of switchgrass (0.2%), which has been considered as a promising crop for biomass fuel [567], and also the value in the one-compartment cell (0.25%) [543].

When surface-modified $BiVO_4$ with iron(III) oxide(hydroxide) (FeO(OH)) and Co^{II}(Ch) were employed as a water oxidation catalyst in the photoanode and as an O_2 reduction catalyst in the cathode, respectively, the highest solar energy conversion efficiency was obtained as 6.6% under simulated solar illumination adjusted to 0.05 sun after one hour of photocatalytic reaction (0.89% under 1 sun illumination) [568].

The chemical energy of H_2O_2 produced by the photocatalytic oxidation of seawater by O_2 in the air was converted to electrical energy through an H_2O_2 fuel cell composed of $Fe^{II}{}_3[Co^{III}(CN)_6]_2$-modified carbon cloth cathode and a nickel mesh anode in a one-compartment cell [563, 569]. The reaction solution (seawater, pH 1.3) containing c. 48 mM of H_2O_2 in the cathode was transferred to the H_2O_2 fuel cell. The cell exhibited an open-circuit potential and a maximum power density of 0.78 V and 1.6 mW cm^{-2}, respectively (Figure 13.3) [563]. The energy conversion efficiency of the H_2O_2 fuel cell was determined to be c. 50% by the measurement of output energy as electrical energy vs. consumed chemical energy H_2O_2, which is comparable to the efficiency of an H_2 fuel cell [563].

14

Photosystem II Mimic

The oxidizing equivalents generated at the donor side of photosystem II (PSII) are used to oxidize water, whereas the reducing equivalents accumulated at the acceptor side of PSII are used to reduce the two quinone molecules, Q_A and Q_B, which act as one-electron and two-electron gates, respectively [570]. Electrons and protons are finally transferred to plastoquinone (PQ) in quinone pool to produce plastoquinol (PQH_2) [571]. The overall solar-driven oxidation of water accompanied by reduction of PQ is given by Eq. (14.1), where PQ is reduced by H_2O to produce O_2 and PQH_2 [571].

$$2H_2O + 2PQ \rightarrow O_2 + 2PQH_2 \tag{14.1}$$

Photodriven oxidation of water accompanied by reduction of *p*-benzoquinone derivatives (X-Q), which is a mimic of Eq. (14.1), has been made possible by using a non-heme iron (II) complex, $[(N4Py)Fe^{II}]^{2+}$ (N4Py = *N*,*N*-bis(2-pyridylmethyl)-*N*-bis(2-pyridyl)methylamine), as a water oxidation catalyst (Eq. (14.2)) [572]. Photoirradiation of an MeCN solution of

$$2H_2O + 2X - Q \rightarrow O_2 + 2PQH_2 \tag{14.2}$$

2,3-dichloro-5,6-dicyano-p-benzoquinone (DDQ) containing $[(N4Py)Fe^{II}]^{2+}$ and H_2O (0.50 M) with a xenon lamp resulted in O_2 evolution as shown in Figure 14.1, where the O_2 yield reaches nearly 100% based on the initial amount of DDQ. DDQ was reduced by H_2O to produce $DDQH_2$ [572]. At prolonged photoirradiation time, the O_2 yield decreased because of the oxidation of $DDQH_2$ by O_2 to produce H_2O_2, accompanied by regeneration of DDQ (Eq. (14.3)) [572]. When DDQ was replaced by *p*-benzoquinone (Q), *p*-chloranil (CA), and

$$DDQH_2 + O_2 \rightarrow DDQ + H_2O_2 \tag{14.3}$$

2,5-dimethyl-*p*-benzoquinone (PXQ), the O_2 yields were 90% [572]. In the case of tetramethyl-*p*-benzoquinone (DQ), however, the O_2 yield became smaller (c. 50%) [572].

The catalytic mechanism of the photodriven water oxidation by X-Q with $[(N4Py)Fe^{II}]^{2+}$ is shown in Scheme 11.5, where DDQ is employed as an oxidant [572]. Photoexcitation of DDQ results in the formation of $^3DDQ^*$ via intersystem crossing from the singlet excited state [573]. Electron transfer from $[(N4Py)Fe^{II}]^{2+}$ to $^3DDQ^*$ occurs to produce $[(N4Py)Fe^{III}]^{3+}$ and $DDQ^{\cdot-}$ with the rate constant of 1.0×10^{10} M^{-1} s^{-1} at 298 K. $[(N4Py)Fe^{III}]^{3+}$ reacts with water to

Electron Transfer: Mechanisms and Applications, First Edition. Shunichi Fukuzumi.

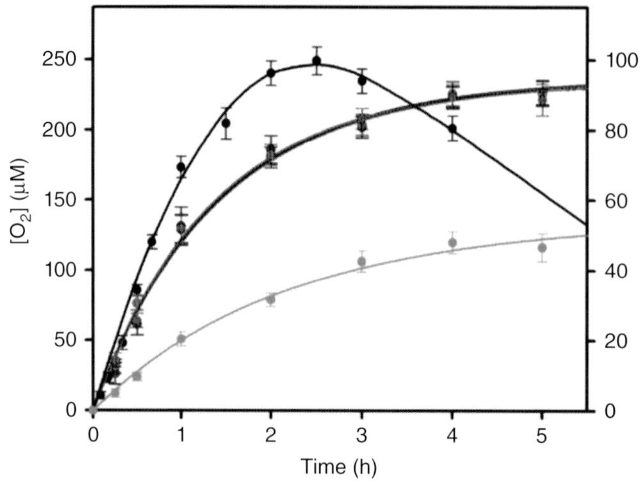

Figure 14.1 Time courses of O_2 evolution by X-Q [0.50 mM; DDQ (black), BQ (second line), CA (second line), DQ (grey), and PXQ (second line)] with $[(N_4Py)Fe^{II}]^{2+}$ (0.20 mM) under photoirradiation (white light) in a deaerated MeCN solution containing H_2O (0.50 M) at 298 K. Source: Hong et al. 2019 [572]. Reproduced with permission of American Chemical Society.

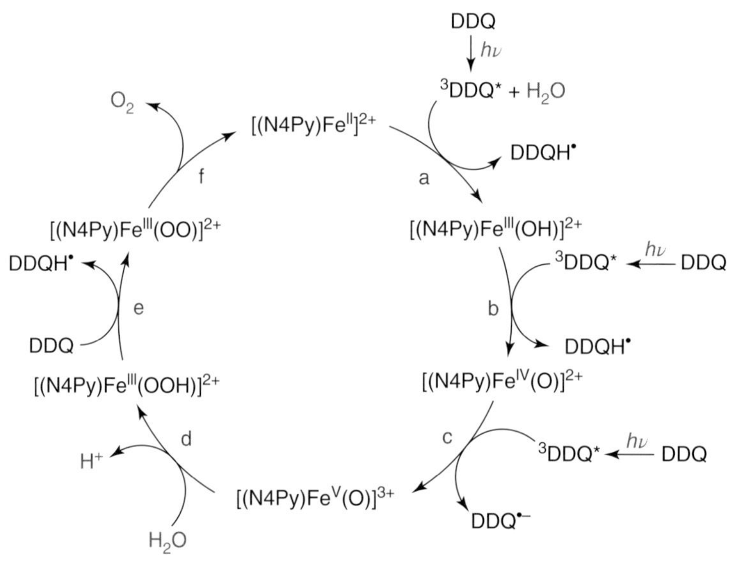

Scheme 14.1 Proposed mechanism of the photodriven water oxidation by DDQ with $[(N4Py)Fe^{II}]^{2+}$. Source: Hong et al. 2019 [572]. Reproduced with permission of American Chemical Society.

produce the Fe(III)-hydroxo complex, $[(N4Py)Fe^{III}(OH)]^{2+}$ (Scheme 14.1, reaction pathway a), whereas $DDQ^{\cdot-}$ is protonated to afford $DDQH^{\cdot}$. Then, electron transfer from $[(N4Py)Fe^{III}(OH)]^{2+}$ to $^3DDQ^*$ occurs to produce the Fe(IV)-oxo complex, $[(N4Py)Fe^{IV}(O)]^{2+}$, and $DDQH^{\cdot}$ (Scheme 14.1, reaction pathway b) [572]. The formation of $[(N4Py)Fe^{IV}(O)]^{2+}$ was confirmed by the absorption spectrum (λ_{max} = 690 nm). $[(N4Py)Fe^{IV}(O)]^{2+}$ is further oxidized by electron transfer from $[(N4Py)Fe^{IV}(O)]^{2+}$ to $^3DDQ^*$ to produce the Fe(V)-oxo complex, $[(N4Py)Fe^{V}(O)]^{3+}$, and $DDQ^{\cdot -}$ with the rate constant of $9.4 \times 10^9\ M^{-1}\ s^{-1}$ at 298 K (Scheme 14.1, reaction pathway c) [572]. $[(N4Py)Fe^{V}(O)]^{3+}$ reacts rapidly with H_2O to produce the Fe(III)-hydroperoxo complex, $[(N4Py)Fe^{III}(OOH)]^{2+}$, and H^+ (Scheme 14.1, reaction pathway d). $DDQ^{\cdot -}$ is protonated to $DDQH^{\cdot}$, which disproportionates to yield DDQ and $DDQH_2$. $[(N4Py)Fe^{III}(OOH)]^{2+}$ is oxidized by DDQ via hydrogen atom transfer to afford $[(N4Py)Fe^{III}(O_2^{\cdot -})]^{2+}$ (Scheme 14.1, reaction pathway e), followed by a rapid release of O_2 from $[(N4Py)Fe^{III}(O_2^{\cdot -})]^{2+}$ to regenerate $[(N4Py)Fe^{II}]^{2+}$ (Scheme 14.1, reaction pathway f) [572]. When $[(N4Py)Fe^{III}(OOH)]^{2+}$ was independently prepared by the reaction of $[(N4Py)Fe^{II}]^{2+}$ and H_2O_2 [574], $[(N4Py)Fe^{III}(OOH)]^{2+}$ was oxidized by DDQ thermally to yield O_2 with 100% yield based on DDQ concentration [572].

15

Conclusion and Perspective

As demonstrated in this book, bioinspired approaches are important in order to design artificial electron-transfer systems including artificial photosynthesis and respiration. The strategies to construct artificial electron-transfer systems are not necessarily the same as those employed in natural systems, although the basic principle to control the electron-transfer processes is the same in light of the Marcus theory of electron transfer. In artificial photosynthesis, simple molecular dyads such as 9-mesityl-10-methylacridininum ion (Acr^+–Mes) have been shown to be capable of fast charge separation but extremely slow charge recombination with minimized energy loss, whereas a significant amount of energy loss is required for the long-range charge separation in the multistep electron-transfer processes in natural photosynthetic reaction centers. Acr^+–Mes can be successfully applied to construct efficient photocatalytic systems including hydrogen evolution as well as organic solar cells by using the Marcus inverted region where the larger the driving force is, the longer the lifetime of the charge-separated state is. Gaseous hydrogen can be stored in the liquid form such as formic acid by efficient interconversion between hydrogen and formic acid under normal pressure and at room temperature. Electron-transfer processes of donor–acceptor dyads containing quinones can be finely modulated by coordination of metal ions to the carbonyl groups. The rate constants of metal ion-coupled electron transfer from electron donors to acceptors increased, exhibiting first-order, second-order, and third-order dependence with respect to concentration of metal ions, when the number of metal ions bound to acceptor radical anions is one, two, and three, respectively. The control of electron-transfer processes by coordination of metal ions to the dyads has led to developing a unique fluorescence sensor for Y^{3+} ion.

A strong binding of one-electron reduced species with metal ion plays an important role in the metal ion-coupled electron-transfer (MCET) reduction of O_2 to control the electron-transfer reactivity by the Lewis acidity of metal ions. In particular, the formation of the μ-peroxo Co(III)–O_2–Co(III) complex is essential for the catalytic four-electron reduction of O_2. The oxidation capability of a nonheme Mn(IV)-oxo complex is much enhanced by binding triflic acid and scandium triflate to the oxo moiety, when epoxidation of styrene, sulfoxidation of thioanisole, and hydroxylation of toluene derivatives by the acid-bound Mn(IV)-oxo complex occur via outer-sphere electron-transfer pathways, exhibiting singly unified driving force dependence of the rate constants.

Electron Transfer: Mechanisms and Applications, First Edition. Shunichi Fukuzumi.

The production of chemical energy utilizing solar energy and its conversion to electrical energy based on H_2O_2 in seawater can provide a practical solution to the construction of an ideal energy-sustainable society using seawater, which is the most earth-abundant resource. Further improvement of the photocatalytic activity for production of H_2O_2 from seawater and O_2 in the air together with more efficient one-compartment H_2O_2 fuel cells without membranes may be required for development of a personal-based sustainable energy conversion system in future.

A functional model of photosystem II in the photosynthesis has been achieved in the photocatalytic oxidation of water by *p*-benzoquinone derivatives with an iron complex via successive electron transfer from an iron (II) complex to the triplet excited state of *p*-benzoquinone derivatives to produce hydroquinone derivatives and the iron(V)-oxo complex that oxidizes water to evolve O_2. The PSII functional model can be combined with a PSI functional model to develop artificial photosynthetic systems. The scope and the applications of bioinspired electron-transfer systems for photosynthesis and respiration to produce sustainable solar fuels and develop solar fuel cells are expected to expand much further in near future.

References

1 (a) Lewis, N.S. and Nocera, D.G. (2006). *Proc. Natl. Acad. Sci. U.S.A.* 103: 15729. (b) Lewis, N.S. (2016). *Nat. Nanotechnol.* 11: 1010.

2 (a) Fukuzumi, S. (2017). *Joule* 1: 689. (b) Fukuzumi, S., Lee, Y.-M., Ahn, H.S., and Nam, W. (2018). *Chem. Sci.* 9: 6017.

3 Faunce, T.A., Lubitz, W., Rutherford, A.W. et al. (2013). *Energy Environ. Sci.* 6: 695.

4 (a) Nelsen, N. (2011). *Biochim. Biophys. Acta* 1807: 856. (b) Nelson, N. and Junge, W. (2015). *Annu. Rev. Biochem.* 84: 659.

5 (a) Medina, M. (2009). *FEBS J.* 276: 3942. (b) Fukuzumi, S., Lee, Y.-M., and Nam, W. (2018). *ChemPhotoChem* 10: 9.

6 (a) Marcus, R.A. (1964). *Annu. Rev. Phys. Chem.* 15: 155. (b) Marcus, R.A. and Sutin, N. (1985). *Biochim. Biophys. Acta, Rev. Bioenerg.* 811: 265. (c) Marcus, R.A. (1993). *Angew. Chem. Int. Ed. Engl.* 32: 1111.

7 (a) Fukuzumi, S. (2008). *Phys. Chem. Chem. Phys.* 10: 2283. (b) Fukuzumi, S. and Ohkubo, K. (2012). *J. Mater. Chem.* 22: 4575.

8 Fukuzumi, S., Ohkubo, K., and Suenobu, T. (2014). *Acc. Chem. Res.* 79: 1455.

9 (a) D'Souza, F. and Ito, O. (2009). *Chem. Commun.*: 4913. (b) Fukuzumi, S. and Kojima, T. (2008). *J. Mater. Chem.* 18: 1427.

10 (a) Nelson, N. and Yocum, C. (2006). *Annu. Rev. Plant Biol.* 57: 521. (b) Nugent, J.H.A. (1996). *Eur. J. Biochem.* 237: 519. (c) Fukuzumi, S., Lee, Y.-M., and Nam, W. (2018). *Biochem. Soc. Trans.* 46: 1279.

11 Williams, L.C. and Taguchi, A.K.W. (1995). *Anoxygenic Photosynthetic Bacteria* (eds. R.E. Blankenship, M.T. Madigan and C.E. Bauer), 1029. Dordrecht: Kluwer.

12 Kirmaier, C., Holten, D., and Parson, W.W. (1985). *Biochim. Biophys. Acta* 810: 49.

13 Lockhart, D.J., Kirmaier, C., Holten, D., and Boxer, S.G. (1990). *J. Phys. Chem.* 94: 6987.

14 Boxer, S.G. (1990). *Annu. Rev. Biophys. Biophys. Chem.* 19: 267.

15 Deisenhofer, J. and Norris, J.R. (eds.) (1993). *The Photosynthetic Reaction Center*. San Diego, CA: Academic Press.

16 Babcock, G.T. and Wikström, M. (1992). *Nature* 356: 301.

Electron Transfer: Mechanisms and Applications, First Edition. Shunichi Fukuzumi.

17 (a) Ferguson-Miller, S. and Babcock, G.T. (1996). *Chem. Rev.* 96: 2889. (b) Adam, S.M., Wijeratne, G.B., Rogler, P.J. et al. (2018). *Chem. Rev.* 118: 10840.

18 (a) Pereira, M.M., Santana, M., and Teixeira, M. (2001). *Biochim. Biophys. Acta* 1505: 185. (b) Zaslavsky, D. and Gennis, R.B. (2000). *Biochim. Biophys. Acta* 1458: 164.

19 Solis, B.H. and Hammes-Schiffer, S. (2014). *Inorg. Chem.* 53: 6427.

20 Warren, J.J. and Mayer, J.M. (2015). *Biochemistry* 54: 186.

21 Migliore, A., Polizzi, N.F., Therien, M.J., and Beratan, D.N. (2014). *Chem. Rev.* 114: 3381.

22 Fukuzumi, S. and Kotani, H. (2012). *RSC Catal. Ser.* 9: 89.

23 Hammes-Schiffer, S. (2012). *Energy Environ. Sci.* 5: 7696.

24 Fukuzumi, S. (1997). *Bull. Chem. Soc. Jpn.* 70: 1.

25 (a) Fukuzumi, S. and Itoh, S. (2001). *Antioxid. Redox Signal.* 3: 807. (b) Fukuzumi, S. and Ohkubo, K. (2010). *Coord. Chem. Rev.* 254: 372.

26 (a) Fukuzumi, S. (2009). *Prog. Inorg. Chem.* 56: 49. (b) Pegis, M.L., Wise, C.F., Martin, D.J., and Mayer, J.M. (2018). *Chem. Rev.* 118: 2340.

27 (a) Fukuzumi, S., Ohkubo, K., and Morimoto, Y. (2012). *Phys. Chem. Chem. Phys.* 14: 8472. (b) Fukuzumi, S., Ohkubo, K., Lee, Y.-M., and Nam, W. (2015). *Chem. Eur. J.* 21: 17548.

28 (a) Miller, J.R., Calcaterra, L.T., and Closs, G.L. (1984). *J. Am. Chem. Soc.* 106: 3047. (b) Closs, G.L. and Miller, J.R. (1988). *Science* 240: 440.

29 (a) Gould, I.R. and Farid, S. (1996). *Acc. Chem. Res.* 29: 522. (b) McLendon, G. (1989). *Acc. Chem. Res.* 21: 160. (c) Winkler, J.R. and Gray, H.B. (1992). *Chem. Rev.* 92: 369. (d) McLendon, G. and Hake, R. (1992). *Chem. Rev.* 92: 481.

30 Fukuzumi, S. (2000). *The Porphyrin Handbook*, vol. 8 (eds. K.M. Kadish, K. Smith and R. Guilard), 115–151. San Diego, CA: Academic Press.

31 Guldi, D.M. and Fukuzumi, S. (2003). *Fullerenes: From Synthesis to Optoelectronic Properties* (eds. D.M. Guldi and N. Martin), 237–265. Dordrecht: Kluwer.

32 Fukuzumi, S., Ohkubo, K., Imahori, H., and Guldi, D.M. (2003). *Chem. Eur. J.* 9: 1585.

33 (a) Fukuzumi, S. and Imahori, H. (2001). *Electron Transfer in Chemistry*, vol. 2 (ed. V. Balzani), 927–975. Weinheim: Wiley-VCH. (b) Fukuzumi, S. and Guldi, D.M. (2001). *Electron Transfer in Chemistry*, vol. 2 (ed. V. Balzani), 270–337. Weinheim: Wiley-VCH.

34 Imahori, H., Tamaki, K., Guldi, D.M. et al. (2001). *J. Am. Chem. Soc.* 123: 2607.

35 Imahori, H., El-Khouly, M.E., Fujitsuka, M. et al. (2001). *J. Phys. Chem. A* 105: 325.

36 Imahori, H., Tkachenko, N.V., Vehmanen, V. et al. (2001). *J. Phys. Chem. A* 105: 1750.

37 Fukuzumi, S., Imahori, H., Yamada, H. et al. (2001). *J. Am. Chem. Soc.* 123: 2571.

38 Imahori, H., Tamaki, K., Araki, Y. et al. (2002). *J. Am. Chem. Soc.* 124: 5165.

39 Imahori, H., Guldi, D.M., Tamaki, K. et al. (2001). *J. Am. Chem. Soc.* 123: 6617.

40 Marguet, S., Hapiot, P., and Neta, P. (1994). *J. Phys. Chem.* 98: 7136.

41 Ohkubo, K., Reimers, J.R., Fukuzumi, S., and Crossley, M.J. (2007). *Phys. Chem. Chem. Phys.* 9: 5260.

42 Guldi, D.M., Imahori, H., Tamaki, K. et al. (2004). *J. Phys. Chem. A* 108: 541.

43 Helms, A., Heiler, D., and McLendon, G. (1992). *J. Am. Chem. Soc.* 114: 6227.

44 Imahori, H., Sekiguchi, Y., Kashiwagi, Y. et al. (2004). *Chem. Eur. J.* 10: 3184.

45 Kashiwagi, Y., Ohkubo, K., McDonald, J.A. et al. (2003). *Org. Lett.* 5: 2719.

46 Fukuzumi, S., Ohkubo, K., Ou, W.E.Z. et al. (2003). *J. Am. Chem. Soc.* 125: 14984.

47 Ohkubo, K., Kotani, H., Shao, J. et al. (2004). *Angew. Chem. Int. Ed.* 43: 853.

48 Fukuzumi, S., Ohkubo, K., Imahori, H. et al. (2001). *J. Am. Chem. Soc.* 123: 10676.

49 Ohkubo, K., Imahori, H., Shao, J. et al. (2002). *J. Phys. Chem. A* 106: 10991.

50 Tkachenko, N.V., Lemmetyinen, H., Sonoda, J. et al. (2003). *J. Phys. Chem. A* 107: 8834.

51 Kesti, T.J., Tkachenko, N., Vehmanene, V. et al. (2002). *J. Am. Chem. Soc.* 124: 8067.

52 (a) Johnson, D.G., Niemczyk, M.P., Minsek, D.W. et al. (1993). *J. Am. Chem. Soc.* 115: 5692. (b) Sakata, Y., Tsue, H., O'Neil, M.P. et al. (1994). *J. Am. Chem. Soc.* 116: 6904. (c) Tsue, H., Imahori, H., Kaneda, T. et al. (2000). *J. Am. Chem. Soc.* 122: 2279. (d) Kang, Y.K., Rubtsov, I.V., Iovine, P.M. et al. (2002). *J. Am. Chem. Soc.* 124: 8275.

53 (a) Asahi, T., Ohkohchi, M., Matsusaka, R. et al. (1993). *J. Am. Chem. Soc.* 113: 5665. (b) Khundkar, L.R., Perry, J.W., Hanson, J.E., and Dervan, P.B. (1994). *J. Am. Chem. Soc.* 116: 9700. (c) Sessler, J.L., Wang, B., and Harriman, A. (1993). *J. Am. Chem. Soc.* 115: 10418.

54 In liquid crystals, the CR rates were attenuated into the submicrosecond time scale; see: (a) Berman, A., Izraeli, E.S., Levanon, H. et al. (1995). *J. Am. Chem. Soc.* 117: 8252. (b) Berg, A., Shuali, Z., Levanon, H. et al. (2001). *J. Phys. Chem. A* 105: 10060.

55 Imahori, H., Yamada, H., Guldi, D.M. et al. (2002). *Angew. Chem. Int. Ed.* 41: 2344.

56 Okamoto, K. and Fukuzumi, S. (2005). *J. Phys. Chem. B* 109: 7713.

57 Fukuzumi, S., Ohkubo, K., Suenobu, T. et al. (2001). *J. Am. Chem. Soc.* 123: 8459.

58 Kikuchi, K., Sato, C., Watabe, M. et al. (1993). *J. Am. Chem. Soc.* 115: 5180.

59 Ohkubo, K., Suga, K., Morikawa, K., and Fukuzumi, S. (2003). *J. Am. Chem. Soc.* 125: 12850.

60 Fukuzumi, S., Kotani, H., Ohkubo, K. et al. (2004). *J. Am. Chem. Soc.* 126: 1600.

61 Harriman, A. (2004). *Angew. Chem. Int. Ed.* 43: 4985.

62 (a) Benniston, A.C., Harriman, A., Li, P. et al. (2005). *Chem. Commun.*: 2701. (b) Benniston, A.C., Harriman, A., Li, P. et al. (2005). *J. Am. Chem. Soc.* 127: 16054.

63 Guo, X., Gan, Z., Luo, H. et al. (2003). *J. Phys. Chem. A* 107: 9747.

64 Osuka, A., Nakajima, S., Maruyama, K. et al. (1993). *J. Am. Chem. Soc.* 115: 4577.

65 Kotani, H., Ohkubo, K., and Fukuzumi, S. (2005). *Chem. Commun.* 3814.

66 Shida, T. (1988). *Electronic Absorption Spectra of Radical Ions*. Amsterdam: Elsevier.

67 Tsudaka, T., Kotani, H., Ohkubo, K. et al. (2017). *Chem. Eur. J.* 23: 1306.

68 Tsudaka, T., Ohkubo, K., and Fukuzumi, S. (2016). *Chem. Commun.* 52: 6178.

69 Xiang, M., Xin, Z.-K., Chen, B. et al. (2017). *Org. Lett.* 19: 3009.

70 Wanga, S., Liua, J., Niua, L. et al. (2018). *J. Photochem. Photobiol., A* 355: 120.

71 Morawski, O. and Prochorow, J. (1995). *Chem. Phys. Lett.* 242: 253.

72 Hoshino, M., Uekusa, H., Tomita, A. et al. (2012). *J. Am. Chem. Soc.* 134: 4569.

73 Fukuzumi, S., Doi, K., Itoh, A. et al. (2012). *Proc. Natl. Acad. Sci. U.S.A.* 109: 15572.

74 Okamura, M.Y. and Feher, G. (1995). *Anoxygenic Photosynthetic Bacteria* (eds. R.E. Blankenship, M.T. Madigan and C.E. Bauer), 577–594. Dordrecht: Kluwer.

75 Mizoshita, N., Yamanaka, K., Shimada, T. et al. (2010). *Chem. Commun.* 46: 9235.

76 Iijima, S. (1991). *Nature* 354: 56.

77 (a) Guldi, D.M. and Martin, N. (eds.) (2010). *Carbon Nanotubes and Related Structures: Synthesis, Characterization, Functionalization, and Applications*. Wiley-VCH. (b) Harris, P.J.F. (2001). *Carbon Nanotubes and Related Structures: New Materials for the Twenty-First Century*. Cambridge: Cambridge University Press. (c) Ma, Y., Wang, B., Wu, Y. et al. (2011). *Carbon* 49: 4098. (d) Zhou, W., Ding, L., Yang, S., and Liu, J. (2010). *J. Am. Chem. Soc.* 132: 336.

78 (a) Avouris, P. (2002). *Acc. Chem. Res.* 35: 1026. (b) Sgobba, V. and Guldi, D.M. (2008). *J. Mater. Chem.* 18: 153.

79 (a) Schnorr, J.M. and Swager, T.M. (2011). *Chem. Mater.* 23: 646. (b) Bitter, J.H. (2010). *J. Mater. Chem.* 20: 7312. (c) Gorityala, B.K., Ma, J., Wang, X. et al. (2010). *Chem. Soc. Rev.* 39: 2925.

80 O'Connell, M.J., Bachilo, S.M., Huffman, C.B. et al. (2002). *Science* 297: 593.

81 Misewich, J.A., Martel, R., Avouris, P. et al. (2003). *Science* 300: 783.

82 (a) Nigoyi, S., Hamon, M.A., Hu, H. et al. (2002). *Acc. Chem. Res.* 35: 1105. (b) Sun, Y.-P., Fu, K., Lin, Y., and Huang, W. (2002). *Acc. Chem. Res.* 35: 1096.

83 (a) Hirsch, A. (2002). *Angew. Chem. Int. Ed.* 41: 1853. (b) Barh, J.L. and Tour, J.M. (2002). *J. Mater. Chem.* 12: 1952. (c) Banerjee, S., Kahn, M.G.C., and Wong, S.S. (2003). *Chem. Eur. J.* 9: 1898.

84 (a) Tasis, D., Tagmatarchis, N., Georgakilas, V., and Prato, M. (2003). *Chem. Eur. J.* 9: 4001. (b) Dyke, C.A. and Tour, J.M. (2004). *Chem. Eur. J.* 10: 812. (c) Banerjee, S., Hemraj-Benny, T., and Wong, S.S. (2005). *Adv. Mater.*

17: 17. (d) Tasis, D., Tagmatarchis, N., and Bianco, A. (2006). *Chem. Rev.* 106: 1105. (e) Chitta, R. and D'Souza, F. (2008). *J. Mater. Chem.* 18: 1440.
85 (a) Guldi, D.M. (2005). *J. Phys. Chem. B* 109: 11432. (b) Guldi, D.M. (2007). *Phys. Chem. Chem. Phys.* 9: 1400. (c) Guldi, D.M., Rahman, G.M.A., Sgobba, V., and Ehli, C. (2006). *Chem. Soc. Rev.* 35: 471.
86 Guldi, D.M. and Sgobba, V. (2011). *Chem. Commun.* 47: 606.
87 Baskaran, D., Mays, J.M., Zhang, X.P., and Bratcher, M.S. (2005). *J. Am. Chem. Soc.* 127: 6916.
88 D'Souza, F., Sandanayaka, A.S.D., and Ito, O. (2010). *J. Phys. Chem. Lett.* 1: 2586.
89 Oelsner, C., Schmidt, C., Hauke, F. et al. (2011). *J. Am. Chem. Soc.* 133: 4580.
90 (a) Ohtani, M. and Fukuzumi, S. (2009). *Chem. Commun.*: 4997. (b) Ohtani, M., Kamat, P.V., and Fukuzumi, S. (2010). *J. Mater. Chem.* 20: 582.
91 (a) Bachilo, S.M., Balzano, L., Herrera, J.E. et al. (2003). *J. Am. Chem. Soc.* 125: 11186. (b) Hata, K., Futaba, D.N., Mizuno, K. et al. (2004). *Science* 306: 1362.
92 (a) Chattopadhyay, D., Lastella, S., Kim, S., and Papadimitrakopoulos, F. (2002). *J. Am. Chem. Soc.* 124: 728. (b) Heller, D.A., Mayrhofer, R.M., Baik, S. et al. (2004). *J. Am. Chem. Soc.* 126: 14567.
93 Endo, M., Kim, Y.A., Hayashi, T. et al. (2002). *Appl. Phys. Lett.* 80: 1267.
94 (a) Kim, C., Kim, Y.J., Kim, Y.A. et al. (2004). *J. Appl. Phys.* 96: 5903. (b) Endo, M., Kim, Y.A., Ezaka, M. et al. (2003). *Nano Lett.* 3: 723.
95 Hasobe, T., Fukuzumi, S., and Kamat, P.V. (2006). *Angew. Chem. Int. Ed.* 45: 755.
96 (a) Hasobe, T., Fukuzumi, S., and Kamat, P.V. (2006). *J. Phys. Chem. B* 110: 25477. (b) Chitta, R., Sandanayaka, A.S.D., Schmacher, A.L. et al. (2007). *J. Phys. Chem. C* 111: 6947.
97 Saito, K., Ohtani, M., Sakata, T. et al. (2006). *J. Am. Chem. Soc.* 128: 14216.
98 Saito, K., Ohtani, M., and Fukuzumi, S. (2007). *Chem. Commun.*: 55.
99 Ohtani, M., Saito, K., and Fukuzumi, S. (2009). *Chem. Eur. J.* 15: 9160.
100 (a) D'Souza, F. and Ito, O. (2012). *Chem. Soc. Rev.* 41: 86. (b) D'Souza, F. and Ito, O. (2005). *Coord. Chem. Rev.* 249: 1410. (c) Fukuzumi, S., Honda, T., Ohkubo, K., and Kojima, T. (2009). *Dalton Trans.*: 3880.
101 Fukuzumi, S., Saito, K., Ohkubo, K. et al. (2011). *Phys. Chem. Chem. Phys.* 13: 17019.
102 (a) D'Souza, F., Maligaspe, E., Ohkubo, K. et al. (2009). *J. Am. Chem. Soc.* 131: 8787. (b) El-Khouly, M.E., Ju, D.K., Kay, K.-Y. et al. (2010). *Chem. Eur. J.* 16: 6193.
103 Kojima, T., Honda, T., Ohkubo, K. et al. (2008). *Angew. Chem. Int. Ed.* 47: 6712.
104 Fukuzumi, S., Ohkubo, K., Saito, K. et al. (2011). *J. Porphyrins Phthalocyanines* 15: 1292.
105 (a) Blanco, M.-J., Consuelo Jimenez, M., Chambron, J.-C. et al. (1999). *Chem. Soc. Rev.* 28: 293. (b) Flamigni, L., Heitz, V., and Sauvage, J.-P. (2006). *Struct. Bond.* 121: 217. (c) Flamigni, L., Talarico, A.M., Chambron, J.-C. et al. (2689). *Chem. Eur. J.* 2004: 10.

106 Beletskaya, I., Tyurin, V.S., Tsivadze, A.Y. et al. (2009). *Chem. Rev.* 109: 165.

107 Katayev, E.A., Ustynyuk, Y.A., and Sessler, J.L. (2006). *Coord. Chem. Rev.* 250: 3004.

108 Kim, S.K. and Sessler, J.L. (2010). *Chem. Soc. Rev.* 39: 3784.

109 Bianchi, A., Bowman-James, K., and García-España, E. (eds.) (1997). *Supramolecular Chemistry of Anions*. New York: Wiley-VCH.

110 Piepenbrock, M.-O.M., Lloyd, G.O., Clarke, N., and Steed, J.W. (2010). *Chem. Rev.* 110: 1960.

111 (a) Hua, Y. and Flood, A.H. (2010). *Chem. Soc. Rev.* 39: 1262. (b) Gamez, P., Mooibroek, T.J., Teat, S.J., and Reedijk, J. (2007). *Acc. Chem. Res.* 40: 435. (c) Amendola, V., Esteban-Gomez, D., Fabbrizzi, L., and Licchelli, M. (2006). *Acc. Chem. Res.* 39: 343.

112 (a) Beer, P.D. and Gale, P.A. (2001). *Angew. Chem. Int. Ed.* 40: 486. (b) Gale, P.D. (2010). *Chem. Soc. Rev.* 39: 3746.

113 (a) Juwarker, H., Suk, J.-M., and Jeong, K.-S. (2009). *Chem. Soc. Rev.* 38: 3316. (b) Chang, K.-J., Kang, B.N., Lee, M.-H., and Jeong, J.S. (2005). *J. Am. Chem. Soc.* 127: 12214.

114 (a) Sessler, J.L. and Davis, J.M. (2001). *Acc. Chem. Res.* 34: 989. (b) Nielsen, K.A., Sarova, G.H., Martín-Gomis, L. et al. (2008). *J. Am. Chem. Soc.* 130: 460. (c) Nielsen, K.A., Cho, W.-S., Lyskawa, J. et al. (2006). *J. Am. Chem. Soc.* 128: 2444.

115 (a) Nishiyabu, R., Palacios, M.A., Dehaen, W., and Anzenbacher, P. Jr., (2006). *J. Am. Chem. Soc.* 128: 11496. (b) Kang, S.K., Powell, D., and Bowman-James, K. (2005). *J. Am. Chem. Soc.* 127: 13478. (c) Kano, K., Kitagishi, H., Tamura, S., and Yamada, A. (2004). *J. Am. Chem. Soc.* 126: 15202.

116 (a) Yoon, D.-W., Gross, D.E., Lynch, V.M. et al. (2008). *Angew. Chem. Int. Ed.* 47: 5038. (b) Gross, D.E., Yoon, D.-W., Lynch, V.M. et al. (2010). *J. Inclusion Phenom. Macrocyclic Chem.* 66: 81. (c) Hay, B.P. (2010). *Chem. Soc. Rev.* 39: 3700.

117 Sessler, J.L., Karnas, E., Kim, S.K. et al. (2008). *J. Am. Chem. Soc.* 130: 15256.

118 (a) Valik, M., Král, V., Herdtweck, E., and Schmidtchen, F.P. (2007). *New J. Chem.* 31: 703. (b) Jadhav, V.D., Herdtweck, E., and Schmidtchen, F.P. (2008). *Chem. Eur. J.* 14: 6098.

119 Fukuzumi, S., Satoh, N., Yuasa, Y., and Ohkubo, K. (2004). *J. Am. Chem. Soc.* 126: 7585.

120 Honda, T., Nakanishi, T., Ohkubo, K. et al. (2010). *J. Am. Chem. Soc.* 132: 10155.

121 (a) Senge, M.O., Forsyth, T.P., Nguyen, L.T., and Smith, K.M. (1994). *Angew. Chem. lnt. Ed. Engl.* 33: 2485. (b) Cheng, B., Munro, O.Q., Marques, H.M., and Scheidt, W.R. (1997). *J. Am. Chem. Soc.* 119: 10732. (c) Gárate-Morales, J.L., Tham, F.S., and Reed, C.A. (2007). *Inorg. Chem.* 46: 1514.

122 (a) Stone, A. and Fleischer, E.B. (1968). *J. Am. Chem. Soc.* 90: 2735. (b) Barakigia, K.M. and Fajer, J. (1995). *Acta Crystallogr.* C51: 511. (c) Medforth, C.J., Haddad, R.E., Muzzi, C.M. et al. (2003). *Inorg. Chem.* 42: 2227.

123 Mizuno, Y., Aida, T., and Yamaguchi, K. (2000). *J. Am. Chem. Soc.* 122: 5278.

124 Honda, T., Kojima, T., and Fukuzumi, S. (2009). *Chem. Commun.*: 4994.
125 Kojima, T., Nakanishi, T., Harada, R. et al. (2007). *Chem. Eur. J.* 11: 8714.
126 Medfortht, C.J., Senge, M.O., Smith, K.M. et al. (1992). *J. Am. Chem. Soc.* 114: 9859.
127 (a) Heitele, H., Pöllinger, F., Häberle, T. et al. (1994). *J. Phys. Chem.* 98: 7402. (b) Fukuzumi, S., Ohkubo, K., Imahori, H. et al. (2001). *J. Am. Chem. Soc.* 123: 10676. (c) Hutchison, J.A., Sintic, P.J., Brotherhood, P.R. et al. (2009). *J. Phys. Chem. C* 113: 11796.
128 (a) Osuka, A., Maruyama, K., Mataga, N. et al. (1990). *J. Am. Chem. Soc.* 112: 4958. (b) Helms, A., Heiler, D., and McLendon, G. (1992). *J. Am. Chem. Soc.* 114: 6221. (c) Pettersson, K., Wiberg, J., Ljungdahl, T. et al. (2006). *J. Phys. Chem. A* 110: 319.
129 (a) Aoyagi, S., Nishibori, E., Sawa, H. et al. (2010). *Nat. Chem.* 4: 678. (b) Aoyagi, S., Sado, Y., Nishibori, E. et al. (2012). *Angew. Chem. Int. Ed.* 51: 3377.
130 Ohkubo, K., Kawashima, Y., and Fukuzumi, S. (2012). *Chem. Commun.* 48: 4314.
131 Kadish, K.M., Rhodes, R.K., and Bottomleym, L.A. (1981). *Inorg. Chem.* 20: 1274.
132 Bill, N.L., Ishida, M., Kawashima, Y. et al. (2014). *Chem. Sci.* 5: 3888–3896.
133 Sun, D., Tham, F.S., Reed, C.A. et al. (2000). *J. Am. Chem. Soc.* 122: 10704.
134 Sun, D., Tham, F.S., Reed, C.A. et al. (2002). *J. Am. Chem. Soc.* 124: 6604.
135 Tanaka, M., Ohkubo, K., Gros, C.P. et al. (2006). *J. Am. Chem. Soc.* 128: 14625.
136 Hernández-Eguía, L.P., Escudero-Adán, E.C., Pintre, I.C. et al. (2011). *Chem. Eur. J.* 17: 14564.
137 Kieran, A.L., Pascu, S.I., Jarrosson, T., and Sanders, J.K.M. (2005). *Chem. Commun.*: 1276.
138 Grimm, B., Schornbaum, J., Cardona, C.M. et al. (2011). *Chem. Sci.* 2: 1530.
139 Hernández-Eguía, L.P., Escudero-Adán, E.C., Pinzón, J.R. et al. (2011). *J. Org. Chem.* 76: 3258.
140 Samanta, S.K. and Schmittel, M. (2013). *Org. Biomol. Chem.* 11: 3108.
141 Börjesson, K., Woller, J.G., Parsa, E. et al. (2012). *Chem. Commun.* 48: 1793.
142 Kang, B., Totten, R.K., Weston, M.H. et al. (2012). *Dalton Trans.* 41: 12156.
143 Fukuzumi, S., Amasaki, I., Ohkubo, K. et al. (2012). *RSC Adv.* 2: 3741.
144 Nobukuni, H., Tani, F., Shimazaki, Y. et al. (2009). *J. Phys. Chem. C* 113: 19694.
145 Saeki, A., Koizumi, Y., Aida, T., and Seki, S. (2012). *Acc. Chem. Res.* 45: 1193.
146 Hoofman, R.J.O.M., de Hass, M.P., Siebbeles, L.D.A., and Warman, J.M. (1998). *Nature* 392: 54.
147 Choi, J.H., Honda, T., Seki, S., and Fukuzumi, S. (2011). *Chem. Commun.* 47: 11213.
148 Frankevich, E., Maruyama, Y., and Ogata, H. (1993). *Chem. Phys. Lett.* 214: 39.
149 Nobukuni, H., Shimazaki, Y., Uno, H. et al. (2010). *Chem. Eur. J.* 16: 11611.
150 Kamimura, T., Ohkubo, K., Kawashima, Y. et al. (2013). *Chem. Sci.* 4: 1451.

151 Kawashima, Y., Ohkubo, K., and Fukuzumi, S. (2012). *J. Phys. Chem. A* 116: 8942.
152 D'Souza, F., Smith, P.M., Gadde, S. et al. (2004). *J. Phys. Chem. B* 108: 11333.
153 Armaroli, N., Diederich, F., Echegoyen, L. et al. (1999). *New J. Chem.*: 77.
154 Da Ros, T., Prato, M., Guldi, D.M. et al. (1999). *Chem. Commun.*: 635.
155 Hauke, F., Swartz, A., Guldi, D.M., and Hirsch, A. (2002). *J. Mater. Chem.* 12: 2088.
156 Takai, A., Chkounda, M., Eggenspiller, A. et al. (2010). *J. Am. Chem. Soc.* 132: 4477.
157 Spur, M., El-Khouly, M.E., Seok, J.H., Kay, K.-Y., and Fukuzumi, S. (2011). *J. Phys. Chem. A* 115: 14430.
158 Schuster, D.I., Loa, K., Guldi, D.M. et al. (2007). *J. Am. Chem. Soc.* 129: 15973.
159 Takai, A., Gros, C.P., Guilard, R., and Fukuzumi, S. (2009). *Chem. Eur. J.* 15: 3110.
160 Takai, A., Gros, C.P., Barbe, J.-M., and Fukuzumi, S. (2011). *Chem. Eur. J.* 17: 3420.
161 Takai, A., Gros, C.P., Barbe, J.-M., and Fukuzumi, S. (2010). *Phys. Chem. Chem. Phys.* 12: 12160.
162 (a) Solladié, N., Hamel, A., and Gross, M. (2000). *Tetrahedron Lett.* 41: 6075. (b) Aziat, F., Rein, R., Peon, J. et al. (2008). *J. Porphyrins Phthalocyanines* 12: 1232.
163 (a) Fujitsuka, M., Hara, M., Tojo, S. et al. (2005). *J. Phys. Chem. B* 109: 33. (b) Fujitsuka, M., Cho, D.W., Solladié, N. et al. (2007). *J. Photochem. Photobiol., A* 188: 346.
164 Saito, K., Troiani, V., Qiu, H. et al. (2007). *J. Phys. Chem. C* 111: 1194.
165 Nielsen, K.A., Cho, W.-S., Jeppesen, J.O. et al. (2004). *J. Am. Chem. Soc.* 126: 16296.
166 Tennyson, A.G., Ono, R.J., Hudnall, T.W. et al. (2010). *Chem. Eur. J.* 16: 304.
167 Park, J.S., Karnas, E., Ohkubo, K. et al. (2010). *Science* 329: 1324.
168 Kadish, K.M. and Ruoff, R.S. (eds.) (2000). *Chemistry, Physics, and Technology*, 1–51. New York: Wiley-Interscience.
169 Yamada, M., Ohkubo, K., Shionoya, M., and Fukuzumi, S. (2014). *J. Am. Chem. Soc.* 136: 13240.
170 Sgobba, V. and Guldi, D.M. (2009). *Chem. Soc. Rev.* 38: 165.
171 (a) Nielsen, K.A., Cho, W.-S., Sarova, G. et al. (2006). *Angew. Chem. Int. Ed.* 45: 6848. (b) Nielsen, K.A., Martín-Gomis, L., Sarova, G.H. et al. (2008). *Tetrahedron* 64: 8449.
172 Fukuzumi, S., Ohkubo, K., Kawashima, Y. et al. (2011). *J. Am. Chem. Soc.* 133: 15938.
173 Bar-Haim, A., Klafter, J., and Kopelman, R. (1997). *J. Am. Chem. Soc.* 119: 6197.
174 (a) Yeow, E.K.L., Ghiggino, K.P., Reek, J.N.H. et al. (2000). *J. Phys. Chem. B* 104: 2596. (b) Gilat, S.L., Adronov, A., and Frechet, J.M.J. (1999). *Angew. Chem. Int. Ed.* 38: 1422.
175 (a) Li, W.-S. and Aida, T. (2009). *Chem. Rev.* 109: 6047. (b) Jiang, D.-L. and Aida, T. (1998). *J. Am. Chem. Soc.* 120: 10895.

176 Huang, C.-Y. and Su, Y.O. (2010). *Dalton Trans.* 39: 8306.
177 Hasobe, T., Kashiwagi, Y., Absalom, M.A. et al. (2004). *Adv. Mater.* 16: 975.
178 D'Souza, F., Deviprasad, G.R., Zandler, M.E. et al. (2002). *J. Phys. Chem. A* 106: 3243.
179 Hasobe, T., Kashiwagi, Y., Absalom, M.A., Sly, J., Hosomizu, K., Crossley, M.J., Imahori, H., Kamat, P.V., and Fukuzumi, S. (2004). *Adv. Mater.* 16: 975.
180 Fukuzumi, S., Saito, K., Ohkubo, K. et al. (2011). *Chem. Commun.* 47: 7980.
181 (a) Hagfeldt, A. and Grätzel, M. (2000). *Acc. Chem. Res.* 33: 269. (b) Grätzel, M. (2001). *Nature* 414: 338.
182 Bignozzi, C.A., Argazzi, R., and Kleverlaan, C.J. (2000). *Chem. Soc. Rev.* 29: 87.
183 Imahori, H. and Fukuzumi, S. (2001). *Adv. Mater.* 13: 1197.
184 Hasobe, T., Imahori, H., Kamat, P.V., and Fukuzumi, S. (2003). *J. Am. Chem. Soc.* 125: 14962.
185 Imahori, H., Arimura, M., Hanada, T. et al. (2001). *J. Am. Chem. Soc.* 123: 335.
186 (a) Imahori, H., Kashiwagi, Y., Endo, Y. et al. (2004). *Langmuir* 20: 73. (b) Imahori, H., Kashiwagi, Y., Hanada, T. et al. (2003). *J. Mater. Chem.* 13: 2890. (c) Fukuzumi, S., Endo, Y., Kashiwagi, Y. et al. (2003). *J. Phys. Chem. B* 107: 11979.
187 Hasobe, T., Imahori, H., Kamat, P.V. et al. (2005). *J. Am. Chem. Soc.* 127: 1217.
188 Kamat, P.V., Barazzouk, S., Hotchandani, S., and Thomas, K.G. (2000). *Chem. Eur. J.* 6: 3914.
189 Hasobe, T., Kamat, P.V., Troiani, V. et al. (2005). *J. Phys. Chem. B* 109: 19.
190 Hasobe, T., Saito, K., Kamat, P.V. et al. (2007). *J. Mater. Chem.* 17: 4160.
191 Hasobe, T., Kamat, P.V., Absalom, M.A. et al. (2004). *J. Phys. Chem. B* 108: 12865.
192 Ohkubo, K., Kawashima, Y., Sakai, H. et al. (2013). *Chem. Commun.* 49: 4474.
193 Hasobe, T., Imahori, H., Fukuzumi, S., and Kamat, P.V. (2003). *J. Phys. Chem. B* 107: 12105.
194 Fukuzumi, S. (2001). *Electron Transfer in Chemistry*, vol. 4 (ed. V. Balzani), 3–67. Weinheim: Wiley-VCH.
195 Fukuzumi, S. and Itoh, S. (1998). *Advances in Photochemistry*, vol. 25 (eds. D.C. Neckers, D.H. Volman and G. von Bünau), 107–172. New York: Wiley.
196 Sankaralingam, M., Lee, Y.-M., Pineda-Galvan, Y. et al. (2019). *J. Am. Chem. Soc.* 141: 1324.
197 Fukuzumi, S., Kuroda, S., and Tanaka, T. (1985). *J. Am. Chem. Soc.* 107: 3020.
198 (a) Fukuzumi, S., Okamoto, T., and Otera, J. (1994). *J. Am. Chem. Soc.* 116: 5503. (b) Fukuzumi, S., Satoh, N., Okamoto, T. et al. (2001). *J. Am. Chem. Soc.* 123: 7756.
199 (a) Fukuzumi, S., Ohkubo, K., and Okamoto, T. (2002). *J. Am. Chem. Soc.* 124: 14147. (b) Fukuzumi, S., Inada, O., Satoh, N. et al. (2002). *J. Am. Chem. Soc.* 124: 9181.

200 Fukuzumi, S. (2006). *Bull. Chem. Soc. Jpn.* 79: 177.
201 Fukuzumi, S. and Ohkubo, K. (2000). *Chem. Eur. J.* 6: 4532.
202 Fukuzumi, S. and Ohkubo, K. (2002). *J. Am. Chem. Soc.* 124: 10270.
203 Okamoto, K., Mori, Y., Yamada, H. et al. (2004). *Chem. Eur. J.* 10: 474.
204 Okamoto, K., Araki, Y., Ito, O., and Fukuzumi, S. (2004). *J. Am. Chem. Soc.* 126: 56.
205 Okamoto, K. and Fukuzumi, S. (2004). *J. Am. Chem. Soc.* 126: 13922.
206 Miura, T., Urano, Y., Tanaka, K. et al. (2003). *J. Am. Chem. Soc.* 125: 8666.
207 Kotani, H., Ohkubo, K., and Fukuzumi, S. (2004). *J. Am. Chem. Soc.* 126: 15999.
208 Fukuzumi, S., Nakanishi, I., and Tanaka, K. (1999). *J. Phys. Chem. A* 103: 11212.
209 Foote, C.S. (1968). *Acc. Chem. Res.* 1: 104.
210 Foote, C.S., Valentine, J.S., Greenberg, A., and Liebman, J.F. (eds.) (1995). *Active Oxygen in Chemistry*, vol. 2. New York: Blackie Academic and Professional.
211 Ogilby, P.R. and Foote, C.S. (1982). *J. Am. Chem. Soc.* 146: 2069.
212 Fukuzumi, S., Fujita, S., Suenobu, T. et al. (2002). *J. Phys. Chem. A* 106: 1241.
213 Ohkubo, K., Nanjo, T., and Fukuzumi, S. (2005). *Org. Lett.* 7: 4265.
214 Ohkubo, K., Nanjo, T., and Fukuzumi, S. (2006). *Catal. Today* 117: 356.
215 Wilson, T. and Schaap, A.P. (1971). *J. Am. Chem. Soc.* 93: 4126.
216 Burns, P.A. and Foote, C.S. (1974). *J. Am. Chem. Soc.* 96: 4339.
217 Lewis, F.D., Bedell, A.M., Dykstra, R.E. et al. (1990). *J. Am. Chem. Soc.* 112: 8055.
218 Ohkubo, K., Mizushima, K., Iwata, R. et al. (2010). *Chem. Commun.* 46: 601.
219 Fukuzumi, S. and Karlin, K.D. (2013). *Coord. Chem. Rev.* 257: 187.
220 Kakuda, S., Peterson, R.L., Ohkubo, K. et al. (2013). *J. Am. Chem. Soc.* 135: 6513.
221 Yamada, Y., Maeda, K., Ohkubo, K. et al. (2012). *Phys. Chem. Chem. Phys.* 14: 9654.
222 Griesbeck, A.G. and Cho, M. (2007). *Org. Lett.* 9: 611.
223 Kotani, H., Ohkubo, K., and Fukuzumi, S. (2008). *Appl. Catal., B* 77: 317–324.
224 Ohkubo, K., Nanjo, T., and Fukuzumi, S. (2006). *Bull. Chem. Soc. Jpn.* 79: 1489.
225 Ohkubo, K., Mizushima, K., and Fukuzumi, S. (2013). *Chem. Res. Intermed.* 39: 205.
226 Ohkubo, K., Fujimoto, A., and Fukuzumi, S. (2011). *Chem. Commun.* 47: 8515.
227 Ohkubo, K. and Hirose, K. (2018). *Angew. Chem. Int. Ed.* 57: 2126.
228 Suzuki, S. (2011). *Angew. Chem. Int. Ed.* 50: 6723.
229 Johansson Seechurn, C.C.C., Kitching, M.O., Colacot, T.J., and Snieckus, V. (2012). *Angew. Chem. Int. Ed.* 51: 5062.
230 Kandepi, V.V.K.M. and Narender, N. (2012). *Synthesis* 44: 15.
231 Podgorsek, A., Zupan, M., and Iskra, J. (2009). *Angew. Chem. Int. Ed.* 48: 8424.

232 Ohkubo, K., Mizushima, K., Iwata, R., and Fukuzumi, S. (2011). *Chem. Sci.* 2: 715.
233 Ohkubo, K., Iwata, R., Miyazaki, S. et al. (2006). *Org. Lett.* 8: 6079.
234 Dubois, D., Moninot, G., Kutner, W. et al. (1992). *J. Phys. Chem.* 96: 7137.
235 Fukuzumi, S., Mori, H., Imahori, H. et al. (2001). *J. Am. Chem. Soc.* 123: 12458.
236 Ohkubo, K., Iwata, R., Yanagimoto, T., and Fukuzumi, S. (2007). *Chem. Commun.*: 3139.
237 Fujitsuka, M., Luo, C., Ito, O. et al. (1999). *J. Phys. Chem. A* 103: 7155.
238 Kitaguchi, H., Ohkubo, K., Ogo, S., and Fukuzumi, S. (2006). *J. Phys. Chem. A* 110: 1718.
239 Ohkubo, K., Yukimoto, K., and Fukuzumi, S. (2006). *Chem. Commun.*: 2504.
240 (a) Kobayashi, K. and Tagawa, S. (2003). *J. Am. Chem. Soc.* 125: 10213. (b) Candeias, L.P. and Steenken, S. (1989). *J. Am. Chem. Soc.* 111: 1094.
241 Hamilton, D.S. and Nicewicz, D.A. (2012). *J. Am. Chem. Soc.* 134: 18577.
242 Nicewicz, D.A. and Nguyen, T.M. (2014). *ACS Catal.* 4: 355.
243 Nguyen, T.M. and Nicewicz, D.A. (2013). *J. Am. Chem. Soc.* 135: 9588.
244 Grandjean, J.-M.M. and Nicewicz, D.A. (2013). *Angew. Chem. Int. Ed.* 52: 3967.
245 Perkowski, A.J. and Nicewicz, D.A. (2013). *J. Am. Chem. Soc.* 135: 10334.
246 Furuya, T., Kamlet, A.S., and Ritter, T. (2011). *Nature* 473: 470.
247 Müller, K., Faeh, C., and Diederich, F. (2007). *Science* 317: 1881.
248 Nagib, D.A., Scott, M.E., and MacMillan, D.W.C. (2009). *J. Am. Chem. Soc.* 131: 10875.
249 Iqbal, N., Jung, J., Park, S., and Cho, E.J. (2014). *Angew. Chem. Int. Ed.* 53: 539.
250 Yasu, Y., Koike, T., and Akita, M. (2013). *Chem. Commun.* 49: 2037.
251 Yasu, Y., Koike, T., and Akita, M. (2012). *Angew. Chem. Int. Ed.* 51: 9567.
252 Wilger, D.J., Gesmundo, N.J., and Nicewicz, D.A. (2013). *Chem. Sci.* 4: 3160.
253 (a) Kerr, R.A. and Service, R.F. (2005). *Science* 309: 101. (b) Baykara, S.Z. (2005). *Int. J. Hydrogen Energy* 30: 545.
254 (a) Lehn, J.-M. and Sauvage, J.-P. (1977). *Nouv. J. Chim.* 1: 449. (b) Kalyanasundaram, K. (1982). *Coord. Chem. Rev.* 46: 159. (c) Darwent, J.R., Douglas, P., Harriman, A. et al. (1982). *Coord. Chem. Rev.* 44: 83.
255 Laane, C., Willner, I., Otvos, J.W., and Calvin, M. (1981). *Proc. Natl. Acad. Sci. U.S.A.* 78: 5928.
256 (a) Kiwi, J. and Grätzel, M. (1979). *J. Am. Chem. Soc.* 101: 7214. (b) Grätzel, M. (1981). *Acc. Chem. Res.* 14: 376. (c) Brugger, P.-A., Cuendet, P., and Grätzel, M. (1981). *J. Am. Chem. Soc.* 103: 2923.
257 (a) Chan, S.-F., Chou, M., Creutz, C. et al. (1981). *J. Am. Chem. Soc.* 103: 369. (b) Krishnan, C.V., Brunschwig, B.S., Creutz, C., and Sutin, N. (1985). *J. Am. Chem. Soc.* 107: 2005.
258 Du, P., Schneider, J., Jarosz, P., and Eisenberg, R. (2006). *J. Am. Chem. Soc.* 128: 7726.
259 (a) Persaud, L., Bard, A.J., Campion, A. et al. (1987). *J. Am. Chem. Soc.* 109: 7309. (b) Yonemoto, E.H., Riley, R.L., Kim, Y.I. et al. (1992). *J. Am. Chem. Soc.* 114: 8081.

260 (a) Okura, I. and Kim-Thuan, N. (1979). *J. Mol. Catal.* 6: 227. (b) Amao, Y., Tomonou, Y., and Okura, I. (2003). *Sol. Energy Mater. Sol. Cells* 79: 103. (c) Okura, I. and Hosono, H. (1992). *J. Phys. Chem.* 96: 4466. (d) Amao, Y. and Okura, I. (2002). *J. Mol. Catal. B: Enzym.* 17: 9.
261 Jiang, D.-L., Choi, C.-K., Honda, K. et al. (2004). *J. Am. Chem. Soc.* 126: 12084.
262 Kotani, H., Ohkubo, K., Takai, Y., and Fukuzumi, S. (2006). *J. Phys. Chem. B* 110: 24047.
263 Kotani, H., Ono, T., Ohkub, K., and Fukuzumi, S. (2007). *Phys. Chem. Chem. Phys.* 9: 1487.
264 Kotani, H., Ohkubo, K., and Fukuzumi, S. (2012). *Faraday Discuss.* 15: 89.
265 Kotani, H., Hanazaki, R., Ohkubo, K. et al. (2777). *Chem. Eur. J.* 2011: 17.
266 Yamada, Y., Miyahigashi, T., Kotani, H. et al. (2011). *J. Am. Chem. Soc.* 133: 16136.
267 Yamada, Y., Miyahigashi, T., Kotani, H. et al. (2012). *Energy Environ. Sci.* 5: 6111.
268 Yamada, Y., Miyahigashi, M., Ohkubo, K., and Fukuzumi, S. (2012). *Phys. Chem. Chem. Phys.* 14: 10564.
269 Yamada, Y., Tadokoro, H., and Fukuzumi, S. (2013). *RSC Adv.* 3: 25677.
270 Yamada, Y., Nomura, A., Ohkubo, K. et al. (2013). *Chem. Commun.* 49: 5132.
271 Yamada, Y., Nomura, A., Tadokoro, H., and Fukuzumi, S. (2015). *Catal. Sci. Technol.* 5: 428.
272 Cao, W., Wu, C., Lei, T. et al. (2018). *Chin. J. Catal.* 39: 1194.
273 Zhang, G., Hu, X., Chiang, C.-W. et al. (2016). *J. Am. Chem. Soc.* 138: 12037.
274 Scholes, G.D., Fleming, G.R., Olaya-Castro, A., and van Grondelle, R. (2011). *Nat. Chem.* 3: 763.
275 Schlapbach, L. and Züttel, A. (2001). *Nature* 414: 353.
276 Irvine, J. (2008). *J. Mater. Chem.* 18: 2295.
277 Eberle, U., Felderhoff, M., and Schüth, F. (2009). *Angew. Chem. Int. Ed.* 48: 6608.
278 Wagemans, R.W.P., van Lenthe, J.H., de Jongh, P.E. et al. (2005). *J. Am. Chem. Soc.* 127: 16675.
279 Orimo, S., Nakamori, Y., Eliseo, J.R. et al. (2007). *Chem. Rev.* 107: 4111.
280 Brayshaw, S.K., Harrison, A., McIndoe, J.S. et al. (2007). *J. Am. Chem. Soc.* 129: 1793.
281 Chen, P., Wu, X., Lin, J., and Tan, K.L. (1999). *Science* 285: 91.
282 Lee, S.M., An, K.H., Lee, Y.H. et al. (2001). *J. Am. Chem. Soc.* 123: 5059.
283 Sun, Q., Wang, Q., Jena, P., and Kawazoe, Y. (2005). *J. Am. Chem. Soc.* 127: 14582.
284 Leonard, A.D., Hudson, J.L., Fan, H. et al. (2008). *J. Am. Chem. Soc.* 131: 723.
285 Rowsell, J.L.C. and Yaghi, O.M. (2005). *Angew. Chem. Int. Ed.* 44: 4670.
286 Rowsell, J.L.C., Millward, A.R., Park, K.S., and Yaghi, O.M. (2004). *J. Am. Chem. Soc.* 126: 5666.
287 Rowsell, J.L.C. and Yaghi, O.M. (2006). *J. Am. Chem. Soc.* 128: 1304.
288 Sun, D., Ma, S., Ke, Y. et al. (2006). *J. Am. Chem. Soc.* 128: 3896.
289 Xiao, B., Wheatley, P.S., Zhao, X. et al. (2007). *J. Am. Chem. Soc.* 129: 1203.

290 Kaye, S.S., Dailly, A., Yaghi, O.M., and Long, J.R. (2007). *J. Am. Chem. Soc.* 129: 14176.

291 Thomas, K.M. (2009). *Dalton Trans.*: 1487.

292 Sculley, J., Yuan, D., and Zhou, H.-C. (2011). *Energy Environ. Sci.* 4: 2721.

293 Sumida, K., Rogow, D.L., Mason, J.A. et al. (2011). *Chem. Rev.* 112: 724.

294 Belof, J.L., Stern, A.C., Eddaoudi, M., and Space, B. (2007). *J. Am. Chem. Soc.* 129: 15202.

295 Dean, D., Davis, B., and Jessop, P.G. (2011). *New J. Chem.* 35: 417.

296 Demirci, U.B. and Miele, P. (2011). *Energy Environ. Sci.* 4: 3334.

297 Dobereiner, G.E., Nova, A., Schley, N.D. et al. (2011). *J. Am. Chem. Soc.* 133: 7547.

298 Bluhm, M.E., Bradley, M.G., Butterick, R. III, et al. (2006). *J. Am. Chem. Soc.* 128: 7748.

299 Himmelberger, D.W., Yoon, C.W., Bluhm, M.E. et al. (2009). *J. Am. Chem. Soc.* 131: 14101.

300 Sebastián, D., Bordejé, E.G., Calvillo, L. et al. (2008). *Int. J. Hydrogen Energy* 33: 1329.

301 Jessop, P. (2009). *Nat. Chem.* 1: 350.

302 Makowski, P., Thomas, A., Kuhn, P., and Goettmann, F. (2009). *Energy Environ. Sci.* 2: 480.

303 Yamaguchi, R., Ikeda, C., Takahashi, Y., and Fujita, K. (2009). *J. Am. Chem. Soc.* 131: 8410.

304 Yamada, Y., Yano, K., Xu, Q., and Fukuzumi, S. (2010). *J. Phys. Chem. C* 114: 16456.

305 Yamada, Y., Yano, K., and Fukuzumi, S. (2012). *Energy Environ. Sci.* 5: 5356.

306 Fukuzumi, S., Suenobu, T., and Ogo, S. (2008). PCT Int. Appl. WO 2008059630, 63pp.

307 Fukuzumi, S. (2008). *Eur. J. Inorg. Chem.*: 1351.

308 Fukuzumi, S., Yamada, Y., Suenobu, T. et al. (2011). *Energy Environ. Sci.* 4: 2754.

309 Enthaler, S. (2008). *ChemSusChem* 1: 801.

310 Joó, F. (2008). *ChemSusChem* 1: 805.

311 Czaun, M., Goeppert, A., May, R. et al. (2011). *ChemSusChem* 4: 1241.

312 Enthaler, S. and Loges, B. (2012). *ChemCatChem* 4: 323.

313 Grasemann, M. and Laurenczy, G. (2012). *Energy Environ. Sci.* 5: 8171.

314 Fukuzumi, S., Kobayashi, T., and Suenobu, T. (2008). *ChemSusChem* 1: 827.

315 Bi, Q.-Y., Du, X.-L., Liu, Y.-M. et al. (2012). *J. Am. Chem. Soc.* 134: 8926.

316 Gu, X., Lu, Z.-H., Jiang, H.-L. et al. (2011). *J. Am. Chem. Soc.* 133: 11822.

317 Tedsree, K., Li, T., Jones, S. et al. (2011). *Nat. Nanotechnol.* 6: 302.

318 O'Neill, B.J., Gürbüz, E.I., and Dumesic, J.A. (2012). *J. Catal.* 290: 193.

319 Bulushev, D.A., Jia, L., Beloshapkin, S., and Ross, J.R.H. (2012). *Chem. Commun.* 48: 4184.

320 Lide, D.R. (ed.) (2008–2009). *Handbook of Chemistry and Physics*, 89e. Boca Raton, FL: CRC Press.

321 Jessop, P.G., Ikariya, T., and Noyori, R. (1995). *Chem. Rev.* 95: 259.

322 Gao, Y., Kuncheria, J.K., Jenkins, H.A. et al. (2000). *J. Chem. Soc., Dalton Trans.*: 3212.

323 Jessop, P.G., Joó, F., and Tai, C.-C. (2004). *Coord. Chem. Rev.* 248: 2425.

324 Federsel, C., Jackstell, R., and Beller, M. (2010). *Angew. Chem. Int. Ed.* 49: 6254.

325 Wang, W., Wang, S., Ma, X., and Gong, J. (2011). *Chem. Soc. Rev.* 40: 3703.

326 Hayashi, H., Ogo, S., and Fukuzumi, S. (2004). *Chem. Commun.*: 2714.

327 Ogo, S., Kabe, R., Hayashi, H. et al. (2006). *Dalton Trans.*: 4657.

328 Himeda, Y., Onozawa-Komatsuzaki, N., Sugihara, H. et al. (2004). *Organometallics* 23: 1480.

329 Fellay, C., Dyson, P.J., and Laurenczy, G. (2008). *Angew. Chem. Int. Ed.* 47: 3966.

330 Loges, B., Boddien, A., Junge, H., and Beller, M. (2008). *Angew. Chem. Int. Ed.* 47: 3962.

331 Boddien, A., Loges, B., Junge, H., and Beller, M. (2008). *ChemSusChem* 1: 751.

332 Fellay, C., Yan, N., Dyson, P.J., and Laurenczy, G. (2009). *Chem. Eur. J.* 15: 3752.

333 Himeda, Y. (2009). *Green Chem.* 11: 2018.

334 Zhang, Z., Hu, S., Song, J. et al. (2009). *ChemSusChem* 2: 234.

335 Federsel, C., Jackstell, R., Boddien, A. et al. (2010). *ChemSusChem* 3: 1048.

336 Enthaler, S., von Langermann, J., and Schmidt, T. (2010). *Energy Environ. Sci.* 3: 1207.

337 Boddien, A., Gärtner, F., Federsel, C. et al. (2011). *Angew. Chem. Int. Ed.* 50: 6411.

338 Himeda, Y., Miyazawa, S., and Hirose, T. (2011). *ChemSusChem* 4: 487.

339 Papp, G., Csorba, J., Laurenczy, G., and Joó, F. (2011). *Angew. Chem. Int. Ed.* 50: 10433.

340 Tanaka, R., Yamashita, M., Chung, L.W. et al. (2011). *Organometallics* 30: 6742.

341 Reutemann, W. and Kieczka, H. (2005). Formic acid. In: *Ullmann's Encyclopedia of Industrial Chemistry, Electronic Release,* 7e, pp. 13–31. Weinheim: Wiley-VCH.

342 Kovács, G., Schubert, G., Joó, F., and Pápai, I. (2006). *Catal. Today* 115: 53.

343 Wheeler, R.A. (2004). Introduction to the molecular bioenergetics of electron, proton, and energy transfer. *ACS Symp. Ser.* 883: 1–6.

344 Dau, H. and Haumann, M. (2008). *Coord. Chem. Rev.* 252: 273.

345 Magnuson, A., Anderlund, M., Johansson, O. et al. (2009). *Acc. Chem. Res.* 42: 1899.

346 Ghirardi, M.L., Dubini, A., Yu, J., and Maness, P.-C. (2009). *Chem. Soc. Rev.* 38: 52.

347 Schut, G.J. and Adams, M.W.W. (2009). *J. Bacteriol.* 191: 4451.

348 Liebgott, P.-P., de Lacey, A.L., Burlat, B. et al. (2010). *J. Am. Chem. Soc.* 133: 986.

349 Amao, Y. (2011). *ChemCatChem* 3: 458.

350 Wagenknecht, P.S., Penney, J.M., and Hembre, R.T. (2003). *Organometallics* 22: 1180.

351 Bastianoni, S. and Marchettini, N. (1996). *Biomass Bioenergy* 11: 411.

352 Gray, K.A., Zhao, L., and Emptage, M. (2006). *Curr. Opin. Chem. Biol.* 10: 141.
353 Agarwal, A.K. (2007). *Prog. Energy Combust. Sci.* 33: 233.
354 Rubin, E.M. (2008). *Nature* 454: 841.
355 Kumar, S., Singh, S.P., Mishra, I.M., and Adhikari, D.K. (2009). *Chem. Eng. Technol.* 32: 517.
356 Matsushika, A., Inoue, H., Kodaki, T., and Sawayama, S. (2009). *Appl. Microbiol. Biotechnol.* 84: 37.
357 Alvira, P., Tomás-Pejó, E., Ballesteros, M., and Negro, M.J. (2010). *Bioresour. Technol.* 101: 4851.
358 Sarkar, N., Ghosh, S.K., Bannerjee, S., and Aikat, K. (2012). *Renewable Energy* 37: 19.
359 Wargacki, A.J., Leonard, E., Win, M.N. et al. (2012). *Science* 335: 308.
360 Ni, M., Leung, D.Y.C., and Leung, M.K.H. (2007). *Int. J. Hydrogen Energy* 32: 3238.
361 Cortright, R.D., Davda, R.R., and Dumesic, J.A. (2002). *Nature* 418: 964.
362 Deluga, G.A., Salge, J.R., Schmidt, L.D., and Verykios, X.E. (2004). *Science* 303: 993.
363 Umegaki, T., Kuratani, K., Yamada, Y. et al. (2008). *J. Power Sources* 179: 566.
364 Bshish, A., Yaakob, Z., Narayanan, B. et al. (2011). *Chem. Pap.* 65: 251.
365 Hayashi, H., Ogo, S., Abura, T., and Fukuzumi, S. (2003). *J. Am. Chem. Soc.* 125: 14266.
366 Maenaka, Y., Suenobu, T., and Fukuzumi, S. (2012). *Energy Environ. Sci.* 5: 7360.
367 Rived, F., Rosés, M., and Bosch, E. (1998). *Anal. Chim. Acta* 374: 309.
368 Ruelle, P., Kesselring, U.W., and Ho, N.-T. (1986). *J. Am. Chem. Soc.* 108: 371.
369 Fukuzumi, S., Kobayashi, T., and Suenobu, T. (2010). *J. Am. Chem. Soc.* 132: 1496.
370 Hull, J.F., Himeda, Y., Wang, W.-H. et al. (2012). *Nat. Chem.* 4: 383.
371 Maenaka, Y., Suenobu, T., and Fukuzumi, S. (2011). *J. Am. Chem. Soc.* 134: 367.
372 Johnson, S.L. and Tuazon, P.T. (1997). *Biochemistry* 16: 1175.
373 Maenaka, Y., Suenobu, T., and Fukuzumi, S. (2012). *J. Am. Chem. Soc.* 134: 9417.
374 Suenobu, T., Shibata, S., and Fukuzumi, S. (2015). *Chem. Commun.* 51: 1670.
375 Nielsen, M., Alberico, E., Baumann, W. et al. (2013). *Nature* 495: 85.
376 Fukuzumi, S., Patz, M., Suenobu, T. et al. (1999). *J. Am. Chem. Soc.* 121: 1605.
377 Bagchi, R.N., Bond, A.M., Scholz, F., and Stösser, R. (1989). *J. Am. Chem. Soc.* 111: 8270.
378 Känzig, W. and Cohen, M.H. (1959). *Phys. Rev. Lett.* 3: 509.
379 Zeller, H.R. and Känzig, W. (1967). *Helv. Phys. Acta* 40: 845.
380 Kasai, P.H. (1965). *J. Chem. Phys.* 43: 3322.
381 Lunsford, J.H. (1973). *Catal. Rev.* 8: 135.
382 Che, M. (1997). *Chem. Rev.* 97: 305.
383 Hou, Z., Fujita, A., Zhang, Y. et al. (1998). *J. Am. Chem. Soc.* 120: 754.
384 Hou, Z. and Wakatsuki, Y. (1997). *Chem. Eur. J.* 3: 1005.

385 Fukuzumi, S. and Okamoto, T. (1993). *J. Am. Chem. Soc.* 115: 11600.
386 Fukuzumi, S., Mochizuki, S., and Tanaka, T. (1989). *Inorg. Chem.* 28: 2459.
387 Sawyer, D.T., Calderwood, T.S., Yamaguchi, K., and Angelis, C.T. (1983). *Inorg. Chem.* 22: 2577.
388 Ohkubo, K., Menon, S.C., Orita, A. et al. (2003). *J. Org. Chem.* 68: 4720.
389 Sato, T., Wakahara, Y., Otera, J. et al. (1991). *J. Am. Chem. Soc.* 113: 4028.
390 An, D.L., Peng, Z., Orita, A. et al. (2006). *Chem. Eur. J.* 12: 1642.
391 Lippard, S.J. (1993). *Science* 261: 699.
392 Mirts, E.N., Bhagi-Damodaran, A., and Lu, Y. (2019). *Acc. Chem. Res.* 52: 935.
393 Wanner, M., Sixt, T., Klinkhammer, K.-W., and Kaim, W. (1999). *Inorg. Chem.* 38: 2753.
394 Schwederski, B., Kasack, V., Kaim, W. et al. (1990). *Angew. Chem. Int. Ed. Engl.* 29: 78.
395 Itoh, S., Kawakami, H., and Fukuzumi, S. (1997). *J. Am. Chem. Soc.* 119: 439.
396 Itoh, S., Kawakami, H., and Fukuzumi, S. (1998). *Biochemistry* 37: 6562.
397 Ernst, S., Hänel, P., Jordanov, J. et al. (1989). *J. Am. Chem. Soc.* 111: 1733.
398 Yuasa, J., Suenobu, T., and Fukuzumi, S. (2006). *ChemPhysChem* 7: 942.
399 Goss, C.A. and Abruna, H.D. (1985). *Inorg. Chem.* 24: 4263.
400 Fukuzumi, S., Koumitsu, S., Hironaka, K., and Tanaka, T. (1987). *J. Am. Chem. Soc.* 109: 305.
401 Fukuzumi, S. and Tokuda, Y. (1992). *J. Phys. Chem.* 96: 8409.
402 Yuasa, J., Suenobu, T., and Fukuzumi, S. (2003). *J. Am. Chem. Soc.* 125: 12090.
403 Tributsch, H. and Pohlmann, L. (1998). *Science* 279: 1891.
404 Gross, E.M., Armstrong, N.R., and Wightman, R.M. (2002). *J. Electrochem. Soc.* 149: E137.
405 Dedeian, K., Djurovich, P.I., Garces, F.O. et al. (1991). *Inorg. Chem.* 30: 1685.
406 Yuasa, J. and Fukuzumi, S. (2004). *Org. Biomol. Chem.* 2: 642.
407 Crano, J.G. and Guglielmetti, R.J. (eds.) (1999). *Organic Photochromic and Thermochromic Compounds*. New York: Kluwer Academic.
408 Day, J.H. (1968). *Chem. Rev.* 68: 649.
409 Ogawa, K., Kasahara, Y., Ohtani, Y., and Harada, J. (1998). *J. Am. Chem. Soc.* 120: 7107.
410 Helton, M.E., Gebhart, N.L., Davies, E.S. et al. (2001). *J. Am. Chem. Soc.* 123: 10389.
411 Adams, D.M., Dei, A., Rheingold, A.L., and Hendrickson, D.N. (1993). *J. Am. Chem. Soc.* 115: 8221.
412 Rall, J., Wanner, M., Albrecht, M. et al. (2802). *Chem. Eur. J.* 1999: 5.
413 Okada, S., Peng, S., Spevak, W., and Charych, D. (1998). *Acc. Chem. Res.* 31: 229.
414 Schenning, A.P.H.J., Kilbinger, A.F.M., Biscarini, F. et al. (2002). *J. Am. Chem. Soc.* 124: 1269.
415 Yuasa, J. and Fukuzumi, S. (2007). *J. Am. Chem. Soc.* 129: 12912.
416 Kaifer, A.E. (1999). *Acc. Chem. Res.* 32: 62.

417 Collin, J.-P., Dietrich-Buchecker, C., Gaviña, P. et al. (2001). *Acc. Chem. Res.* 34: 477.

418 Matyushov, D.V. (2007). *Acc. Chem. Res.* 40: 294.

419 Milischuk, A.A., Matyushov, D.V., and Newton, M.D. (2006). *Chem. Phys.* 324: 172.

420 Yuasa, J., Yamada, S., and Fukuzumi, S. (2008). *Chem. Eur. J.* 14: 1866.

421 Yuasa, J., Yamada, S., and Fukuzumi, S. (2006). *J. Am. Chem. Soc.* 128: 14938.

422 Fukuzumi, S., Nishizawa, N., and Tanaka, T. (1985). *J. Chem. Soc., Perkin Trans. 2*: 371.

423 Eberson, L. (1982). *Adv. Phys. Org. Chem.* 18: 79.

424 Fukuzumi, S., Okamoto, K., and Imahori, H. (2002). *Angew. Chem. Int. Ed.* 41: 620.

425 Okamoto, K., Imahori, H., and Fukuzumi, S. (2003). *J. Am. Chem. Soc.* 125: 7014.

426 Sessler, J.L., Wang, B., Springs, S.L., and Brown, C.T. (1996). Electron- and energy-transfer reactions in noncovalently linked supramolecular model systems. In: *Comprehensive Supramolecular Chemistry*, vol. 4 (ed. Y. Murakami), 311–336. New York: Pergamon.

427 Scheiner, S. (1997). *Hydrogen Bonding*. Oxford: Oxford Press.

428 Nocek, J.M., Zhou, J.S., Forest, S.D. et al. (1996). *Chem. Rev.* 96: 2459.

429 Cukier, R.I. and Nocera, D.G. (1998). *Annu. Rev. Phys. Chem.* 49: 337.

430 Rotello, V.M. (2001). *Electron Transfer in Chemistry*, vol. 4 (ed. V. Balzani), 68–87. Weinheim: Wiley-VCH.

431 Lubitz, W., Lendzian, F., and Bittl, R. (2002). *Acc. Chem. Res.* 35: 313.

432 Chang, C.J., Brown, J.D.K., Chang, M.C.Y. et al. (2001). *Electron Transfer in Chemistry*, vol. 3 (ed. V. Balzani), 409–461. Weinheim: Wiley-VCH.

433 Fukuzumi, S., Honda, T., and Kojima, T. (2012). *Coord. Chem. Rev.* 256: 2488.

434 Okamoto, K., Yoshida, Y., Imahori, H. et al. (2003). *J. Am. Chem. Soc.* 125: 1007.

435 O'Malley, P.J. (1997). *J. Phys. Chem. A* 101: 6334.

436 Okamoto, K., Imahori, H., Araki, Y., and Ito, O. (2002). *J. Am. Chem. Soc.* 124: 6794.

437 Fukuzumi, S., Nakanishi, I., Maruta, J. et al. (1998). *J. Am. Chem. Soc.* 120: 6673.

438 Fukuzumi, S., Nishizawa, N., and Tanaka, T. (1984). *J. Org. Chem.* 49: 3571.

439 Bunting, J.W. (1991). *Bioorg. Chem.* 19: 456.

440 He, G.-X., Blasko, A., and Bruice, T.C. (1993). *Bioorg. Chem.* 21: 423.

441 Fukuzumi, S. (1992). *Advances in Electron Transfer Chemistry* (ed. P.S. Mariano), 67–175. Greenwich, CT: JAI Press.

442 Fukuzumi, S. and Tanaka, T. (1988). *Photoinduced Electron Transfer*, Part C, Chapter 10 (eds. M.A. Fox and M. Chanon), pp. 636–687. Amsterdam: Elsevier.

443 Fukuzumi, S., Yuasa, J., and Suenobu, T. (2002). *J. Am. Chem. Soc.* 124: 12566.

444 Yuasa, J., Suenobu, T., Ohkubo, K., and Fukuzumi, S. (2003). *Chem. Commun.*: 1070.

445 Fukuzumi, S., Fujii, Y., and Suenobu, T. (2001). *J. Am. Chem. Soc.* 123: 10191.

446 Ishikawa, M. and Fukuzumi, S. (1990). *J. Chem. Soc., Faraday Trans.* 86: 3531.

447 Kaim, W. and Schwederski, B. (1994). *Bioinorganic Chemistry: Inorganic Elements in the Chemistry of Life*. Chichester: Wiley.

448 Fridovich, I. (1989). *J. Biol. Chem.* 264: 7761.

449 Bertini, I., Banci, L., and Piccioli, M. (1990). *Coord. Chem. Rev.* 100: 67.

450 Cass, A.E.G. (1985). *Superoxide Dismutases in Metalloproteins*, Part 1 (ed. P. Harrison), 121. Weinheim: Verlag Chemie.

451 Fukuzumi, S., Ohtsu, H., Ohkubo, K. et al. (2002). *Coord. Chem. Rev.* 226: 71.

452 Ellerby, L.M., Cabelli, D.E., Graden, J.A., and Valentine, J.S. (1996). *J. Am. Chem. Soc.* 118: 6556.

453 Oberley, L.W. and Buettner, G.R. (1979). *Cancer Res.* 39: 1141.

454 Farmer, K.J. and Sohal, R.S. (1989). *Free Radical Biol. Med.* 7: 23.

455 Strotkamp, K.G. and Lippard, S.J. (1982). *Acc. Chem. Res.* 15: 318.

456 Murphy, B.P. (1993). *Coord. Chem. Rev.* 124: 63.

457 Gartner, A. and Weser, U. (1986). *Top. Curr. Chem.* 132: 1.

458 Ohtsu, H., Shimazaki, Y., Odani, A. et al. (2000). *J. Am. Chem. Soc.* 122: 5733.

459 Ohtsu, H. and Fukuzumi, S. (2001). *Chem. Eur. J.* 7: 4947.

460 Fridovich, I. and Beauchamp, C. (1971). *Anal. Biochem.* 44: 276.

461 Morimoto, Y., Kotani, H., Park, J. et al. (2011). *J. Am. Chem. Soc.* 133: 403.

462 Lee, Y.-M., Kotani, H., Suenobu, T. et al. (2008). *J. Am. Chem. Soc.* 130: 434.

463 Rohde, J.-U., In, J.-H., Lim, M.H. et al. (2003). *Science* 299: 1037.

464 Fukuzumi, S., Morimoto, Y., Kotani, H. et al. (2010). *Nat. Chem.* 2: 756.

465 Swart, M. (2013). *Chem. Commun.* 49: 6650.

466 Prakash, J., Rohde, G.T., Meier, K.K. et al. (2015). *J. Am. Chem. Soc.* 137: 3478.

467 Li, F., Van Heuvelen, K.M., Meier, K.K. et al. (2013). *J. Am. Chem. Soc.* 135: 10198.

468 Thibon, A., England, J., Martinho, M. et al. (2008). *Angew. Chem. Int. Ed.* 47: 7064.

469 Nishida, Y., Lee, Y.-M., Nam, W., and Fukuzumi, S. (2014). *J. Am. Chem. Soc.* 136: 8042.

470 Morimoto, Y., Lee, Y.-M., Nam, W., and Fukuzumi, S. (2013). *Chem. Commun.* 49: 2500.

471 Comba, P., Lee, Y.-M., Nam, W., and Waleska, A. (2014). *Chem. Commun.* 50: 412.

472 Kakuda, S., Rolle, C.J., Ohkubo, K. et al. (2015). *J. Am. Chem. Soc.* 137: 3330.

473 Kakuda, S., Peterson, R.L., Ohkubo, K. et al. (2013). *J. Am. Chem. Soc.* 135: 2825.

474 Fukuzumi, S., Kotani, H., Lucas, H.R. et al. (2010). *J. Am. Chem. Soc.* 132: 6874.

475 Chen, J., Lee, Y.-M., Davis, K.M. et al. (2013). *J. Am. Chem. Soc.* 135: 6388.

476 Yoon, H., Lee, Y.-M., Wu, X. et al. (2013). *J. Am. Chem. Soc.* 135: 9186.

477 Yoon, H., Lee, Y.-M., Nam, W., and Fukuzumi, S. (2014). *Chem. Commun.* 50: 12944.
478 Hong, S., Pfaff, F.F., Kwon, E. et al. (2014). *Angew. Chem. Int. Ed.* 53: 10403.
479 Damatoy, D.B., Shimon, L.J.W., Weiner, L. et al. (2014). *Inorg. Chem.* 53: 1779.
480 Pfaff, F.F., Kundu, S., Risch, M. et al. (2011). *Angew. Chem. Int. Ed.* 50: 1711.
481 Lacy, D.C., Park, Y.J., Ziller, J.W. et al. (2012). *J. Am. Chem. Soc.* 134: 17526.
482 Park, J., Morimoto, Y., Lee, Y.-M. et al. (2014). *Inorg. Chem.* 53: 3618.
483 Parkin, G. (2009). *Acc. Chem. Res.* 42: 315.
484 Chen, J., Yoon, H., Lee, Y.-M. et al. (2015). *Chem. Sci.* 6: 3624.
485 Morimoto, Y., Park, J., Suenobu, T. et al. (2012). *Inorg. Chem.* 51: 10025.
486 Park, J., Morimoto, Y., Lee, Y.-M. et al. (2011). *J. Am. Chem. Soc.* 133: 5236.
487 Park, J., Morimoto, Y., Lee, Y.-M. et al. (2012). *J. Am. Chem. Soc.* 134: 3903.
488 Park, J., Lee, Y.-M., Nam, W., and Fukuzumi, S. (2013). *J. Am. Chem. Soc.* 135: 5052.
489 Park, J., Lee, Y.-M., Ohkubo, K. et al. (2015). *Inorg. Chem.* 54: 5806.
490 Park, J., Morimoto, Y., Lee, Y.-M. et al. (2011). *Inorg. Chem.* 50: 11612.
491 Nehru, K., Seo, M.S., Kim, J., and Nam, W. (2007). *Inorg. Chem.* 46: 293.
492 Wu, X., Seo, M.S., Davis, K.M. et al. (2011). *J. Am. Chem. Soc.* 133: 20088.
493 Lee, D. and Lippard, S.J. (2001). *J. Am. Chem. Soc.* 123: 4611.
494 Lee, Y.-M., Kim, S., Ohkubo, K. et al. (2019). *J. Am. Chem. Soc.* 141: 2614.
495 Lee, Y.-M., Bang, S., Kim, Y.M. et al. (2013). *Chem. Sci.* 4: 3917.
496 Bang, S., Lee, Y.-M., Hong, S. et al. (2014). *Nat. Chem.* 6: 934.
497 Lionetti, D. and Agapie, T. (2014). *Nature* 513: 495.
498 Babcock, G.T. (1999). *Proc. Natl. Acad. Sci. U.S.A.* 96: 12971.
499 Wikström, M., Krab, K., and Saraste, M. (1981). *Cytochrome Oxidase: A Synthesis*. New York: Academic Press.
500 Yoshikawa, S., Shinzawa-Itoh, K., Nakashima, R. et al. (1998). *Science* 280: 1723.
501 Michel, H., Behr, J., Harrenga, A., and Kannt, A. (1998). *Ann. Rev. Biophys. Biomol. Struct.* 27: 329.
502 Collman, J.P., Boulatov, R., and Sunderland, C.J. (2003). *The Porphyrin Handbook*, vol. 11 (eds. K.M. Kadish, K.M. Smith and R. Guilard), 1–49. San Diego, USA: Elsevier Science.
503 Chufán, E.E., Puiu, S.C., and Karlin, K.D. (2007). *Acc. Chem. Res.* 40: 563.
504 Kim, E., Chufn, E.E., Kamaraj, K., and Karlin, K.D. (2004). *Chem. Rev.* 104: 1077.
505 Chufan, E.E., Mondal, B., Gandhi, T. et al. (2007). *Inorg. Chem.* 46: 6382.
506 Collman, J.P., Boulatov, R., Sunderland, C.J., and Fu, L. (2004). *Chem. Rev.* 104: 561.
507 Anson, F.C., Shi, C.N., and Steiger, B. (1997). *Acc. Chem. Res.* 30: 437.
508 Ogura, T. and Kitagawa, T. (2004). *Biochim. Biophys. Acta* 1655: 290.
509 Fukuzumi, S., Okamoto, K., Gros, C.P., and Guilard, R. (2004). *J. Am. Chem. Soc.* 126: 10441.
510 Fukuzumi, S., Mochizuki, S., and Tanaka, T. (1989). *Chem. Lett.*: 27.
511 Deng, Y., Chang, C.J., and Nocera, D.J. (2000). *J. Am. Chem. Soc.* 122: 410.

512 Bolze, F., Gros, C.P., Drouin, M. et al. (2002). *J. Organomet. Chem.* 643–644: 89.

513 Chang, C.J., Baker, E.A., Pistorio, B.J. et al. (2002). *Inorg. Chem.* 41: 3102.

514 Goto, Y., Watanabe, Y., Fukuzumi, S. et al. (1998). *J. Am. Chem. Soc.* 120: 10762.

515 Goto, Y., Matsui, T., Ozaki, S. et al. (1999). *J. Am. Chem. Soc.* 121: 9497.

516 Fukuzumi, S., Mochizuki, S., and Tanaka, T. (1990). *Inorg. Chem.* 29: 653.

517 Fukuzumi, S., Okamoto, K., Tokuda, Y. et al. (2004). *J. Am. Chem. Soc.* 126: 17059.

518 Fukuzumi, S., Tokuda, Y., Kitano, T. et al. (1993). *J. Am. Chem. Soc.* 115: 8960.

519 Fukuzumi, S., Ohkubo, K., Tokuda, Y., and Suenobu, T. (2000). *J. Am. Chem. Soc.* 122: 4286.

520 Wayner, D.D.M., McPhee, D.J., and Griller, D. (1998). *J. Am. Chem. Soc.* 110: 132.

521 Takada, N., Itoh, S., and Fukuzumi, S. (1996). *Chem. Lett.*: 1103.

522 Murakami, M., Hong, D., Suenobu, T., and Fukuzumi, S. (2011). *J. Am. Chem. Soc.* 133: 11605.

523 Nakazono, T., Parent, A.R., and Sakai, K. (2015). *Chem. Eur. J.* 21: 6723.

524 Dogutan, D.K., McGuire, R. Jr., and Nocera, D.G. (2011). *J. Am. Chem. Soc.* 133: 9178.

525 Wasylenko, D.J., Palmer, R.D., Schott, E., and Berlinguette, C.P. (2012). *Chem. Commun.* 48: 2107.

526 Wang, D. and Groves, J.T. (2013). *Proc. Natl. Acad. Sci. U.S.A.* 110: 15579.

527 Wang, L., Duan, L., Wang, Y. et al. (2014). *Chem. Commun.* 50: 12947.

528 Neudeck, S., Maji, S., López, I. et al. (2014). *J. Am. Chem. Soc.* 136: 24.

529 Bozoglian, F., Romain, S., Ertem, M.Z. et al. (2009). *J. Am. Chem. Soc.* 131: 15176.

530 Sala, X., Maji, S., Bofill, R. et al. (2014). *Acc. Chem. Res.* 47: 504.

531 Berardi, S., Drouet, S., Francàs, L. et al. (2014). *Chem. Soc. Rev.* 43: 7501.

532 Nakazono, T., Parent, A.R., and Sakai, K. (2013). *Chem. Commun.* 49: 6325.

533 Ishizuka, T., Watanabe, A., Kotani, H. et al. (2016). *Inorg. Chem.* 55: 1154.

534 Yamada, Y., Fukunishi, Y., Yamazaki, S., and Fukuzumi, S. (2010). *Chem. Commun.* 46: 7334–7336.

535 Hatay, I., Su, B., Li, F. et al. (2009). *J. Am. Chem. Soc.* 131: 13453.

536 Mase, K., Ohkubo, K., and Fukuzumi, S. (2015). *Inorg. Chem.* 54: 1808.

537 Mase, K., Ohkubo, K., and Fukuzumi, S. (2013). *J. Am. Chem. Soc.* 135: 2800.

538 Honda, T., Kojima, T., and Fukuzumi, S. (2012). *J. Am. Chem. Soc.* 134: 4196.

539 Su, B., Hatay, I., Trojanek, A. et al. (2010). *J. Am. Chem. Soc.* 132: 2655.

540 Kadish, K.M., Shen, J., Fremond, L. et al. (2008). *Inorg. Chem.* 47: 6726.

541 Qu, J., Shen, Y., Qu, X., and Dong, S. (2004). *Electroanalysis* 16: 1444–1450.

542 El-Deab, M.S., Othman, S.H., Okajima, T., and Ohsaka, T. (2008). *J. Appl. Electrochem.* 38: 1445.

543 Kato, S., Jung, J., Suenobu, T., and Fukuzumi, S. (2013). *Energy Environ. Sci.* 6: 3756.

544 Das, A., Joshi, V., Kotkar, D. et al. (2001). *J. Phys. Chem. A* 105: 6945.

545 Wua, X., Li, F., Zhanga, B., and Sun, L. (2015). *J. Photochem. Photobiol., C* 25: 71–89.
546 Fukuzumi, S., Jung, J., Yamada, Y. et al. (2016). *Chem. Asian J.* 11: 1138.
547 Okamura, M. and Masaoka, S. (2015). *Chem. Asian J.* 10: 306.
548 Fukuzumi, S. and Hong, D. (2014). *Eur. J. Inorg. Chem.* 2014: 645.
549 Yin, Q., Tan, J.M., Besson, C. et al. (2010). *Science* 328: 342.
550 Lv, H., Song, J., Geletii, Y.V. et al. (2014). *J. Am. Chem. Soc.* 136: 9268.
551 Fukuzumi, S., Hong, D., and Yamada, Y. (2013). *J. Phys. Chem. Lett.* 4: 3458.
552 Parent, A.R. and Sakai, K. (2014). *ChemSusChem* 7: 2070.
553 Blakemore, J.D., Crabtree, R.H., and Brudvig, G.W. (2015). *Chem. Rev.* 115: 12974.
554 Hong, D., Murakami, M., Yamada, Y., and Fukuzumi, S. (2012). *Energy Environ. Sci.* 5: 5708.
555 Yamada, Y., Oyama, K., Gates, R., and Fukuzumi, S. (2015). *Angew. Chem. Int. Ed.* 54: 5613.
556 Shiraishi, Y., Kanazawa, S., Kofuji, Y. et al. (2014). *Angew. Chem. Int. Ed.* 53: 13454.
557 Hong, D., Yamada, Y., Nagatomi, T. et al. (2012). *J. Am. Chem. Soc.* 134: 19572.
558 Isaka, Y., Kato, S., Hong, D. et al. (2015). *J. Mater. Chem. A* 3: 12404.
559 Isaka, Y., Yamada, Y., Suenobu, T., and Fukuzumi, S. (2016). *Catal. Sci. Technol.* 6: 681.
560 Isaka, Y., Yamada, Y., Ohkubo, K. et al. (2016). *RSC Adv.* 6: 42041.
561 Xu, P., Feng, J., Fang, T. et al. (2016). *RSC Adv.* 6: 9905.
562 (a) Park, Y., McDonald, K.J., and Choi, K.-S. (2013). *Chem. Soc. Rev.* 42: 2321. (b) Luan, P. and Zhang, J. (2019). *ChemElectroChem* https://doi.org/10.1002/celc.201900398.
563 Mase, K., Yoneda, M., Yamada, Y., and Fukuzumi, S. (2016). *Nat. Commun.* 7: 11470.
564 Chen, Z., Concepcion, J.J., Song, N., and Meyer, T.J. (2014). *Chem. Commun.* 50: 8053.
565 Huang, L., Li, R., Chong, R. et al. (2014). *Catal. Sci. Technol.* 4: 2913.
566 Fukuzumi, S., Lee, Y.-M., and Nam, W. (2017). *ChemSusChem* 10: 4264.
567 Melis, A. (2009). *Plant Sci.* 177: 272.
568 Mase, K., Yoneda, M., Yamada, Y., and Fukuzumi, S. (2016). *ACS Energy Lett.* 1: 913.
569 Fukuzumi, S. and Yamada, Y. (2016). *ChemElectroChem* 3: 1978.
570 (a) Najafpour, M.M., Renger, G., Hołyńska, M. et al. (2016). *Chem. Rev.* 116: 2886. (b) Shen, J.-R. and Annu, J.-R. (2015). *Rev. Plant Biol.* 66: 23.
571 (a) Zobnina, V., Lambreva, M.D., Rea, G. et al. (2017). *Photosynth. Res.* 131: 15. (b) Müh, F., Glöckner, C., Hellmich, J., and Zouni, A. (2012). *Biochim. Biophys. Acta* 1817: 44–65.
572 Hong, Y.H., Jung, J., Nakagawa, T. et al. (2019). *J. Am. Chem. Soc.* 141: 6748.
573 (a) Ohkubo, K., Fujimoto, A., and Fukuzumi, S. (2013). *J. Am. Chem. Soc.* 135: 5368–5371. (b) Ohkubo, K., Hirose, K., and Fukuzumi, S. (2015). *Chem. Eur. J.* 21: 2855–2861.
574 Hong, S., Lee, Y.-M., Shin, W. et al. (2009). *J. Am. Chem. Soc.* 131: 13910.

Index

Electron Transfer: Mechanisms and Applications, First Edition. Shunichi Fukuzumi.